Modeling Human and Organizational Behavior

APPLICATION TO MILITARY SIMULATIONS

Richard W. Pew and Anne S. Mavor, editors

Panel on Modeling Human Behavior and Command Decision Making: Representations for Military Simulations

Commission on Behavioral and Social Sciences and Education

National Research Council

NATIONAL ACADEMY PRESS
Washington, D.C. 1998

NATIONAL ACADEMY PRESS • 2101 Constitution Avenue, NW • Washington, D.C. 20418

NOTICE: The project that is the subject of this report was approved by the Governing Board of the National Research Council, whose members are drawn from the councils of the National Academy of Sciences, the National Academy of Engineering, and the Institute of Medicine. The members of the committee responsible for the report were chosen for their special competences and with regard for appropriate balance.

This study was supported by Technical Support Services Contract DACW61-96-D-0001 between the National Academy of Sciences and the Defense Modeling and Simulation Office of the U.S. Department of Defense. Any opinions, findings, conclusions, or recommendations expressed in this publication are those of the author(s) and do not necessarily reflect the view of the organizations or agencies that provided support for this project.

Library of Congress Cataloging-in-Publication Data

Modeling human and organizational behavior : application to
military simulations / Richard W. Pew and Anne S. Mavor, editors.
p. cm.
"Panel on Modeling Human Behavior and Command Decision Making: Representations for Military Simulations, Commission on Behavioral and Social Sciences and Education, National Research Council."
Includes bibliographical references and index.
ISBN 0-309-06096-6
1. Psychology, Military. 2. Human behavior—Simulation methods. 3. Decision-making. 4. Command of troops. I. Pew, Richard W. II. Mavor, Anne S. III. National Research Council (U.S.). Panel on Modeling Human Behavior and Command Decision Making: Representations for Military Simulations.
U22.3 .M58 1998
355′.001′9—ddc21

98-19705

Additional copies of this report are available from:

National Academy Press
2101 Constitution Avenue, N.W.
Washington, D.C. 20418
Call 800-624-6242 or 202-334-3313 (in the Washington Metropolitan Area).

This report is also available online at **http://www.nap.edu**

Printed in the United States of America

PANEL ON MODELING HUMAN BEHAVIOR AND COMMAND DECISION MAKING: REPRESENTATIONS FOR MILITARY SIMULATIONS

RICHARD W. PEW (*Chair*), BBN Technologies, GTE Internetworking, Cambridge, MA
JEROME BUSEMEYER, Psychology Department, Indiana University
KATHLEEN M. CARLEY, Department of Social and Decision Sciences, Carnegie Mellon University
TERRY CONNOLLY, Department of Management and Policy and College of Business and Public Administration, University of Arizona, Tucson
JOHN R. CORSON, JRC Research and Analysis, L.L.C., Williamsburg, VA
KENNETH H. FUNK, II, Industrial and Manufacturing Engineering, Oregon State University, Corvallis
BONNIE E. JOHN, Human-Computer Interaction Institute, Carnegie Mellon University
RICHARD M. SHIFFRIN, Psychology Department, Indiana University, Bloomington
GREG L. ZACHARIAS, Charles River Analytics, Cambridge, MA

ANNE S. MAVOR, *Study Director*
JERRY S. KIDD, *Senior Adviser*
SUSAN R. McCUTCHEN, *Senior Project Assistant*

COMMITTEE ON HUMAN FACTORS

WILLIAM C. HOWELL (*Chair*), Arizona State University, Tempe
TERRY CONNOLLY, Department of Management and Policy and College of Business and Public Administration, University of Arizona, Tucson
COLIN G. DRURY, Industrial Engineering Department, University of Buffalo, New York
MARTHA GRABOWSKI, Rensselaer Polytechnic and LeMoyne College, New York
DANIEL R. ILGEN, Department of Psychology and Department of Management, Michigan State University
RICHARD J. JAGACINSKI, Department of Psychology, Ohio State University, Columbus
LAWRENCE R. JAMES, Department of Management, University of Tennessee
BONNIE E. JOHN, Human-Computer Interaction Institute, Carnegie Mellon University
TOM B. LEAMON, Liberty Mutual Insurance Co. and Liberty Mutual Research Center for Safety and Health, Hopkinton, MA
DAVID C. NAGEL, AT&T Laboratories, Basking Ridge, NJ
KARLENE ROBERTS, Haas School of Business, University of California, Berkeley
LAWRENCE W. STARK, School of Optometry, University of California, Berkeley
KIM J. VICENTE, Department of Mechanical and Industrial Engineering, University of Toronto, Canada
EARL L. WIENER, Department of Management Science, University of Miami
GREG L. ZACHARIAS, Charles River Analytics, Cambridge, MA

ANNE S. MAVOR, *Director*
JERRY S. KIDD, *Senior Adviser*
SUSAN R. McCUTCHEN, *Senior Project Assistant*

Contents

The National Academy of Sciences is a private, nonprofit, self-perpetuating society of distinguished scholars engaged in scientific and engineering research, dedicated to the furtherance of science and technology and to their use for the general welfare. Upon the authority of the charter granted to it by the Congress in 1863, the Academy has a mandate that requires it to advise the federal government on scientific and technical matters. Dr. Bruce M. Alberts is president of the National Academy of Sciences.

The National Academy of Engineering was established in 1964, under the charter of the National Academy of Sciences, as a parallel organization of outstanding engineers. It is autonomous in its administration and in the selection of its members, sharing with the National Academy of Sciences the responsibility for advising the federal government. The National Academy of Engineering also sponsors engineering programs aimed at meeting national needs, encourages education and research, and recognizes the superior achievements of engineers. Dr. William A. Wulf is president of the National Academy of Engineering.

The Institute of Medicine was established in 1970 by the National Academy of Sciences to secure the services of eminent members of appropriate professions in the examination of policy matters pertaining to the health of the public. The Institute acts under the responsibility given to the National Academy of Sciences by its congressional charter to be an adviser to the federal government and, upon its own initiative, to identify issues of medical care, research, and education. Dr. Kenneth I. Shine is president of the Institute of Medicine.

The National Research Council was organized by the National Academy of Sciences in 1916 to associate the broad community of science and technology with the Academy's purposes of furthering knowledge and advising the federal government. Functioning in accordance with general policies determined by the Academy, the Council has become the principal operating agency of both the National Academy of Sciences and the National Academy of Engineering in providing services to the government, the public, and the scientific and engineering communities. The Council is administered jointly by both Academies and the Institute of Medicine. Dr. Bruce M. Alberts and Dr. William A. Wulf are chairman and vice chairman, respectively, of the National Research Council.

Preface

This report is the work of the Panel on Modeling Human Behavior and Command Decision Making: Representations for Military Simulations. The panel was established by the National Research Council (NRC) in 1996 in response to a request from the Defense Modeling and Simulation Office of the U.S. Department of Defense. The charge to the panel was to review the state of the art in human behavior representation as applied to military simulations, with emphasis on the challenging areas of cognitive, team, and organizational behavior. The panel formed to meet these goals included experts in individual behavior, organizational behavior, decision making, human factors, computational modeling, and military simulations.

The project extended over an 18-month period. At the end of the first phase, in February 1997, the panel published an interim report (Pew and Mavor, 1997) that argued for the need for models of human behavior, summarized a methodology for ensuring the development of useful models, and described selected psychological process models that have the potential to improve the realism with which human-influenced action is represented. In the second phase of the project, the panel conducted an in-depth analysis of the theoretical and applied research in human behavior modeling at the individual, unit, and command levels. The result of that analysis is presented in this final report.

This report is intended not only for policy makers in the Defense Modeling and Simulation Office and the military services, but also for the broader behavioral science community in the military, other government agencies, industry, and universities, whose modeling efforts can contribute to the development of more realistic and thus more useful military simulations.

Many individuals have made a significant contribution to the panel's thinking and to various sections of the report by serving as presenters, consultants, and reviewers. Although all of these individuals provided valuable information, a few played a more direct role in developing this manuscript and deserve special mention. First, we extend our gratitude to Eva Hudlicka of Psychometrix Associates for her substantial contribution to the chapters on situation awareness and behavior moderators; in the latter chapter she provided draft material on modeling the effects of emotion on the cognitive activities of command decision makers. Next, we extend our gratitude to John Anderson of Carnegie Mellon University for his contributions to the discussion of ACT-R, to Stephen Grossberg of Boston University for his contribution on adaptive resonance theory, and to Stephen Deutsch of BBN Technologies, GTE Internetworking, for his work on OMAR. Finally, we offer a special thank you to David Kieras of the University of Michigan for his important insights as a member of the panel through its first phase and as a contributor of key information on EPIC for this volume.

Other individuals who provided important information and help include: Laurel Allender, Army Research Laboratory, Human Research and Engineering Directorate; Susan Archer, Micro Analysis and Design; Floyd Glenn, CHI Systems; Paul Lehner, MITRE Corporation; John Laird, University of Michigan; Ron Laughery, Micro Analysis and Design; John Lockett, Army Research Laboratory, Human Research and Engineering Directorate; Commander Dennis McBride, Office of Naval Research; James L. McClelland, Center for the Neural Basis of Cognition; H. Kent Pickett, TRADOC Analysis Center; Douglas Reece, Science Applications International Corporation; Gerard Rinkus, Charles River Analytics; Jay Shively, NASA Ames; Barry Smith, NASA Ames; Magnus Snorrason, Charles River Analytics; and Dave Touretzky, Carnegie Mellon University.

To our sponsors, the Defense Modeling and Simulation Office, we are most grateful for their interest in the topic of this report and their many useful contributions to the panel's work. We particularly thank Judith Dahmann, James Heusmann, Ruth Willis, and Major Steve Zeswitz, USMC. We also extend our thanks to Lieutenant Colonel Peter Polk for his support and encouragement during the projects first phase.

In the course of preparing this report, each member of the panel took an active role in drafting chapters, leading discussions, and reading and commenting on successive drafts. Jerome Busemeyer provided material on learning and decision making; Kathleen Carley drafted chapters on command and control at the unit level and on information warfare; Terry Connolly provided sections on decision making; John Corson provided expertise and drafted material on military needs and operations, Kenneth Funk took the major responsibility for coordinating and drafting material on integrative architectures and on multitasking; Bonnie John contributed significantly to the chapter on integrative architectures; Richard Shiffrin drafted sections on attention and memory; and Greg Zacharias drafted material on situation awareness and planning. We are deeply indebted to the panel members for their broad scholarship, their insights, and their cooperative spirit. Truly, our report is the product of an intellectual team effort.

This report has been reviewed by individuals chosen for their diverse perspectives and technical expertise, in accordance with procedures approved by the NRC's Report Review Committee. The purpose of this independent review is to provide candid and critical comments that will assist the authors and the NRC in making the published report as sound as possible and to ensure that the report meets institutional standards for objectivity, evidence, and responsiveness to the study charge. The content of the review comments and draft manuscript remain confidential to protect the integrity of the deliberative process.

We thank the following individuals for their participation in the review of this report: Ruzena Bajcsy, Department of Computer and Information Science, University of Pennsylvania; Kevin Corker, NASA Ames Research Center, Moffett Field, California; Scott Gronlund, Department of Psychology, University of Oklahoma; William Howell, American Psychological Association, Washington, D.C.; John F. Kihlstrom, Department of Psychology, University of California at Berkeley; R. Duncan Luce, Institute for Mathematical Behavioral Science, University of California at Irvine; Krishna Pattipati, Department of Electrical and Systems Engineering, University of Connecticut; Paul S. Rosenbloom, Department of Computer Science, University of Southern California; Anne Treisman, Department of Psychology, Princeton University; and Wayne Zachary, CHI Systems, Lower Gwynedd, Pennsylvania.

Although the individuals listed above provided many constructive comments and suggestions, responsibility for the final content of this report rests solely with the authoring panel and the NRC.

Staff of the National Research Council made important contributions to our work in many ways. We extend particular thanks to Susan McCutchen, the panel's senior project assistant, who was indispensable in organizing meetings, arranging travel, compiling agenda materials, coordinating the sharing of information among panel members, and managing the preparation of this report. We are also indebted to Jerry Kidd, who provided help whenever it was needed and who made significant contributions to the chapter on the behavior moderators. Finally, we thank Rona Briere, whose editing greatly improved the report.

Richard W. Pew, *Chair*
Anne S. Mavor, *Study Director*
Panel on Modeling Human Behavior and Command Decision Making: Representations for Military Simulations

Modeling Human and Organizational Behavior

Executive Summary

This report represents the findings of an 18-month study conducted by the Panel on Modeling Human Behavior and Command Decision Making: Representations for Military Simulations. For this study, the panel, working within the context of the requirements established by military simulations, reviewed and assessed the state of the art in human behavior representation—or modeling of the processes and effects of human behavior—at the individual, unit, and command levels to determine what is required to move military simulations from their current limited state to incorporate realistic human and organizational behavior.

The need to represent the behavior of individual combatants as well as teams and larger organizations has been expanding as a result of the increasing use of simulations for training, systems analysis, systems acquisition, and command decision aiding. Both for training and command decision aiding, the behaviors that are important to represent realistically are those that can be observed by the other participants in the simulation, including physical movement and detection and identification of enemy forces. It is important that observable actions be based on realistic decision making and that communications, when they originate with a simulated unit, be interpretable as the result of sensible plans and operations. A team should manifest a range of behaviors consistent with the degree of autonomy it is assigned, including detection of and response to expected and unexpected threats. It should be capable of carrying out actions on the basis of communications typically received from its next-highest-echelon commander.

In the panel's view, achieving realism with respect to these observable outcomes requires that the models of human behavior employed in the simulation be based on psychological, organizational, and sociological theory. For individual combatants, it is important to represent the processes underlying the observable behavior, including attention and multitasking, memory and learning, decision making, perception and situation awareness, and planning. At the unit level it is important to represent the command and control structure, as well as the products of that structure. Added realism can also be achieved by representing a number of behavior moderators at the individual and organizational levels. Moderators at the individual level, such as workload and emotional stress, serve to enhance or degrade performance, as reflected in the speed and accuracy of performance. Moderators at the organizational level, including the average level of training, whether standard operating procedures are followed, the level and detail of those procedures, and the degree of coupling between procedures, all affect performance. In each of these essential areas, this report presents the panel's findings on the current state of knowledge, as well as goals for future understanding, development, and implementation. The goals found at the end of each chapter are presented as short-, intermediate-, and long-term research and development needs. The report also provides descriptions of integrative architectures for modeling individual combatants. Overall conclusions and recommendations resulting from the study are presented as well. This summary presents the panel's overall recommendations in two broad areas: a framework for the development of models of human behavior, and infrastructure and information exchange. Detailed discussion of these recommendations is provided in Chapter 13 of this report.

A FRAMEWORK FOR THE DEVELOPMENT OF MODELS OF HUMAN BEHAVIOR

The panel has formulated a general framework that we believe can guide the development of models of human behavior for use in military simulations. This framework reflects the panel's recognition that given the current state of model development and computer technology, it is not possible to create a single integrative model or architecture that can meet all the potential simulation needs of the services. The framework incorporates the elements of a plan for the Defense Modeling and Simulation Office (DMSO) to apply in pursuing the development of models of human behavior to meet short-, intermediate-, and long-term goals. For the short term, the panel believes it is important to collect real-world, war-game, and laboratory data in support of the development of new models and the development and application of human model accreditation procedures. For the intermediate term, we believe DMSO should extend the scope of useful task analysis and encourage sustained model development in focused areas. And for the long term, we believe DMSO should advocate theory development and behavioral research that can lead to future generations of models of human and organizational behavior. Work on achieving these short-, intermediate-, and long-term goals should begin concurrently. We recommend that these efforts be focused on four themes, in the following order of priority: (1) collect and disseminate human performance data, (2) develop accreditation procedures for models of human behavior, (3) support sustained model development in focused areas, and (4) support theory development and basic research in relevant areas.

Collect and Disseminate Human Performance Data

The panel has concluded that all levels of model development depend on the sustained collection and dissemination of human behavior data. Data needs extend from the kind of real-world military data that reflect, in context, the way military forces actually behave, are coordinated, and communicate, to laboratory studies of basic human capacities. Between these extremes are data derived from high-fidelity simulations and war games and from laboratory analogs to military tasks. These data are needed for a variety of purposes: to support the development of measures of accreditation, to provide benchmark performance for comparison with model outputs in validation studies, to help set the parameters of the actual models of real-world tasks and test and evaluate the efficacy of those models, and to challenge existing theory and lead to new conceptions that will provide the grist for future models. In addition to the collection of appropriate data, there must be procedures to ensure that the data are codified and made available in a form that can be utilized by all the relevant communities—from military staffs who need to have confidence in the models to those in the academic sphere who will develop the next generation of models. It is important to note that clear measures of performance for military tasks are needed. Currently, these measures are poorly defined or lacking altogether.

Create Accreditation Procedures for Models of Human Behavior

The panel has observed very little quality control among the models that are used in military simulations today. DMSO should establish a formal procedure for accrediting models to be used for human behavior representation. One component needed to support robust accreditation procedures is quantitative measures of human performance. In addition to supporting accreditation, such measures would facilitate evaluation of the cost-effectiveness of alternative models so that resource allocation judgments could be made on the basis of data rather than opinion. The panel does not believe that the people working in the field are able to make such judgments now, but DMSO should promote the development of simulation performance metrics that could be applied equivalently to live exercises and simulations. The goal would be to create state-of-health statistics that would provide quantitative evidence of the payoff for investments in human behavior representation.

There are special considerations involved in human behavior representation that warrant having accreditation procedures specific to this class of behavioral models. The components of accreditation should include those described below.

Demonstration/Verification

Provide proof that the model actually runs and meets the design specifications. This level of accreditation is similar to that for any other model, except that verification must be accomplished with human models in the loop, and to the extent that such models are stochastic, will require repeated runs with similar but not identical initial conditions to verify that the behavior is as advertised.

Validation

Show that the model accurately represents behavior in the real world under at least some conditions. Validation with full generality is not possible for models of this complexity; rather, the scope and level of the required validation should be very focused and matched closely to the intended uses of each model. One approach to validation is to compare model outputs with data collected during prior live simulations conducted at various military training sites (e.g., the National Training Center, Red Flag, the Joint Readiness Training Center). Another approach is to compare model outputs with data derived from laboratory experiments or various archival sources. The panel suggests that to bring objectivity and specialized knowledge to the validation process, the validation team should include specialists in modeling and validation who have not participated in the actual model development. For those areas in which the knowledge base is insufficient and the costs of data collection are too high, it is suggested that the developers rely on expert judgment. However, because of the subjectiveness of such views, we believe that judgment should be the alternative of last resort.

Analysis

Describe the range of predictions that can be generated by the model. This information is necessary to define the scope of the model; it can also be used to link this model with others. Analysis is hampered by the complexity of these models, which makes it difficult to extract the full range of behavior covered. Thus investment in analysis tools is needed to assist in this task.

Documentation Requirements

The accreditation procedures should include standards for the documentation that explains how to run and modify the model and a plan for maintaining and upgrading the model. Models will be used only if they are easy to run and modify to meet the changing needs of the user organization. Evaluation of the documentation should include exercising specific scenarios to ensure that the documentation facilitates the performance of the specified modeling tasks.

Summary

As a high priority, the panel recommends that the above accreditation procedures be applied to military models of human behavior that are either currently in use or being prepared for use, most of which have not had the benefit of rigorous quantitative validation, and that the results of these analyses be used to identify high-payoff areas for improvement. Significant improvements may thereby be achievable relatively quickly for a small investment.

Provide Support for Sustained Model Development in Focused Areas

Several specific activities are associated with model development. They include the following:

- **Develop task analysis and structure.** Researchers and model users must continue and expand the development of detailed descriptions of military contexts—the tasks, procedures, and structures that provide the foundation for modeling of human behavior at the individual, unit, and command levels.
- **Establish model purposes.** The modeler must establish explicitly the purpose(s) for which a model is being developed and apply discipline to enhance model fidelity only to support those purposes.
- **Support focused modeling efforts.** Once high-priority modeling requirements have been established, we recommend sustained support in focused areas for human behavior model development that is responsive to the methodological approach outlined in Chapter 12 of this report.
- **Employ interdisciplinary teams.** It is important that model development involve interdisciplinary teams composed of military specialists and researchers/modelers with expertise in cognitive psychology, social psychology, sociology, organizational behavior, computer science, and simulation technology.
- **Benchmark.** Periodic modeling exercises should be conducted throughout model development to benchmark the progress being made and to enable a focus on the most important shortfalls of the prototype models. These exercises should be scheduled so as not to interfere with further development advances.
- **Promote interoperability.** In concert with model development, DMSO should evolve policy to promote interoperability among models representing human behavior. Although needs for human behavior representation are common across the services, it is simplistic to contemplate a single model of human behavior that could be used for all military simulation purposes, given the extent to which human behavior depends on both task and environment.

- **Employ substantial resources.** Improving the state of human behavior representation will require substantial resources. Even when properly focused, this work is at least as resource demanding as environmental representation. Further, generally useful unit-level models are unlikely to emerge simply through minor adjustments in integrative individual architectures.

In the course of this study, the panel examined the current state of integrated computational models of human behavior and human cognitive processes that might lead to improved models of the future. However, the current state of the art offers no single representation architecture that is suited to all individual human or organizational modeling needs. Each integrated model we reviewed implies its own architecture, and the chapters of this report on particular cognitive content areas each suggest specific alternative modeling methodologies. It is not likely, even in the future, that any single architecture will address all modeling requirements.

On the other hand, we recognize the value of having a unitary architecture. Each new architecture requires an investment in infrastructure beyond the investment in specific models to be built using that architecture. Having an architecture that constrains development can promote interoperability of component modeling modules. As applications are built with a particular architecture, the infrastructure can become more robust, and some applications can begin to stand on the shoulders of others. Development can become synergistic and therefore more efficient.

At this point in the maturity of the field, it would be a mistake for the military services to make a choice of one or another architecture to the exclusion of others. Therefore, we recommend that the architectures pursued within the military focus initially on the promising approaches identified in Chapter 3 of this report. This recommendation is especially important because the time scale for architecture development and employment is quite long, and prior investment in particular architectures can continue to produce useful payoffs for a long time after newer and possibly more promising architectures have appeared and started to undergo development. On the other hand, this recommendation is in no way meant to preclude exploration of alternative architectures. Indeed, resources need to be devoted to the exploration of alternative architectures, and in the medium and especially long terms, such research will be critical to continued progress.

Support Theory Development and Basic Research in Relevant Areas

There is a need for continued long-term support of theory development and basic research in areas such as decision making, situation awareness, learning, and organizational modeling. It would be short-sighted to focus only on the immediate payoffs of modeling; support for future generations of models needs to be sustained as well. It might be argued that the latter is properly the role of the National Science Foundation or the National Institutes of Health. However, the kinds of theories needed to support human behavior representation for military situations are not the typical focus of these agencies. Their research tends to emphasize toy problems and predictive modeling in restricted experimental paradigms for which data collection is relatively easy. To be useful for the representation of military human behavior, the research needs to be focused on the goal of integration into larger military simulation contexts and on specific military modeling needs.

RECOMMENDATIONS FOR INFRASTRUCTURE AND INFORMATION EXCHANGE

The panel has identified a set of actions we believe are necessary to build consensus more effectively within the Department of Defense modeling and simulation community on the need for and direction of human performance representation within military simulations. The focus is on near-term actions DMSO can undertake to influence and shape modeling priorities within the services. These actions are in four areas: collaboration, conferences, interservice communication, and education/training.

Collaboration

The panel believes it is important in the near term to encourage collaboration among modelers, content experts, and behavioral and social scientists, with emphasis on unit/organizational modeling, learning, and decision making. It is recommended that specific workshops be organized in each of these key areas.

Conferences

The panel recommends an increase in the number of conferences focused on the need for and issues associated with human behavior representation in military models and simulations. The panel believes the previous biennial conferences on computer-generated forces and behavioral representation have been valuable, but could be made more useful through changes in organization and structure. We recommend that external funding be provided for these and other conferences and that papers be submitted in advance and refereed. The panel believes organized sessions and tutorials on human behavior representation, with invited papers by key contributors in the various disciplines associated with the field, can provide important insights and direction. Conferences also provide a proactive stimulus for the expanded interdisciplinary cooperation the panel believes is essential for success in this arena.

Expanded Interservice Communication

There is a need to actively promote communication across the services, model developers, and researchers. DMSO can lead the way in this regard by developing a clearinghouse for human behavior representation, perhaps with a base in an Internet web site, with a focus on information exchange. This clearinghouse might include references and pointers to the following:

- Definitions
- Military task descriptions
- Data on military system performance
- Live exercise data for use in validation studies
- Specific models
- Resource and platform descriptions
- DMSO contractors and current projects
- Contractor reports
- Military technical reports

Education and Training

The panel believes opportunities for education and training in the professional competencies required for human behavior representation at a national level are lacking. We recommend that graduate and postdoctoral fellowships in human behavior representation and modeling be provided. Institutions wishing to offer such fellowships would have to demonstrate that they could provide interdisciplinary education and training in the areas of human behavior representation, modeling, and military applications.

A FINAL THOUGHT

The modeling of cognition and action by individuals and groups is quite possibly the most difficult task humans have yet undertaken. Developments in this area are still in their infancy. Yet important progress has been and will continue to be made. Human behavior representation is critical for the military services as they expand their reliance on the outputs from models and simulations for their activities in management, decision making, and training. In this report, the panel has outlined how we believe such modeling can proceed in the short, medium, and long terms so that DMSO and the military services can reap the greatest benefit from their allocation of resources in this critical area.

1

Introduction

The Panel on Modeling Human Behavior and Command Decision Making: Representations for Military Simulations was formed by the National Research Council in response to a request from the Defense Modeling and Simulation Office (DMSO). The charge to the panel was to review the state of the art in human behavior representation as applied to military simulations, with emphasis on the challenging areas of cognitive, team, and organizational behavior.

This report represents the findings of an 18-month study in which the panel, working within the context of the requirements established for military simulations, reviewed and assessed the processes and effects of human behavior at the individual, unit, and command levels to determine what is required to move the application of these kinds of models from their current, limited state to the inclusion of realistic human and organizational behavior. Based on the results of these efforts, the panel is convinced that (1) human behavior representation is essential to successful applications in both wargaming and distributed interactive simulation; (2) current models of human behavior can be improved by transferring what is already known in the behavioral science, social science, cognitive science, and human performance modeling communities; and (3) great additional progress can be expected through the funding of new research and the application of existing research in areas the panel explored.

In addition to summarizing the current state of relevant modeling research and applications, this report recommends a research and development agenda designed to move the representation of humans in military simulations forward in a systematic and integrated manner. Both the review of the state of the art and the panel's recommendations are intended to offer guidance to researchers and practitioners who are developing military simulations, as well as to those who are responsible for providing the research and development framework for future military simulation activities.

STUDY APPROACH AND SCOPE

In the first phase of the study, several panel members attended workshops and conferences sponsored by DMSO and the Simulation, Training and Instrumentation Command (STRICOM) at which leading military contractors described their efforts to model human behavior for a variety of military simulations. The panel heard a review of modeling requirements, the state of military modeling in general, and current initiatives from representatives of DMSO. Selected presentations were obtained from specialists in the modeling community. An interim report reflecting this first phase of the study was produced in March 1997 (Pew and Mavor, 1997). During the second phase of the study, the panel held more extensive discussions with military modelers and others involved in human and organizational modeling and, taking advantage of the expertise within its membership, explored the scientific domain of human behavior to identify those areas in the literature that are pertinent to military modeling problems. The panel conducted a thorough review and analysis of selected theoretical and applied research on human behavior modeling as it applies to the military context at the individual, unit, and command levels.

It should be noted that discussion among the experts working in the domain of human behavior representation ranges much more broadly than is represented by the charge of this panel. Our focus was on the technology and knowledge available for developing useful and usable models of human behavior, from the individual combatant to the highest levels of command and control. Because they are important to the generation and success of such models, we also addressed the front-end analysis required as a prerequisite for model development and the verification and validation needed to ensure that models meet their stated requirements. The state of the art in the management of simulation and modeling processes, including scenario generation mechanisms and human interfaces to the models themselves, was considered outside the scope of the panel's work. Moreover, because the panel was charged to emphasize cognitive, team, and organizational behavior, computer science and artificial intelligence models that are not associated with behavioral organizational theories were not pursued, nor did the panel focus on theories and research related to sensory and motor behavior.

WHAT IS HUMAN BEHAVIOR REPRESENTATION?

The term *model* has different meanings for different communities. For some, a model is a physical replica or mock-up; for others, a model can be a verbal/analytical description or a block diagram with verbal labels. For the panel, use of the term implies that human or organizational behavior can be represented by computational formulas, programs, or simulations. A *simulation* is a method, usually involving hardware and software, for implementing a model to play out the represented behavior over time. The term *human behavior representation* has been coined by the Department of Defense (DoD) modeling and simulation community to refer to the modeling of human behavior or performance that needs to be represented in military simulations. In this report we use the term *human behavior representation* to denote a computer-based model that mimics either the behavior of a single human or the collective action of a team of humans. The term may be used in the context of a self-contained *constructive computer simulation* that is used to simulate a battle and is run once or many times to produce outputs that reflect the battle outcomes, either individually or statistically. Or it may be used in the context of a *distributed simulation* of the behavior of selected battlefield elements that can be viewed by real crews performing in other battlefield element simulators, such as squads of individual soldiers, ground vehicles, or aircraft, so that the battle can be played out in the simulated world interactively.

Today's military services use human behavior representation for many different purposes. The main beneficiaries of improved behavior representations are the end-user communities for whom simulation has become an important tool in support of their activities. *Training simulation users* are instructors and trainees who use simulations for individual or team instruction. *Mission rehearsal simulation users* are members of operational forces who use simulations to prepare for specific missions. *Analysis simulation users* employ their simulations to evaluate alternative weapon systems, staffing requirements, doctrine, and tactics. *Acquisition simulation users* are those who use simulations to support acquisition decisions based on the anticipated performance of weapons systems. *Joint force analysis simulation users* address questions associated with improving the command, control, and communications interoperability of joint forces. In all of these domains, it has become valuable to include human behavior representation in the simulations. Of course, the scientists and engineers who will implement the models also stand to benefit from the availability of improved representations.

As the armed forces look to the future, they are attempting to identify and assess ways of effectively applying information technology, employing smart precision munitions, and integrating joint and combined operations to enhance military operations. These factors, coupled with the vision of employing military forces in an uncertain quasi-battle environment that requires information dominance to build the correct military response, add new dimensions to future battle actions. Greater importance will be placed on the ability of commanders to exercise command and control and make more precise battlefield decisions. In addition, there is increased ambiguity surrounding decisions about what military weapon systems should be developed, what joint scenarios and battle contingencies should be trained, and what doctrine or rules of engagement should be employed.

In the face of an increasing number of military contingency missions, military planners must develop a better understanding of a broader range of force employment and potential battle outcomes for which the military services have no solid basis in experience. All of these factors lead to the conclusion that in the future, models and simulations used to train military forces, develop force structures, and design and develop weapon systems must be able to create more realistic representations of the command and control process and the impact of command decisions on battle outcomes. The representations needed are ones that more accurately reflect the impact of human behavior and the decision process of friendly and enemy leaders at multiple levels of command within real-time constraints.

In constructive simulation, it is no longer sufficient simply to use the relative strength of opposing forces, together with their fire power, to represent battle outcomes. As suggested in the *Annual Report of Army-After-Next* (U.S. Army, 1997)—a forward look at the implications of the Army of 2025—future battles, fought with the benefit of all the information technology now under development, will not necessarily be won by the side with the greatest fire power. To model and predict the outcomes of future wars, it will be necessary to consider information warfare as well. This implies a need for much greater emphasis on realistic modeling of the human element in battle because the human battle participants are the focus of information utilization.

The armed services are also increasingly using distributed simulation in support of technology design and evaluation, military planning, and training goals. As suggested above, in such simulations individuals participate in war games involving multiple players, each at a simulated workstation, each acting as if he or she were taking part in a real battle with views of the other participants not unlike those that would exist on a real battlefield. In this domain, human behavior representation is used to simulate the behavior of enemy forces or collateral friendly forces when there are not enough individuals available to represent all the needed players. There is also an interest in simulating the behavior of higher echelons in the command structure regarding their orders and reactions to the progress of the battlefield operations.

The rapidly changing state of the technology poses an additional challenge. Improvements in military technology—new kinds of decision aids and automation—will change the nature of the tasks to be modeled. Not only is the state of modeling technology changing, but the behavior that is to be modeled and reflected on the battlefield will change as well.

Two primary critics will view the outputs of human behavior representation and judge how successful they are. First, players in non-real-time constructive battlefield war games will observe only the resulting movements of troops and units, attrition results, and battle outcomes. Second, participants in real-time distributed interactive battlefield simulations will see the performance of individual soldiers and higher-level units in terms of the individual and unit behavior they exhibit, the execution of plans they formulate, and the battle outcomes that result. Although explanations of how the behavior comes about may be useful for after-action reviews, they are not needed during simulation execution. Only the outcomes need to meet the expectations of the audiences that will observe them. Similarly, detailed rationales for how groups accomplish tasks are generally irrelevant. What is important is that the group behavior mirror that which is expected in the real world.

When viewed from the perspective of the simulation user (exclusive of developers), the characteristics of behavior that are visible and interpretable to the users of a simulation depend on the level of aggregation at which the behavior is presented. We consider first the individual players, either dismounted or associated with a vehicle. These individuals may be the individual combatants, ground vehicle or air system commanders, squad or platoon leaders, or commanders at a higher level. They may observe units at different levels of aggregation as well.

The most obvious behavior to be observed is the *physical movement* in the battlespace. It must be at an appropriate speed, and the path followed must make sense in light of the current situation and mission.

The *detection and identification* of enemy or friendly individual units in the human behavior representation must appear reasonable to the observer (see also Chapter 7). The visual search should depend on situation awareness; prior knowledge of the participant; current task demands; and external environmental factors, such as field of view, distance, weather, visibility, time of day, and display mode (unaided vision versus night vision goggles).

Decision-making outcomes should reflect situation awareness and real environmental conditions (see also Chapter 6). The decisions concern such observations as which way to move given the plan and the situation presented by the opposing forces; they also concern whether to shoot, seek cover (evade in the case of aircraft or ship), or retreat. Movement decisions should be consistent and coordinated with the behavior of others in the same unit. Decisions should be consistent with the currently active goals. Ideally, individuals will exhibit behavior that reflects rational analysis and evaluation of alternative courses of action, including evaluation of alternative enemy actions, given the context. In practice, in time-critical, high-stakes situations, individual decisions are more likely to be "recognition-primed," that is, made on the basis of previously successful actions in similar situations. For example, Klein et al. (1986) show how experienced fire team commanders used their expertise to characterize a situation and generate a "workable" course of action without explicitly generating multiple options for comparative evaluation and selection. In more recent work, Kaempf et al. (1996) describe how naval air defense officers spent most of their time deciding on the nature of the situation; when decisions had to be made about course-of-action plans, fewer than 1 in 20 decisions focused on option evaluation.

Representation of *communication processes* also depends on the specific purposes of the simulation, but should follow doctrine associated with the particular element. Communication needs to be represented only when it is providing relevant objective status, situation assessment, or unit status information that will affect action at the level of the unit being represented. Communication may take several forms and employ several modes, including direct verbal communication, hand gestures, radio communication, and data link. High-resolution models of small teams may require explicit representation of message content, form, and mode.

THE ROLE OF PSYCHOLOGICAL AND ORGANIZATIONAL SCIENCE

The panel believes that movement from the current state of human behavior representation to the achievement of higher levels of realism with respect to observable outcomes requires significant understanding and application of psychological and organizational science. Scientific psychology has more than a century's accumulation of data, theory, and experience in research concerning basic human abilities. Many of these results are so useful in practical domains that they have disappeared from psychology and become integrated into technology. For example, the design of high-fidelity audio equipment is based on precise measurements of human auditory abilities collected many years ago (Lindsey and Norman, 1977). Similarly, a number of other practical domains have been utilizing various aspects of psychological research. The development of practical human behavior representations for military simulations is especially intriguing because it presents an opportunity to construct and apply comprehensive models of human abilities that span the various subareas of psychology. The resulting synthesis of results and theory will not only be practically useful, but also serve as a stimulus for a broader and deeper theoretical integration that is long overdue.

In addition to a century of research, psychology also has about three decades of experience with computational theories of human ability. Prior to this time, most psychological theory was expressed as verbal descriptions of mental processes whose implications were difficult to define because of a lack of precision. The rise of information processing theory in psychology after World War II helped considerably by applying a metaphor: humans process information in a manner analogous to that of computer systems. Information is acquired, manipulated, stored, retrieved, and acted on in the furtherance of a given task by distinct mechanisms. The metaphor was taken further in the 1970s with theories and models of mental processes being expressed in terms of computer programs. By writing and running the programs, researchers could explore the actual implications of a theoretical idea and generate quantitative predictions from the theory.

Sociology and organizational science also have accumulated almost a century of data, theory, and experience in research concerning the behavior of groups of humans. As with psychology, many of these results are useful in practical domains and so have disappeared from these fields, in this case becoming integrated into operations research techniques and best management practices. For example, shop floor allocation procedures were derived from early work on scientific management. The development of practical applications of human behavior representations is exciting in this context because it presents an opportunity to construct and apply comprehensive models of units that span distributed artificial intelligence, organizational science, sociology, small-group psychology, and political science studies of power.

Following from early work in cybernetics, an information processing tradition emerged within sociology and organizational science. This movement arose more or less in parallel with that in psychology. However, unlike the movement in psychology, which focused on how the individual human acquires, manipulates, stores, retrieves, and acts on information, the movement in sociology and organizational science concentrated on how cognitive, temporal, physical, and social constraints limit the acquisition of information and the consequent actions taken by individuals and groups. The part of this tradition that focused on temporal, physical, and social constraints became known as structural theory. Sociologists and organizational theorists found further that people's opinions, attitudes, and actions are affected by whom they know and interact with and by what they believe others think of them. Social information processing theory and the various mathematical models of exchange and influence grew out of this research.

Many social and organizational theories are expressed as verbal descriptions of institutional, social, and political processes. As with such descriptions of psychological theories, the implications of these processes are difficult to determine, particularly for dynamic behavior. The primary reason it is difficult to derive a consistent set of predictions for dynamic behavior from these verbal models is that the behavior of units is extremely nonlinear, involves multiple types of feedback, and requires the concurrent interaction of many adaptive agents. Humans, unassisted by a computer, are simply not good at thinking through the implications of such complexity.

In addition to almost a century of research, sociology and organizational science have about four decades of experience with computational modeling of unit-level behavior. Most of these computational models grew out of work in information processing, social information processing, and structural theory. By writing and running these computational programs, researchers can explore the actual implications of theoretical ideas and generate quantitative predictions for unit-level behavior. Also, such models can be used to examine the impact of alterations in group size and composition on the resultant outcomes.

THE CHALLENGE

To review the state of the art in human performance modeling with specific focus on potential military applications under the purview of DMSO is especially challenging because the way the models will be used differs substantially from the goals and purposes of typical academic researchers studying and modeling human performance. Most academic researchers concerned with human performance are interested in the interplay between empirical data (experimental, field, or archival) and theory. They implement their theories through executable models so the associated detailed assumptions will be revealed, and so they can validate and evaluate the implications of those theories. Their theories are typically about specific human performance capacities and limitations, such as attention, decision making, and perceptual-motor performance. Rarely do these researchers articulate a comprehensive model of human performance that will in the aggregate reflect the behavior of real humans. Nevertheless, this is the challenge presented by the requirements of military simulations.

At the unit level, theories are typically about group performance and how it is affected by the communication and interaction among group members, procedures, command and control structures, norms, and rules. Many of these theories can be articulated as computational models. These models often illustrate the potential impact of an isolated change in procedures or structures, but they are not typically simulation models in the sense that they generate observable outputs.

The panel has been challenged by the need to focus on behavioral outcomes and to connect knowledge and theory of human behavior with realistic behavioral outcomes, rather than becoming bogged down in details of theory. However, it is our underlying belief that achieving the desired outcomes with realism and with generality requires models that are based on the best psychological and sociological theory available. In fact, the lack of such a theoretical foundation is a limitation of the current modeling efforts the panel reviewed. In the absence of theory the models are "brittle" in the sense that mild deviations from the conditions under which they were created produce unrealistic behavior and simplistic responses that do not correspond to the behavior of real individual soldiers or units. To avoid this brittleness and lack of correspondence between the model and real behavior, it is necessary to approximate the underlying structure correctly.

An example will illustrate this point. In a simulation of the behavior of a flight of attacking helicopters, the helicopters were moving out to attack. One was designated the scout and moved ahead out of sight while the others hovered, waiting for a report. The scout was shot down. Having no further instructions, the others continued hovering until they ran out of fuel. The model could obviously be fixed to eliminate this specific bug in the program by concatenating further if-then rules. However, what is really needed is a more general decision-making process for the lead pilot that can select among alternative courses of action when expected information does not become available as needed. Existing theory is the best means of defining this structure. Furthermore, as one attempts to aggregate forces and model larger units, the unique processes and theory become even more important.

The panel also examined models of learning as frameworks within which to build specific behavioral models. These models of learning are often not based on an explicit theory and use a representational framework that is broader than the specific behavior to be simulated. However, operating within such a framework may be helpful and important in minimizing brittleness.

SETTING EXPECTATIONS IN THE USER COMMUNITY

In the panel's discussions with various representatives of the user community, it became clear that there is wide variation in users' expectations of what is possible with regard to generating human behavior that is doctrinal, realistic, creative, and/or adaptive. We suspect that what can be achieved in the near term is much more limited than some of these users expect. One purpose of this study was to elaborate those aspects of model theory and implementation the panel believes are achievable now and those aspects that require significant translation of scientific theory and principles before being developed as components of computer-based behavioral models, as well as those aspects of behavior for which the behavioral/social science community has inadequate knowledge for use in developing realistic models in the near future. In addition to presenting an approach to modeling methodology, including both model development and model validation, the panel's goal was to set forth in general terms the theoretical and operating principles of models that are applicable to human behavior representation, to describe specific applications of these theories and principles, and to identify the most promising paths to pursue in each modeling area. Much work remains to be done. There is an enormous gap between the current state of the art in human and organizational modeling technology on the one hand and the military needs on the other.

Subsequent chapters examine the potential psychological and sociological underpinnings of extensions to both the approaches and content of such models. These chapters represent the collective judgment of the panel concerning representative promising areas and approaches for expanding the models' behavioral content. Given the scope of psychological and sociological inquiry, it is likely that another panel at another time would put forth an equally appropriate, overlapping but different set of areas and approaches. What is presented here reflects this panel's expertise and collective judgment.

A fundamental problem that faces the human behavior representation community is how to determine which of the many modeling requirements will make a difference in the resultant quality of the models, based on the intended use of the simulation. As the panel deliberated, it became clear that a consultative behavioral science panel cannot set these priorities without much more experience in dealing with the specific concerns of the military community. The panel may be able to say which requirements will produce models that behave more like real humans, but this is a different set of priorities from the requirements that will produce models likely to be perceived by simulation users as being more like real individual combatants and military units. If one asks program managers or subject matter experts, they will say they need all the fidelity they can get, but this is not a helpful response in an environment of limited resources where design decisions involve tradeoffs among a set of desirable simulation goals. It is just not known which of the many improvements in human behavior representation will really make a difference in the way a modeled combatant will be viewed as regards meeting the expectancies of the opposing force and minimizing the ability to "game" the simulation. This issue is analogous to the traditional problem of simulation fidelity. Analysts would like to have high fidelity only where it matters, but no one has shown clearly just what aspects of fidelity matter. The situation is no different with human behavioral representation.

ORGANIZATION OF THE REPORT

Chapter 2 characterizes the current and future modeling and simulation requirements of the military and reviews several existing military modeling efforts. The central portion of the report comprises chapters that focus on the science and technology of human behavior representation. Chapter 3 provides a general review of several integrative architectures for modeling the individual combatant. Each review includes a brief discussion of how the architecture has been applied in the military context. Chapters 4 through 9 are devoted to an analysis of the theory, data, and state of modeling technology of individual human behavior in six key areas: attention and multitasking (Chapter 4), memory and learning (Chapter 5), decision making (Chapter 6), situation awareness (Chapter 7), planning (Chapter 8), and behavior moderators (Chapter 9). The emphasis in these chapters is on the current state of research and development in the field under review. To the extent possible, the focus is on the behavioral theory on which understanding of cognitive mechanisms is based. However, in the chapters on situation awareness and planning, where less behavioral theory exists, we focus on the strengths and weakness of current modeling approaches. Chapters 10 and 11 address issues and modeling efforts at the organizational level: Chapter 10 covers command, control, and communications (C^3), whereas Chapter 11 deals with belief formation and diffusion, topics of particular interest in the context of information warfare. Chapters 3 through 11 each conclude by presenting conclusions and goals for the short, intermediate, and long terms in their respective areas. Chapter 12 presents general methodological guidelines for the development, instantiation, and validation of models of human behavior. Finally, Chapter 13 provides the panel's conclusions and recommendations regarding a programmatic framework for research, development, and implementation and for infrastructure/information exchange.

2

Human Behavior Representation: Military Requirements and Current Models

This chapter examines the military requirements for human behavior representation and reviews current military modeling efforts in light of those requirements. The first section identifies key military modeling requirements in terms of levels of aggregation. These requirements are then illustrated and elaborated by a vignette describing a tank platoon in a hasty defense. This is followed by a discussion of the types of military simulations and their uses. The final section reviews selected military models, focused on providing computer generated forces. The annex to this chapter reviews other current models used by the various services that are briefly referred to in the chapter.

MILITARY/MODELING REQUIREMENTS

The Under Secretary of Defense for Acquisition and Technology has set as an objective to "develop authoritative representations of individual human behavior" and to "develop authoritative representations of the behavior of groups and organizations" (U.S. Department of Defense, 1995:4-19 to 4-21). Yet presentations made at three workshops held by DMSO, formal briefings to the panel, and informal conversations among panelists, Department of Defense (DoD) representatives, and DoD contractor personnel all suggested to the panel that users of military simulations do not consider the current generation of human behavior representations to be reflective of the scope or realism required for the range of applications of interest to the military. The panel interprets this finding as indicating the need for representation of larger units and organizations, as well as for better agreement between the behavior of modeled forces (individual combatants and teams) and that of real forces; for less predictability of modeled forces, to prevent trainees from gaming the training simulations; for more variability due not just to randomness, but also to reasoned behavior in a complex environment, and for realistic individual differences among human agents; for more intelligence to reflect the behavior of capable, trained forces; and for more adaptivity to reflect the dynamic nature of the simulated environment and intelligent forces.

Authoritative behavioral representations are needed at different levels of aggregation for different purposes. At various times, representations are needed for the following:

- Individual combatants, including dismounted infantry
- Squad, platoon, and/or company
- Individual combat vehicles
- Groups of combat vehicles and other combat support and combat service support
- Aircraft
- Aircraft formations
- The output of command and control elements
- Large units, such as Army battalions, brigades, or divisions; Air Force squadrons and wings; and Navy battle groups

Representations are needed for OPFOR (opposing forces or hostiles), BLUFOR (own forces or friendlies) to represent adjacent units, and GRAYFOR (neutrals or civilians) to represent operations other than war and the interactions among these forces.

EXAMPLE VIGNETTE: A TANK PLATOON IN THE HASTY DEFENSE

To illustrate the scope of military model requirements, the panel prepared a vignette describing the typical activities of an Army platoon leader preparing for and carrying out what is referred to as a "hasty defense"—a basic military operation in which a small unit of 16 soldiers manning four M1 tanks is participating as a part of a larger force to defend a tactically important segment of the battlefield. The platoon leader's tasks include planning the defense, making decisions, rehearsing the mission, moving to and occupying the battle positions, and conducting the defense. This vignette is realistic as regards what is required of such an Army unit. Clearly none of the currently known modeling technologies or methods for representing human behavior can even come close to mimicking all the behaviors exhibited in this vignette, nor would they be expected to do so. The vignette is intended to provide an overview for the reader who is unfamiliar with the kinds of military operations that are typically trained. Annotations appearing throughout link various elements of the vignette to the discussion in later chapters of the report.

While the planning actions described do not need to be represented in detail, the results of those actions do—since they lead the platoon leader to decide where to locate his forces and what specific orders to give. Command and control models will require the generation of such a set of orders (see Chapter 10). Similarly, the detailed analysis of specified and implied tasks and the evaluation of courses of action do not typically need to be modeled directly, but the outcomes of these activities—the actual behavior of the individual soldiers responding to the resulting orders—do need to be represented. These observations may help clarify why the panel believes that at some level of detail, the actual behavioral and organizational processes underlying the platoon leader's orders and the soldiers' behavior must be addressed if any degree of realism in the observable results is to be achieved. Finally, some of the activities described in the execution phase of the battle are reactive in that the platoon leader and elements of his platoon change their behavior in response to what they see and hear during contact with the enemy. It is this kind of reactive behavior that cannot be modeled unless behavioral and organizational processes are represented at a deeper level than a script to be carried out by each soldier.

It is important to note that this vignette touches on only a fraction of the command and control decisions that occur throughout the levels of command as a battle starts to unfold. Enemy identification, fratricide, casualty evacuation, target identification between units, target handoff, the effects of stress and fatigue, the level of destruction desired, and civilian casualties are but a few of the issues that affect combat decisions and the actions of military decision makers.

Enemy and Friendly Situation

Enemy Situation

A small country friendly to the United States is under the threat of attack from its northern neighbor. An enemy armored brigade has deployed along the country's current northern border. The lead battalion of an enemy force has pushed armored scout elements across the border into the country's territory to gain intelligence on its expected defensive strong points. Enemy forces have the capability to attack with one full-strength armored brigade and two motorized infantry brigades at any time.

Friendly Situation

In support of the friendly country, a U.S. armor brigade from the 1st U.S. Armor Division has occupied assembly areas south of a proposed line of defense and is preparing to occupy a series of defensive positions to stop a threatened attack by the enemy forces. The 3rd Battalion 33rd Armor has the northernmost defensive strong point in the country.

Planning the Defense

Initial Planning

The planning process described below parallels the planning considerations presented in Chapter 8 of this report. After the 1st platoon leader, A Company, 3-33 Armor, receives his unit's mission from the company commander, he begins his planning process with an analysis of the mission and situation. He starts with a basic mission assessment process that looks at mission, enemy, terrain, troops, and time available (METT-T). METT-T is a standardized thought process taught to leaders as a way to facilitate the planning process. [*Potentially useful theories and models of planning are discussed in Chapter 8.*]

Mission A forward defensive position along the expected avenue of enemy attack is defended by A Company. 1st platoon is to engage the enemy lead armor elements from a forward battle position, then withdraw to the main line of defense occupied by A Company (-). 1st platoon is to occupy its forward battle position (BP1) under cover of darkness.

Enemy 1st platoon is expected to encounter an enemy force of up to 10 T-80 tanks, possibly reinforced with a mechanized infantry platoon of 30 soldiers.

Terrain The ground is wooded, rolling terrain and provides good concealment with folds in the earth. Long-range fields of fire are possible from high ground at the edge of the wood lines. Trafficability is high, forestation does not prevent tank and BMP[1] movement off road, and all streams in the area are fordable. [*The terrain configuration affects intervisibility between friendly and enemy forces. Realistic models require representation of what each soldier, vehicle commander, and driver can detect and identify as the battlefield is scanned, as discussed in Chapter 7.*]

Troops 1st platoon is at 100 percent strength with experienced tank commanders, morale is high, troops are rested, and all four M1 tanks/systems are functioning. [*Fatigue affects the overall level of performance of the troops. See Chapter 9 for a discussion of this and other variables that moderate behavior.*]

Time 1st platoon has little time after moving to its forward position before the expected attack by the enemy. The platoon has no time for an actual reconnaissance and will do a thorough map reconnaissance of the terrain instead.

---

[1]BMP = boyevaga mashina pakhoty, a Soviet armored personnel carrier.

Initial Schedule

The 1st platoon leader develops an initial schedule that maps out the time he has available. His tentative schedule is to issue a warning order to his tank commanders immediately and issue the platoon operations order after finishing the leader rehearsal with the company commander. At 1900 hours, the platoon leader must be ready for the company commander's rehearsal with a proposed platoon plan. He will let the platoon continue its sleep plan (one-third of the platoon awake on guard, the other two-thirds asleep) until after the commander's rehearsal. He formulates a tentative plan based on his METT-T analysis and his knowledge of the platoon's capabilities. Upon returning to the platoon, he briefs his platoon sergeant, issues a warning order to his tank commanders, and begins to prepare his plan. [*See Chapter 8 for a discussion of potential planning models.*]

Tentative Plan and Schedule

To complete his tentative plan, the platoon leader conducts a detailed mission analysis. In this analysis, he assesses his mission and the intent of the commander two levels higher. He determines the platoon's purpose and mission, assesses the unit's mission and its execution to identify specified and implied tasks, assesses time constraints and limitations, and makes a tentative time schedule. [*Planning models are discussed in Chapter 8.*]

Estimate of the Situation (Situation Awareness)

The use of an estimate of the situation reflects the platoon leader's need to gain situation awareness. Gaining situation awareness includes performing assessments of terrain and weather and the enemy and friendly situations. [*The impact of situation awareness is a significant cognitive factor and has a major effect on decision making. Situation awareness is discussed in Chapter 7.*]

Assessment of Terrain and Weather This assessment includes observation and fields of fire, cover and concealment, obstacles, key terrain, and avenues of approach (OCOKA), as well as weather.

- **Observation and Fields of Fire.** The rolling nature of the ground and the presence of the trees limits observation from the tops of the hills. There will be no time to clear fields of fire. Optical sights in the tanks will allow infrared detection of the enemy, but the enemy will not necessarily move on the open trail. The terrain and trafficability increase security concerns for the platoon and accomplishment of the mission. Local security will be more difficult if the platoon is split into two sections in different locations. Dismounted listening posts/observation posts will be necessary during hours of darkness. The platoon leader makes a note to request a squad of infantrymen for dismounted security. [*This element of the vignette highlights some of the complexities of military planning that will be challenging to model.*]

• **Cover and Concealment.** Cover and concealment, represented by the rolling terrain and the trees, affect both sides equally. The enemy is likely to approach the platoon's position without realizing the platoon is there. However, dismounted enemy soldiers will also be able to approach the platoon with little chance of detection if the platoon reveals its position. The platoon leader makes a note that if he is not given infantry support, he will have to dismount loaders to man listening posts during hours of darkness and turn off the tank engines once in the forward BP.

• **Obstacles.** There are no major obstacles in the defensive sector. Some limited obstacles may be forward of BP1 (see Figure 2.1). Initially, the platoon will move on a trail and then travel cross country to the forward BP. The enemy could use the main road that runs from the northeast into BP1 or the trail running from the north, or move in the cross-country compartments. Natural impediments in the form of creek beds and fallen trees should not constitute real obstacles. The trees are far enough apart to allow the tanks to move between and around them, and the creeks are fordable to wheels and tracks. Cross-country movement would reduce the speed of movement of either side. [*Military units coordinate their movement based on their organizational structure. It is not sufficient to model the behavior of each individual soldier. Rather, a coherent model of the larger unit-level behavior is required, and it will be governed by unit orders, rules, and procedures. See Chapter 10.*]

• **Key Terrain.** BP1 is north of a road and trail intersection and includes a series of hills that allow long-range observation. The main road and the trail are two likely enemy avenues of approach. The high ground along and on either side of these features is key terrain.

• **Avenues of Approach.** The road and trail that intersect south of BP1 are the two obvious avenues of approach for both sides. The enemy may fear land mines on the trail and road and therefore may avoid using them. On the other hand, the enemy forces will be reacting to indirect fires and can be expected to use the fastest route to their objective. [*Models of the red forces will need to be responsive to actions of the blue forces. Modeling interactive forces is a much greater challenge than simply scripting the behavior of individual combatants independently of what the other side is doing.*]

• **Weather.** The weather forecast is continued good visibility with 10 percent chance of rain. Visibility could be restricted during periods of heavy rain. In addition, heavy rain would make off-road movement more difficult and prolong the length of time required for the platoon to reach the BP and/or slow the movement of the enemy.

Assessment of Enemy Situation The enemy tank elements are expected to attempt to fight through any resistance. They will try to attack with overwhelm-

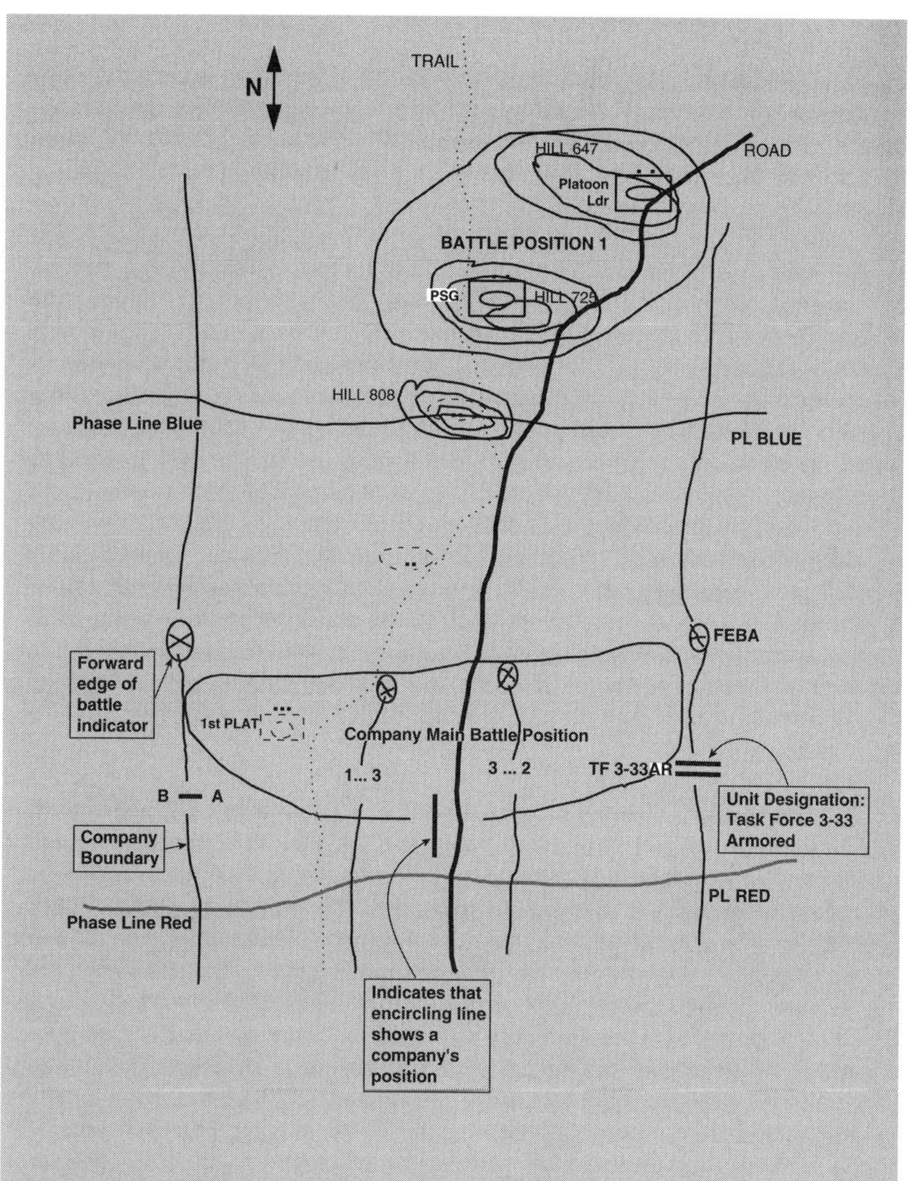

FIGURE 2.1 Scenario map. See text for discussion.

ing firepower. Other company-sized enemy elements will simultaneously be attacking the other friendly forces. Initially, the platoon can expect an enemy tank platoon of three T-80 tanks, followed by a second and third tank platoon and possibly a mechanized infantry platoon of approximately 30 men in three BMP personnel carriers. [*See Chapter 6 for a discussion of relevant modeling efforts.*]

Assessment of Friendly Situation The task force is defending with two armored companies and one mechanized company. It has priority of fires from the brigade's direct support artillery battalion and can expect up to six attack helicopters in support. The brigade has an attack aviation battalion under its control and will use it to reinforce forward ground units once the location of the main enemy has been determined. The best use of attack helicopters is against enemy armor. The brigade commander wants to use the attack helicopters to engage the enemy as early and as far forward as possible and at long range. Sending a tank platoon forward of the main battle area will provide early warning and give some protection to the helicopters. During the conduct of the defense, helicopters could be forced to leave because of fuel requirements or enemy ground fire. The task force commander wants a platoon from A Company to be pushed forward (north) to support brigade air attack plans and to cover the northern avenues of approach leading into the task force sector. [*These are examples of some of the principles and rules that could be used to guide models of planning.*]

Platoon Leader's Concerns from OCOKA and Enemy/Friendly Assessments The platoon leader is concerned about the friendly attack helicopters. His platoon will initially be forward of the main battle areas. He is fearful that if the enemy attacks during hours of darkness, the attack choppers could become disoriented and mistake his unit for an enemy force. The company commander has already voiced this concern at the battalion operation order (OPORD) and has returned with coordinating instructions for the platoon leader. The platoon leader is going to mark his vehicles with laser-reflecting tape that will allow his unit to be differentiated from the enemy through heat signatures. He is also concerned with the possibility that the helicopters will be pursuing the enemy and will literally run into his force when he is attempting to engage the enemy. He has been given the radio frequency for coordination with the attack helicopters, and a "no-fire line" has been coordinated through artillery channels forward of BP1. The no-fire line is not easily identified from the air. Finally, the platoon leader is concerned that the enemy may attack in greater strength than expected. If this is true, it may be difficult to maintain needed maneuver flexibility, and his ammunition supply could become a problem. His platoon will have to protect itself while operating forward of the company. Once the enemy starts to close on BP1, the enemy's position will be identified and targeted. The decision point for when to initiate the unit maneuver to the rear after enemy engagement is critical. The platoon leader is not comfortable with the length of time his elements can and should stay in their initial positions. He decides to limit the rate of ammunition expenditure within the platoon during the initial phase of the enemy engagement. [*This element of the vignette illustrates the complexity of battlefield decision making. See Chapter 6 for a discussion of possible modeling frameworks and behavioral decision-making phenomena that need to be taken into account.*]

Development of Courses of Action

Now the platoon leader must develop potential courses of action. He has three basic courses of action in mind that he will discuss with the company commander.

Course of Action #1—One Battle Position Occupied in a Diamond Formation The platoon will occupy a BP in a diamond formation, with the guns of the four tanks pointed in all four cardinal directions, covering all potential avenues until the enemy approaches. Based on the direction of the enemy advance, one tank will engage immediately, and two of the tanks will relocate as necessary to engage. The fourth tank will provide rear security and will overwatch the withdrawal of the other tanks after they engage the enemy. The platoon will move from this position after destroying the lead tank platoon and causing the enemy force to deploy, and then maneuver to the company main BP.

Course of Action #2—Two Battle Positions Occupied by Sections Covering Two Avenues of Approach The platoon will split into two sections—one with the platoon leader and the other with the platoon sergeant. The platoon leader will occupy the north edge of BP1 along the most likely avenue of approach. The platoon sergeant will occupy BP1 farther south and southwest of the platoon leader's section to cover a second avenue (the trail) and support the platoon leader's section by fire. Once the enemy is engaged and deploys, the two sections will support each other as they maneuver to the rear to reach the company main BP.

Course of Action #3—One Platoon Battle Position on Line The platoon will occupy its position on line, with all four tanks in a platoon BP to protect the most likely avenue of approach. However, once the enemy is engaged, the platoon will displace by sections, covering each other until they all reach a position at the center of the company main BP. [*The formulation and evaluation of alternative courses of action are generally considered part of the human decision-making process. They are sometimes undertaken formally, but more often are accomplished only informally or within the head of the commander and/or his staff.*]

The Decision Process

[The initial decision process during the planning of the defense reflects, in a simple form, the types of considerations and theories presented in Chapter 6 of this report. The challenge in military simulations is to bring the effects of realistic leader planning and execution decisions into complex models in which multiple levels of decision outcomes are embedded.]

Analysis of Courses of Action

Course of Action #1 This course of action is flexible and can accomplish the mission of blocking the initial enemy attack regardless of the direction from which the enemy approaches. It offers the potential for engaging the enemy before reaching BP1 regardless of what route the enemy chooses. At the very least it guarantees early warning for the task force elements in the rear. This course of action relies on the notion that the enemy will use any one of three avenues, including the open road and trail, in advancing toward their objectives. This course of action leaves the platoon in one BP for support; however, two of the tanks could require repositioning to bring effective fires on the enemy, depending on the actual direction of enemy approach. The deployment may not allow maximum fires at first contact, but it does support flexibility.

Course of Action #2 The enemy is most likely to move on one of the high-speed avenues of approach, with no bounding maneuver until contact is made. This course of action ensures that two tanks can immediately engage any enemy forces that approach on the road (the most likely avenue of approach) and on the trail (the other high-speed approach). It also provides overwatch to the forward tank section. This course of action provides less local security than course of action #1 since the sections are split, and the separate sections are thus more open to dismounted attack.

Course of Action #3 This course of action focuses all the platoon's combat power on the road, the most likely high-speed approach. The platoon is susceptible to flank attack in its exposed position; however, all tanks are mutually supporting. Being in one forward position requires a planned movement by the platoon at a critical time to ensure that it is not decisively engaged. It also requires selection of alternate positions for the platoon to cover the trail and potential cross-country approaches.

Comparison of Courses of Action

The platoon leader decides the most important criteria for choosing a

TABLE 2.1 Course of Action (COA) Rating Chart

Criterion	Rating COA 1	COA 2	COA 3
1 Long-range enemy engagement	3	2	1
2 Survivability of the platoon	2	1	3
3 Simplicity of executing actions	3	2	1
4 Security	1	3	2
5 Flexibility	2	1	3
Total Score	11	9	10

course of action are (1) engagement of the enemy at long range, (2) survivability of the platoon when engaged, (3) simplicity of executing battle actions, (4) security of the platoon prior to attack, and (5) flexibility of the dispositions. He compares each course of action against these criteria. He adopts a simple rating of 1 for the best course of action for a criterion, 2 for the next best, and 3 for the least desirable, as shown in Table 2.1.

Selecting a Course of Action

The platoon leader decides to go with course of action #2—which has the lowest total score (Table 2.1)—for the commander's rehearsal. He brings to the rehearsal questions regarding coordination with the attack helicopters, his ability to move in and around BP1, and the availability of infantry support.

The Final Plan

After the company commander's rehearsal, the platoon leader consults with his platoon sergeant and tank commanders. He has learned that there is no infantry support available for his mission and that he is confined in his movement until the platoon arrives in the vicinity of the company's battle position. He is informed that the attack helicopters are aware of the no-fire line and will not engage the enemy south of that line. He is still uncomfortable about the attack helicopters operating so close to the platoon. After incorporating the new information from the company commander's rehearsal, he finalizes the plan. He then issues the platoon OPORD to his tank commanders and gunners.

Mission Rehearsal

In his timeline the platoon leader has roughly 1.5 hours for rehearsal. He plans to rehearse right up to the moment the platoon marshals for movement to the start point. This is not very much time. He sets priorities for the rehearsal:

- Movement to the battle position
- Actions on enemy contact en route to BP1
- Occupation of the BP—emphasis on security and camouflage
- Actions on enemy contact in the BP
- Planned maneuvers to the rear after enemy engagement

[The behavior of the unit can be expected to change as a result of this rehearsal. Learning will take place. Representing such changes will require models of the learning process, such as those discussed in Chapter 5 for the individual combatant and in Chapter 10 for the unit.]

Movement to and Occupation of BP1

The movement phase begins with the platoon marshaling to move to the start point. Movement to the company release point near the company main BP will take about 2 hours. 1st platoon moves last in the A Company column. In the vicinity of the main company BP, 1st platoon receives a radio message to move forward to BP1. The platoon leader radios his tank commanders to turn left in a column formation, with the platoon leader's tank in the lead. He observes the area to his front with a night vision sight for signs of the enemy as they move. His gunner observes the terrain to the left with a night thermal sight, and the following tank observes to the right. As the terrain opens up, the platoon leader radios his tank commanders to shift to a wedge formation. This maneuver will bring more firepower to the front and still provide some flank security within the platoon. The platoon leader knows the platoon is proficient in the wedge formation, but at night it is more difficult to maintain this formation, and their rate of movement is slower. Possible enemy contact en route is the platoon leader's primary concern with this phase of the operation. His mission requires establishment of a blocking position at BP1, so it will not be enough to encounter enemy resistance and simply break contact. In the reaction drills in his rehearsal, the platoon leader emphasized fighting through any enemy.

As the platoon approaches the southern edge of BP1, the platoon leader reports to the company commander, "Crossing phase line blue with negative enemy contact." On the move, he directs his platoon sergeant and his tank section to occupy fighting positions on the high ground oriented to the trail to the north, but positioned with observation to the northeast. The platoon leader now moves astride the road to the northeast, with his wingman covering his left rear. As the platoon leader continues his move to the northeast, his platoon sergeant signals "set in position." As the platoon leader approaches the piece of high ground he has planned to occupy to block the road, he orders his wingman to stop while he makes a reconnaissance of the area. He edges forward to observe for any enemy before breaking the crest of the hill. On foot, he locates the exact positions that give his section cover to the northeast, but long-range fields of fire over the road. He directs the section into position—"set." The tanks are now in hide positions. They can observe any enemy approach at long range, but they are not exposed.

Now comes the battle of nerves. The platoon leader uses the loader from each tank as forward security, observing for dismounted infantry. The gunner mans the night thermal sight, scanning at long range for signs of enemy movement. The platoon leader, tank commanders, and drivers improve their positions by camouflaging their tanks and locating alternate firing positions. The wait continues toward dawn, with situation reports to the company every 30 minutes. Where is the enemy? [*Intervisibility, detection, and identification are paramount here, but the parameters of a model that includes these elements for nighttime conditions will be quite different than those for a daytime engagement. The way the platoon leader allocates his attention and time to these tasks and the tasks associated with conducting the defense are covered in Chapter 4, on attention and multitasking.*]

Conduct of the Defense: The Battle Decision Process

[*Execution of the battle plan represents another human decision-making process. This process differs from the deliberate decision making discussed earlier. It is rushed and filled with rapid cognitive perceptions by the decision maker trying to sense rapidly changing conditions that must be confronted. This process does not always result in logical or sound decisions. This process must also be reflected within military simulations—with the same models as those presented in Chapter 6.*]

In 1st platoon's battle plan, the intent of the platoon leader at BP1 is to attrite the enemy at long range using indirect fires and attack helicopters first and then long-range direct fires. He has instructed his tank commanders to engage enemy tanks first and then enemy personnel carriers. From their hide positions, the tank commanders will pop up into firing positions and then relocate to alternate hide or firing positions. It is now 0535 hours and starting to become light. A radio message is received from the platoon sergeant: "Enemy tanks! Three, no! six T-80s approaching along the trail in column at grid NA 943672" (4.5 kilometers to the sergeant's front)." The platoon leader searches to his front—no signs of the enemy on the road. The enemy is coming directly from the north, not along the road! The platoon leader calls the spot report to company and requests artillery fires and attack helicopters. The platoon leader and his section must relocate. He radios to his platoon sergeant, "Am moving to your location! Will go into overwatch 500 meters to your south VIC Hill 808. Over!" His platoon sergeant answers, "Roger Plan B!"

Now everything starts to happen rapidly. In minutes, artillery is pounding the grid. The platoon leader's section is racing to positions on Hill 808. Two gun ships arrive, hovering behind Hill 808, and come up on the platoon's radio frequency. They have the enemy in sight. The platoon leader's section moves into hide positions forward of the crest on Hill 808. One enemy tank is destroyed by artillery. The attack birds pop up, and two Longbow missiles streak toward the first two enemy tanks. Two enemy tanks are destroyed! The platoon sergeant reports six more T-80s moving through the woods east of the trail. The enemy is now 3,000 meters from the platoon sergeant's tank section. The tank commanders pick their targets, and the section moves forward to fire. Two more enemy tanks are destroyed. The platoon leader orders the platoon sergeant to relocate to the high ground south of Hill 808. The helicopters report to the platoon leader that they have spotted four BMPs and three more tanks west of the trail 3,000 meters from Hill 808, and they see signs of a large force deploying farther to the north. The choppers pop up once more and fire two Longbows and then duck down and race toward the south. The platoon leader reports a total of five enemy tanks destroyed and calls for smoke to cover the movement of the platoon sergeant's section. Soon he will be the front line. He hears the platoon sergeant's section racing past Hill 808. He spots two enemy tanks coming out of a wood line 2,000 meters to his front. He alerts the gunner closest to this position and radios to his wingman. They pull up, fire, and then immediately pull back. The platoon leader orders his wingman to withdraw, calls for the platoon sergeant to cover his move, and radios to company: "Two more enemy tanks destroyed at grid 938626! Enemy forces deploying to my north. Request permission to withdraw to main BP." [*The stresses of too much workload and time pressure both tend to degrade performance from ideal levels. In modeling, these stresses might be treated as moderator variables (see Chapter 9) or, where sufficient theory exists, incorporated into models of situation awareness and decision making (see Chapters 7 and 6, respectively).*]

With permission granted, the platoon leader requests artillery at the enemy locations and requests smoke on the northern slope of Hill 808. In minutes, his section is racing toward the platoon sergeant's overwatching section. He has destroyed the initial forces sighted and has caused the enemy to deploy. The platoon will move by echelon to supplementary positions and potentially experience more engagements before reaching the company main BP. As the platoon moves toward the company main BP, it faces another critical set of events—recognition and passage. With communications and planned signals in place, the platoon's final movement can be covered by the other platoons as they approach the main BP. This could have become a chaotic situation if the platoon and company had not trained for such an operation. When the platoon reaches the main BP, another set of decisions and actions awaits. The platoon leader and platoon sergeant must set the platoon in its positions, check its ammunition and fuel status, check for battle damage, report to company, and be ready for the enemy's next action.

This is the end of a small unit leader tactical vignette, but it represents the beginning of the process of describing models and promising research for more accurately depicting human actions and command and control decision outcomes from the tactical, to the operational, to the strategic level within military simulations.

MILITARY SIMULATIONS: TYPES AND USES

As noted in the introduction, there are at least three major categories of simulation users in the military—those that are training individual combatants or leaders and teams; those that are performing analyses of systems, doctrine, and tactics for purposes of acquisition and advanced development; and those that are addressing questions associated with improving command and control and the interoperability of joint forces. Because training of the force and its leaders is the largest mission of the military in peacetime, it has received the most attention and the highest level of funding. We begin this section with some brief definitions of the types of simulations used in the military services. We then discuss the above three major areas of application. In the final section we examine efforts to create models that can be used to introduce human behavior into military simulations.

Definitions

As noted in Chapter 1, we define a model as a physical, mathematical, or otherwise logical representation of a system, entity, phenomenon, or process and a simulation as a method for implementing a model over time. The widely used military taxonomy for classifying types of simulation includes three categories: live, virtual, and constructive. Depending on the type of simulation, human behavior may be introduced by computational models or by actual human participants. Since humans play a central role in the outcomes of combat engagements, the panel believes that when human roles are involved and human participants are not used, it is important to provide models of human behavior in order to ensure realistic outcomes for training and analysis.

The traditional approach to maintaining readiness and testing new employment concepts has been to conduct field exercises or live maneuver exercises, also referred to as *live simulations*. Live simulations are least dependent on accurate models of human behavior because in these exercises, real humans operate real equipment. The aspects that are simulated involve weapon firing and how hits of different types of ordnance are measured and recorded. All of the services have large investments in major training centers that are designed to exercise battalion/squadron and larger formations in a realistic combat environment. Exercises conducted at major training centers are often joint exercises that include participation by multiple services.

Limited physical maneuver space and the cost of transporting units and operating major weapon systems constrain the number of units that can participate in live simulations and the duration of such exercises. Additionally, the time between exercises/rotations is too long for units to maintain the required level of combat readiness. In the future, there will be a continuing need to use live simulations. However, the above shortcomings in effectively meeting military readiness needs with regard to physical space, cost, and time considerations will continue to become more limiting.

Virtual simulations are characterized by real humans operating simulated equipment in simulated environments. These simulations use a full range of multimedia technology and computer software to replicate weapon and system functions in real time with a suitable level of fidelity for interactive training with humans in the loop. The requirement for models of human behavior arises from the need to represent intelligent opponents and collateral forces. In some cases, opposing forces are played by real humans; in other cases, a human controller represents the decision making and tactics of enemy commanders and the execution of battle movements, and engagement behavior is simulated to represent doctrinal maneuvers. Virtual simulations are used for individual combatant, team, and leader training. Networked virtual simulations (sometimes referred to as distributed interactive simulations [DIS]) are used to engage trainees located in physically remote distributed sites in a common combat exercise. In these cases, simulators representing units are linked electronically. Although these simulations hold the promise of creating highly effective training environments, they are pushing the limits of available simulation technology, and simulators that apply this technology are costly.

The Army's close combat tactical trainer (CCTT), discussed in detail later in the chapter, is an example of a virtual simulation. The CCTT program includes company and battalion sets of M1 tank simulators, M2 Bradley fighting vehicle simulators, and AH64 Apache helicopter simulators, and now has semiautomated force representations of infantry soldiers/dismounted units. CCTTs can be networked from multiple remote locations to execute an integrated battle scenario with effective simulation of maneuvers and battle engagements, command and control, and logistics support. They are operated as human-in-the-loop, fully crewed systems.

Finally, *constructive simulations* involve simulated people operating simulated equipment, usually not in real time. Real people provide inputs that set limiting values on parameters, but they are not involved in determining the outcomes. All human behavior in the event flow comes from models. A few simulation systems have a feature that allows those running the simulation to freeze the action periodically in order to make fine adjustments to the dynamics. Variability in the behavior of systems and humans can be introduced by stochastic processes. The level of human behavior representation in these systems varies widely, but even the best of them assume ideal human behavior according to doctrine that will be carried out literally, and rarely take account of the vagaries of human performance capacities and limitations.

Within the military today, constructive simulations are the largest and most broadly applied type of simulation. Each of the services has developed several simulations of this type, many of which support multiple purposes, including planning, training, force development, organizational analysis, and resource assessments within the services. Examples of multipurpose constructive simulations include JANUS (Army), naval simulation system (NSS) (Navy), advanced air-to-air system performance evaluation model (Air Force), and joint warfare system (JWARS) (Joint Services). These and other selected models and simulations are briefly described and categorized in the addendum to this chapter.

The linking of constructive and virtual reality simulations into federated networks is the newest and most rapidly expanding application of simulations in the DoD arena. Advances in this area were born of the necessity to reduce costs associated with the development of training and weapon systems. All of the services are seeking means of maintaining the fighting readiness of units and large combat organizations that cannot be deployed or exercised in peacetime to the degree required because of environmental and cost constraints. In addition, in the joint and combined arena, the military services and joint service commanders have a range of requirements that involve the development of command and control doctrine, force readiness, and procedures for joint service task forces and combined international coalition forces that may best be accomplished with the support of simulation.

Training Applications

Individual Combatant, Unit, and Leader Training

As noted above, training is a central concern for the military services. Key objectives of the Army's use of simulations in this arena have been to decrease training time, decrease the cost of training, and increase the realism of training events. The Army uses constructive simulations such as JANUS and Corps Battle Simulation (CBS) to train leaders and other decision makers. In the last decade, with assistance from the Defense Advanced Research Projects Agency (DARPA), the Army has developed virtual system simulators for use in local and distributed computer networks. For example, systems such as simulation network (SIMNET) and its successor, CCTT, were designed to provide training in crew maneuver and fighting skills, as well as to provide unit-level training up to the battalion task force level for organizations equipped with M1 tanks, M2 Bradleys, and AH64 Apache helicopters. Recent training applications have been focused on attempts to integrate constructive simulations with virtual and live simulations.

The Navy has developed an array of simulations to support team/crew, classroom, and shipboard training requirements. Simulation and simulator training requirements are focused on ship systems and fleet operations. Simulations of Navy command and control task force and fleet operations have been developed within all of the unified commands, including European, Atlantic, and Pacific.

These large force constructive simulations, such as NSS, are used primarily for command and staff training. They are constructive two-sided simulations that have red team opposition force players and blue force staff and commander players.

The Air Force Air Combat Command has used airframe-specific flight simulators for decades in transition training and at the squadron and wing levels for sustainment training. The advent of virtual simulations that replicate system-specific performance functions with good fidelity offers the possibility of reduced training costs and increased levels of individual and unit proficiency. These simulations may allow the development of more highly refined coordination procedures and tactics that have not previously been possible in a peacetime environment.

Joint and Combined Training

An area of expanding growth in the number, size, and capability of models and simulations is Joint Chiefs of Staff and major theater command simulations. The Joint Chiefs of Staff Joint Warfighting Center, located at Fort Monroe, Virginia, conducts joint staff training exercises for the unified commands around the world. These exercises operate with a number of joint and service simulations, some of which are described in the annex to this chapter. A consistent direction in the joint training area has been a growing need for DIS to allow the conduct of realistic joint training exercises with increased fidelity and significantly decreased costs. The synthetic theater of war (STOW) is an example of the blending of constructive and virtual simulations for joint and combined training. STOW-Europe was the first operational demonstration of the concept of linked constructive simulations, with Army ground force players at remote sites being linked to the Air Force Air Warrior Center. Incorporation of intelligent forces into STOW is discussed later in the chapter.

Analysis Applications

Research, Development, and Acquisition

The thrust of simulations for research, development, and acquisition is to determine system capabilities and levels of performance that can be expected from new systems. Simulations are applied to address human engineering concerns, the design of systems and their interoperability with other services or multinational forces, and option prioritization and risk assessment decisions, as well as to examine survivability, vulnerability, reliability, and maintainability. All of the military services use simulations in their respective research, development, and acquisition processes; however, they do not have comparable organizations for the development of systems, and, therefore do not have a common set of processes for the development or application of such simulations. Training and Doctrine Command (TRADOC) is responsible for the development of operational system requirements, training, and doctrine for the U.S. Army. Once a weapon system has been approved for development, it is assigned to a program office headed by a program manager and supported by a staff that coordinates all aspects of the development. The various offices are grouped by type of system (e.g., electronic, combat, combat support, combat service support) and are controlled by program executive officers on the Army staff. They are reinforced by proponent organizations from TRADOC and by material developers from the U.S. Army Materiel Command. The most commonly used simulations in support of Army research, development, and acquisition are of the constructive type.

The Naval Systems Command, organized into air, surface, and subsurface systems, oversees the system requirements process, but actual development is placed in the hands of a system program office. Each such office is organized with all of the resources and organizational structure needed to control and execute program development and acquisition. Programs are organized around the fighting system, such as the Aegis class cruiser or a carrier-based attack aircraft. Typically, the system program offices develop unique simulation capabilities for their system for both analysis and training applications.

The Air Force Systems Command is organized into system program offices similar to those of the Navy and has operational commands, such as the Air Combat Command, that provide doctrine and training requirements. The Air Force acquisition process is built entirely around the weapon system, and the system program office has control of all of the development processes, including simulation and simulator development.

Advanced Concepts and Requirements

Analysis of advanced concepts and requirements is focused on the development and evaluation of doctrine and organizations at a stage that precedes system development. The thrust is on predicting and assessing the impact of force effectiveness outcomes that will result from new technology, changes in force structure mixes, or the integration of joint or combined forces. Users of simulations in this advanced-technology area include DARPA, the U.S. Army Concepts Analysis Agency (CAA), the Center for Naval Analyses (CNA), several Air Force analysis organizations, such as the RAND Corporation, and DoD joint task force study organizations, all of which research new defense threat issues or future service structures/requirements. While such organizations may develop new simulations through civilian contracts, they use or adapt many of the same constructive simulations used in the research, development, and acquisition area.

Development of Doctrine, Techniques, and Procedures

The services use both constructive and virtual simulations to support doctrine development, assess battle tactics for new systems and organizations, and refine techniques for system and unit employment. Creating new or revised doctrine and procedures is a function of the introduction of new technology, new systems, or new missions. Many of the simulations used to analyze advanced development and acquisition issues are also used to examine doctrine and procedures.

Command Decision Making and Interoperability— Joint/Combined Forces

Interoperability of joint and combined forces has been less emphasized than changes in the world situation and U.S. military commitments around the globe would suggest. Joint training has seen an ever-increasing level of activity with the development of joint/combined forces organizations, such as the Joint Warfighting Center and a number of joint models and simulations developed by the unified commands. But these training simulations and the standardization of procedures incorporating potential coalition forces and the differences in respective decision processes have not received the same level of attention as other development areas.

CURRENT MILITARY MODELS OF HUMAN BEHAVIOR AND THEIR LIMITATIONS

This section reviews existing military models that have been developed to represent human elements in computer-generated forces. Table 2.2 lists these models and shows the application areas from the previous section (training, analysis, interoperability) to which each relates. The models with an asterisk are those discussed in this section; the other models listed in the table are those discussed in the annex to this chapter and only referred to briefly in the chapter.

Computer-Generated Forces

Computer-generated forces are computer representations of vehicles and humans for use in simulations. The goal is to develop these representations to the point that they will act automatically and with some realism without human intervention. Computer-generated forces populate both constructive and virtual simulations. They are used in combination with human players as a semiautomated force in the DIS context to represent opposing forces and collateral friendly forces. There are specific programs devoted to creating these entities. The most widely used is modular semiautomated forces (ModSAF), which plays in a number of applications within the services and in joint force models; however, several other models are also used, such as CCTT semiautomated force

TABLE 2.2 Key Military Models and Simulations

Model/ Simulation	Training		Analysis			Interoperability
	Individual/ Leader	Joint and Combined	Research, Development, Acquisition	Advanced Concepts	Doctrine Development	
ARMY						
CAST-FOREM	X		X	X		
CBS	X	X			X	
CCTT SAF*	X				X	
CSSTSS	X			X	X	
EADSIM		X		X		
Eagle		X		X		X
JANUS	X		X		X	
ModSAF*	X	X				
RWA-Soar*	X					
VIC			X	X	X	
WARSIM						
NAVY AND MARINE CORPS						
MCSF*	X					
MTWS	X					
NSS	X		X	X	X	
SUTT*	X				X	
AIR FORCE						
AASPEM	X		X		X	
FWA-Soar*	X					
TAC-BRAWLER	X					
JOINT SERVICE						
CFOR*	X	X				
JCATS	X	X			X	
JCM	X	X				X
JTLS	X	X				
JWARS		X		X	X	X
STOW	X	X				

*Indicates those models and simulations used for human behavior representation.

(SAF); computer-controlled hostiles, intelligent forces (IFOR); and command forces (CFOR) models. Both IFOR and CFOR models have been developed in the context of the synthetic theater of war (STOW), an advanced-concept technology demonstration jointly sponsored by the U.S. Atlantic Command and DARPA. A key element of STOW is the representation of both fighting forces and their commanders in software.

Modular Semiautomated Forces (ModSAF)

ModSAF is an open architecture for modeling semiautomated forces, developed by the Army's Simulation, Training, and Instrumentation Command (STRICOM). It provides a set of software modules for constructing advanced distributed simulation and computer-generated force applications. Semiautomated forces are virtual simulations of multiple objects that are under the supervisory control of a single operator. According to Downes-Martin (1995), the human behaviors represented in ModSAF include move, shoot, sense, communicate, tactics, and situation awareness. The authoritative sources of these behaviors are subject matter experts and doctrine provided by the Army Training and Doctrine Command (TRADOC). Task-based explicit behaviors are enumerated by a finite-state machine that represents all the behavior and functionality of a process for a limited number of states. The finite-state machine includes a list of states, a list of commands that can be accepted while in each state, a list of actions for each command, and a list of state conditions required for an action to be triggered.

In ModSAF there is no underlying model of human behavior, so that any behavior representation must be coded into the finite-state machine. As a result, it is impractical to use ModSAF to construct general-purpose behavioral or learning models. Typically, ModSAF models are employed to represent individual soldiers or vehicles and their coordination into orderly-moving squads and platoons, but their tactical actions as units are planned and executed by a human controller. Ceranowicz (1994) states that because the human agents in ModSAF are not intelligent enough to respond to commands, it is necessary for the human in command to know everything about the units under his/her control. This would not be the case in a real-world situation. ModSAF supports the building of models of simple behaviors at the company level and below.

The initial ModSAF (developed by the Army) has been adopted by the other services—Navy, Marine Corps, and Air Force. Each of these services now has its own ModSAF source and capability tailored to its own needs. In a recent STOW exercise, four types of ModSAFs were used. The scenarios for this exercise were preset and the role of the human controller minimized.

Close Combat Tactical Trainer Semi-Automated Force (CCTT SAF)

CCTT SAF, also under development by STRICOM, is designed to simulate actual combat vehicles, weapon systems, and crew and platoon commander elements. The approach used to model human behavior is rule-based knowledge representation. The source of the knowledge is doctrine provided by natural-language descriptions of tactical behavior incorporated into combat instruction sets developed by subject matter experts. Combat instruction sets contain the following elements: a behavior description, a sequence of actions taken in the behavior, initial conditions, input data, terminating conditions, and situational interrupts. For action selection, blue force behaviors are grouped into categories of move, shoot, observe, or communicate; opposing force actions are listed in order, but not catalogued.

According to Downes-Martin (1995), the CCTT SAF human behavior representations work reasonably well for small-unit tactics, but because there is no underlying model of human behavior, it is difficult to increase the size or scope of the simulated operations. Like ModSAF, the system requires hand coding of rules, and therefore its models can be generalized only to groups of similar size and complexity. It should also be noted that behavior in CCTT SAF is based solely on following doctrine and does not allow for intelligent human responses. According to Kraus et al. (1996), work is under way on using CCTT combat instruction sets as inputs to ModSAF.

Marine Corps Synthetic Forces

Marine Corps Synthetic Forces (MCSF) is a DARPA-supported project conducted by Hughes Research Laboratory. The simulation provides a computer-generated force representation for the Marine Corps that is intended to model individual fire-team members, fire-team leaders, and squad leaders for the purpose of training their respective superiors (Hoff, 1996). MCSF is based on ModSAF. Individual behavior is controlled by unit-level (i.e., higher-echelon) tasks. Units and individuals perform doctrinally correct sequences of actions that can be interrupted in limited ways by reactions to events. In MCSF, units plan assaults and other tasks using fuzzy decision tables. These decision tables encode a subject matter expert's ranking of decision alternatives for a number of key decisions, such as attack or abort, avoid contact, select assault point, and select route.

Ruleset antecedents are essentially features of the tactical situation (e.g., enemy posture is dug in), which are obtained directly by querying the ModSAF module(s) responsible for maintaining the state of the simulation. No attempt is made to model the individual's cognitive functions of information gathering, perception, correlation, or situation assessment.

Computer-Controlled Hostiles for the Small Unit Tactical Trainer[2]

The small unit tactical trainer (SUTT) is under development for the Marine Corps by the Naval Air Warfare Center, Training Systems Division (NAWCTSD). It is a virtual world simulator for training Marine rifle squads in military operations in urban terrain (MOUT) clearing (Reece, 1996). The virtual world in the trainer is populated by computer-controlled hostiles (CCHs) developed by the Institute for Simulation and Training at the University of Central Florida. The CCHs behave as smart adversaries and evade and counterattack friendly forces. The key state variables include low-level facts (e.g., soldier position, heading, posture), situation knowledge (e.g., threats, current target), and the current task. Perception is based on simple visual and aural detection and identification of enemy forces. Situation awareness involves both visible and remembered threats.

Most of the activity is reflexive. The model has an action selection component based on hierarchical task decomposition, starting with top-level-tasks, e.g., engage enemy, seek cover, watch threat, look around, and run away. Situation dependent rules propose tasks to perform. Rule priorities allow specification of critical, normal, and default behavior. Tasks may be deliberately proposed with the same priority level, allowing random selection to choose a task according to predefined weights. This mechanism supports variation in behavior that obeys a prior probability distribution (e.g., probability of fight or flight reaction can be set for a CCH). A new task may be selected each time frame as the situation changes. Proposed, but not implemented, architecture extensions would allow the selection of tasks that would serve more than one top-level goal or task simultaneously. Other models under development as part of this program include one for visual detection and another for hearing (Reece and Kelly, 1996).

Intelligent Forces (IFOR) Models

IFOR models have been created using the Soar architecture to model the combat behavior of fixed- and rotary-wing pilots in combat and reconnaissance missions. (Details on the Soar architecture are presented in Chapter 3.) IFORs are very large expert systems—they use encoded knowledge as a basis for action and problem solving. The Soar architecture was originally devised to support the study of human problem-solving behavior. It incorporates many of the concepts of artificial intelligence, and its main feature is the use of production rules as the means to link an initial condition (a stimulus) to a particular response. Although Soar is capable of learning, this function has not been exercised.

To meet the objectives of simulating intelligent forces, specific contexts were needed. In the IFOR framework, adaptations for both fixed-wing attack (FWA)-Soar and rotary-wing attack (RWA)-Soar air operations were developed.

---

[2]Previously called the team target engagement simulator.

FWA-Soar This adaptation of IFOR (formerly known as tactical air [TacAir]-Soar) can simulate a fleet of up to 500 fixed-wing aircraft in a context that includes such elements as enemy aircraft, surface-to-air missiles, and enemy radar installations. It is usable by Army, Navy, Marine, and Air Force units. Its most extensive and intensive use recently was as the primary source of air missions for STOW-97. Soar's participation in this exercise was initially intended to serve as a demonstration of the capabilities of FWA-Soar; however, a training mission was added to the exercise.

FWA-Soar presents an excellent example of the adjustments often required when a software program of major dimensions (5,200 production rules) is taken from the laboratory into an operational environment. Specifically, it was clear early in the exercise that some real-world behavioral complexities had not been captured. Some of these were relatively trivial, such as the real-world behavior of pilots in avoiding known sites of ground-to-air missile defenses. Others were more serious in the sense that they highlighted questions about costs and benefits related to the extent to which functions could or should be automated. For example, the translation of an air tasking order into a detailed mission plan/profile was a new requirement presented by STOW that was not provided for in the original software. During the exercise, human operator intervention was needed via a computer subsystem known as the exercise editor to achieve the translation.

The net effect of the postexercise evaluation of FWA-Soar was to encourage the possible addition of computational capabilities. However, the proposed additions were not tied to modeling of human behavior at the basic level, but were focused on developing a simulation that was capable of executing subtasks omitted from the original program (Laird et al., 1997).

RWA-Soar The development of RWA-Soar includes representations of helicopter company-level anti-armor attacks and multicompany operations in air transport (the movement of foot soldiers) and escort missions. In addition to its uses in the Army, RWA-Soar has been used to model Marine transport and escort missions. At present, RWA-Soar resides in a smaller computer program than that of FWA-Soar. However, effort toward the implementation of RWA-Soar is closely integrated with the work on the Soar version of CFOR (discussed below).

Command Forces (CFOR) Simulations

The CFOR project is part of the STOW program. Its focus has been on providing a framework for the simulation of behavior at the command and control level. The program has three components: (1) an architecture in which simulations of command and control interact with the simulated battlefield, (2) a command language for information exchange between simulated and real commanders (the command and control simulation interface language, or CCSIL), and (3) a strategy that provides for the integration of command entities developed by different contractors and services. Each of the services has its CFOR requirements definitions and development efforts (Hartzog and Salisbury, 1996).

One example of a development effort for the Army in this domain is Soar/CFOR. At present, this effort encompasses the command functions of a helicopter battalion-to-company operation. RWA-Soar provides the simulation of the individual helicopter pilots in this setting (vehicle dynamics are provided by ModSAF). The entity simulated by Soar/CFOR is the company commander. In exercises, a live human represents the battalion commander, who generates orders that are rendered in computer language by CCSIL. From that point forward in the exercise, all actions are computer formulated. A critique by one of the developers (Gratch, 1996) stipulates that the most difficult features to put into the computer program are responses to unanticipated contingencies. As in FWA-Soar, the main developmental thrust in trying to overcome such a difficulty is to add subtask elements to the computer program so that every contingency can be met at the automated level. Development of an approach that would solve the general problem—such as a truly adaptive model of human problem solving—has not been actively pursued.

Human Behavior in Constructive Wargaming Models

In the domain of constructive wargaming models, human behavior typically is not represented explicitly at all, for either blue or red forces. When human decisions are called for, a doctrinally based decision rule is inserted that reflects what the individual ought to do in the ideal case. Human performance capacities and limitations are ignored. For example, in the NSS, a Monte Carlo simulation, plans are executed based on decision tables developed by subject matter experts. This approach does not provide the fidelity of battle outcomes on a real battlefield, where the number of casualties or weapon system losses depends on real human strengths and frailties and varies significantly from unit to unit based on leadership, stress, consistency of tactical decisions, and effectiveness of training. This lack of human performance representation in models becomes more significant as the size, scope, and duration of wargaming simulations continue to grow. In the future, these limitations will become more noticeable as greater reliance is placed on the outcomes of models/simulations to support training and unit readiness, assessments of system performance, and key development and acquisition decisions.

Thus it is fair to say that, in terms of models in active use, the introduction of human behavior into military simulations is in its infancy. However, because of the wide range of potential uses of these kinds of models, it is badly needed to create more realistic and useful evaluations. It is timely, therefore, to consider ways in which human behavior representation can be expanded. The following chapters review the state of the science in several areas of human behavior and suggest goals for incorporation into existing models as well as goals for promising research directions.

ANNEX: CURRENT MILITARY MODELS AND SIMULATIONS

This annex presents a brief characterization of some of the current models and simulations used by the various services. This review is not intended to be exhaustive, but rather to provide a selective sampling of these activities. Table 2.2 shows the uses of each model and simulation described below.

Army Models and Simulations

JANUS

JANUS is a constructive model that provides the battle results of set engagements by forces in contact. It is focused on the levels of squad/team/crew to battalion task force, but has been extended with sacrifices in fidelity to the brigade and division levels. Engagement results are based on a series of mathematical computations with stochastic distributions of the probabilities of (1) detection, based on line of sight; (2) hit, based on the ballistic characteristics of the weapon; and (3) kill, based on the lethality of the firer and protection characteristics of the target. The model requires entry of the capabilities and location of every weapon system on the notional battlefield. Organizations are represented by player cells on the blue and red sides. The model can incorporate such elements as normal staff action drills, experimental weapon characteristics, and new tactics and procedures, depending on the setup and particular version of JANUS employed. Within the Army, the principal users of JANUS are combat development and training organizations of the branch schools within TRADOC; the TRADOC Battle Laboratories; and the TRADOC Analysis Command at White Sands Missile Range, New Mexico, and Fort Leavenworth, Kansas.

Combined Arms Task Force Effectiveness Model

The combined arms task force effectiveness model (CASTFOREM) is a large-scale model focused on tactics and employment of forces at the task force and brigade levels. The model uses mathematical calculations and stochastic distributions, along with subroutines to execute some command and control functions. It uses doctrinal tactics and maneuvers and key characteristics of the weapon systems to determine the battle outcome of a scenario against a postulated threat force used by a red team cell employing threat weapons and maneuver characteristics. The model has been expanded to the division level with some loss in fidelity. The key user of CASTFOREM is TRADOC.

Eagle

Eagle is a high-level constructive model that addresses and assesses force-level requirements at the division and corps levels. Essentially a mathematical model, it provides gross requirement determinations based on the execution of a battle scenario involving up to a two-division corps against semiautomated forces in a combined and/or joint air-land battle. It incorporates logistics requirements based on expected rates of consumption.

Vector-in-Command

Vector-in-command is a constructive model that can be run at the battalion task force to division and corps levels, depending on scenario construction. It is used extensively within TRADOC to assess the performance effectiveness of proposed new weapon systems, conduct assessments of new doctrine, and address the potential impacts of organizational or procedural changes. The model is stochastic and operates according to sets of rules programmed for the scenarios planned for the assessment.

Extended Air Defense Simulation

The extended air defense simulation (EADSIM) is a system-level analysis simulation capable of assessing the effectiveness of theater missile defense and air defense systems against a broad range of extended air defense threats. The model incorporates fixed- and rotary-wing aircraft, ballistic missiles, cruise missiles, radar and infrared sensors, satellites, command and control structures, electronic warfare effects, and fire support. It is a time- and event-stepped two-side reactive model that executes a planned scenario. In its basic configuration it does not operate as a human-in-the-loop simulation. Used primarily by the U.S. Army Space and Strategic Defense Command, it can be confederated with other theater- or campaign-level constructive models, as well as some system virtual simulators.

Corps Battle Simulation

Currently, corps battle simulation (CBS) is the primary simulation used by the Army to train staff officers at the Army's Command and General Staff College at Fort Leavenworth, Kansas. It is employed to support a number of higher-level division and corps exercise programs and has been used extensively to support assessment of the Army's advanced warfighting experiments. CBS is a constructive simulation originally designed to be conducted at the division level, and subsequently revised and expanded for use at the corps level. It interfaces with other models and simulations, such as air warfare simulation (AWSIM), tactical simulation model (TACSIM), combat service support training support simulation (CSSTSS), and Marine air-ground task force tactical warfare system (MTWS). CBS uses human-in-the-loop commanders and staff organizations from brigade-, division-, and corps-level staffs. It executes the battle outcomes of approximately 3 hours of combat based on player inputs that establish unit locations, weapon system status, and intended actions/maneuvers. It computes battle losses and logistics consumption in gross terms down to the company and battalion task force levels, with reports and status given to each command and staff level participating. The simulation requires considerable pre-exercise setup, a significant opposing forces (OPFOR) player cell, and blue forces (BLUFOR) commanders and staff officers for all unit command and staff organizations in the scenario task organization being represented.

Close Combat Tactical Trainer

CCTT is a family of virtual simulations and simulators currently under development by the Army and TRADOC. It is organized into unit formations of battalion task forces equipped with M1 tank devices and M2 Bradley infantry fighting vehicle devices, and supported by AH64 attack helicopter simulators. CCTT is being designed to support brigade-level operations with high task fidelity down to the crew and squad levels.

Combat Service Support Training Support Simulation

The combat service support training support simulation (CSSTSS) is a deterministic logistics requirements and performance model. It is a logistics exercise model and can be used as the logistics driver for operational simulations such as corps battle simulation (CBS). CSSTSS replicates all the automated logistics system functions of the Army's standard Army management information system (STAMIS) systems, with support logistics and personnel operations from the conterminous United States (CONUS) level to the theater, corps, and division levels and below. The level of logistics fidelity is high, but setup time and input operations are extensive.

War Simulation 2000

War Simulation (WARSIM) 2000 is a proposed Army corps- and theater-level simulation that is currently under development as the replacement for CBS, the tactical simulation model (TACSIM), CSSTSS, and the brigade/battalion battle simulation (BBS). It is intended to provide a high-fidelity simulation for operational exercises that will effectively model the new digital command control, Intel, and logistics systems and requirements. As proposed, it will allow interface with a constructive model environment, as well as with virtual simulators/simulations and live simulations.

Navy and Marine Corps Models and Simulations

Naval Simulation System

The naval simulation system (NSS) is an object-oriented, Monte Carlo, multiresolution constructive simulation under development for naval operations—OPNAV N812 and OPNAV N62. It has a planned (future) virtual simulation mode of operation. NSS is being developed for use in simulating naval operations in support of analyses of tactics, decision support applications, and training (Stevens and Parish, 1996). It provides explicit treatment of command structures from the national to the operating unit level, operational plans and tactics, sensors, weapons, and countermeasures. The simulation applies situation awareness based on a commander's perception of the status of friendly, neutral, and hostile forces. Tactics are represented by means of what is termed a decision table, although it might be more properly termed a prioritized production rule system. The simulation normally operates faster than real time. The basic unit is the "entity," typically a ship or aircraft. Explicit representations are included for the "hardware" component, consisting of the environment, sensors, weapons, and communications channels, and the "software" component, consisting of the command structure, the data fusion process, the tactics, and the operational plans. This latter set of components is the key determinant of unit behavior. The data fusion process serves to generate the tactical picture or the assessed situation. This, in combination with the tactics, generates the unit behavior.

Marine Air-Ground Task Force Tactical Warfare System

The Marine air-ground task force tactical warfare system (MTWS) is a commander and staff air and ground training simulation that simulates the ground combat of a task force and its air and indirect fire support elements. It is used as a staff trainer, but not validated for analysis. MTWS unit behavior is not sophisticated, and the simulation does not model individual behavior. Doctrine is parametric, and communications is not represented. The simulation is controlled by human controllers whose skills are critical to the realism of the play.

Air Force Models and Simulations

Advanced Air-to-Air System Performance Evaluation Model

The advanced air-to-air system performance evaluation model (AASPEM) 4.1 is a comprehensive tool for performing air combat analysis in a realistic few-on-few engagement environment of up to 75 aircraft and missile combinations for six different aircraft and six different missiles at a time. AASPEM can be used to explore tactics, maneuver versus detection, launch ranges, flight testing, and mission planning. It has been used for studies of the effectiveness of new aircraft, missiles, countermeasure system designs, and concept development. Aircraft and missiles are explicitly flown incorporating Air Force-specified missile guidance laws and missile propulsion characteristics. AASPEM comprises a family of four integrated submodels: (1) the advanced missile flyout model (AMFOM); (2) the aircraft, missile, avionics performance simulation (AMAPS); (3) the interactive tactical air combat simulation (INTACS); and (4) the automated decision logic tactical air combat simulation (ALTACS). The simulation requires a large amount of time to create realistic scenarios.

Tactical Air Combat Simulator

The tactical air combat simulator (TACBRAWLER) simulates air-to-air combat between multiple flights of aircraft in both visual and beyond-visual ranges. The simulation incorporates user determination of mission and tactical doctrine, pilot aggressiveness, pilot perception of the enemy, reaction time, and quality of the decision made. It models the aircraft's aerodynamics, missiles, radars, communications, infrared situation track (IRST), identification friend or foe (IFF), radar warning receiver (RWR), and missile warning devices. It is an event-store simulation operated on Monte Carlo principles. It requires a large amount of time for initialization and does not simulate terrain.

Joint Service Models and Simulations

Joint Conflict Model

The joint conflict model (JCM) is a deterministic model developed by the Pacific Theater Commander for joint staff-level training with joint task force-level structures and scenarios for the Pacific Theater.

Joint Theater Level Simulation

The joint theater level simulation (JTLS) system is an interactive, multisided analytical wargaming system that models a theater-level joint and coalition force air, land, and naval warfare environment. The system consists of six major programs and a number of smaller support programs used to plan, execute, and analyze plans and operations. The simulation uses Lanchester attrition algorithms; detailed logistics modeling; and explicit movement of air, land, and naval forces.

Synthetic Theater of War

Synthetic theater of war (STOW) is a family of constructive and virtual simulations linked together to support joint training and theater contingency exercises. STOW-Europe (STOW-E) was the first operational demonstration of the concept of linked constructive simulations, with Army ground force players at remote sites being linked to the U.S. Air Force Air Warrior Center. An exercise called STOW-97 was conducted to test new doctrine, tactics, and weapon system concepts. It incorporated Soar-based IFORs to represent fixed-wing aircraft that exhibited human-like behavior. The STOW concept has expanded and evolved to include linking of virtual simulations in both joint service and North Atlantic Treaty Organization scenarios.

Joint Combat Operations

Joint combat operations (JCATS) is being developed by the Conflict Simulation Laboratory at Lawrence Livermore National Laboratory to model joint combat operations, as well as unconventional warfare (e.g., hostage rescue, operations other than war). Coverage is from the individual up to the division level. Currently, no human behavior representation is included, and all tactics and play are specified by the human players. However, efforts are under way to incorporate an optimal route-planning algorithm developed by Lawrence Livermore National Laboratory to support route planning for individual and unit entities. No cognitive architecture has been specified for representing synthetic human players.

Joint Warfare System

The joint warfare system (JWARS) is being developed by the JWARS program office to model joint combat operations (Prosser, 1996a, 1996b). Key command and control, communications, and computers (C^4) intelligence, surveillance, and reconnaissance (ISR) modeling requirements include the following tasks:

- Situation development and assessment
- Intelligence planning and direction
- Command and control decision making

It is not clear whether any attempts are being made to represent the way human decision makers accomplish these tasks. For example, situation development and assessment are modeled by a stepwise process of (1) data fusion, (2) situation map generation, and (3) enemy course-of-action assessment through best-fit matching of library templates to the estimated situation map. Although this approach reflects a plausible process for *automated* situation assessment, whether it provides an appropriate basis for human behavior representation is unclear. Likewise, there is no description of how or whether JWARS models the human planning function for either intelligence planning and collection or operations planning. Additional review is clearly needed, especially in light of the emphasis being placed on JWARS as the keystone tool for joint operations analysis.

3

Integrative Architectures for Modeling the Individual Combatant

We have argued in the introduction that in order to make human behavior representation more realistic in the setting of military simulations, the developers will have to rely more on behavioral and organizational theory. In Chapters 3 through 11 we present scientific and applied developments that can contribute to better models. Chapters 4 through 8 review component human capacities, such as situation awareness, decision making, planning, and multitasking, that can contribute to improved models.

However, to model the dismounted soldier, and for many other purposes where the behavior of intact individuals is required, an integrative model that subsumes all or most of the contributors to human performance capacities and limitations is needed. Although such quantitative integrative models have rarely been sought by psychologists in the past, there is now a growing body of relevant literature. In the few cases in which these integrative models have been applied to military simulations, they have focused on specific task domains. Future integrative model developers will likely be interested in other domains than the ones illustrated here. Because each of our examples embodies a software approach to integrative models, we are referring to these developments as integrative modeling architectures. They provide a framework with behavioral content that shows promise of providing a starting point for model developers who wish to apply it to their domains. Each has its own strengths and weaknesses and we do not explicitly recommend any one. The developers' choices depend entirely on their goals and objectives in integrative model development. In Chapter 13, we do not recommend converging on a single integrative architecture, although we do argue that it is important to adopt modular structures that will allow easier interoperability among developed models.

The chapter begins with a general introduction to integrative architectures that review their various components; this discussion is couched in terms of a stage model of information processing. Next we review 10 such architectures, describing the purpose, assumptions, architecture and functionality, operation, current implementation, support environment, validation, and applicability of each. The following section compares these architectures across a number of dimensions. This is followed by a brief discussion of hybrid architectures as a possible research path. The final section presents conclusions and goals in the area of integrative architectures.

GENERAL INTRODUCTION TO INTEGRATIVE ARCHITECTURES

A general assumption underlying most if not all of the integrative architectures and tools reviewed in this chapter is that the human can be viewed as an information processor, or an information input/output system. In particular, most of the models examined are specific instantiations of a *modified stage model* of human information processing. The modified stage model is based on the classic stage model of human information processing (e.g., Broadbent, 1958). The example given here is adapted from Wickens (1992:17). Such a stage model is by no means the only representation of human information processing as a whole, but it is satisfactory for our purposes of introducing the major elements of the architectures to be discussed.

In the modified stage model, *sensing and perception* models transform representations of external stimulus energy into internal representations that can be operated on by cognitive processes (see Figure 3.1).

Memory consists of two components. *Working memory* holds information temporarily for cognitive processing (see below). *Long-term memory* is the functional component responsible for holding large amounts of information for long periods of time. (See also Chapter 5.)

Cognition encompasses a wide range of information processing functions. *Situation awareness* refers to the modeled individual combatant's state of knowledge about the environment, including such aspects as terrain, the combatant's own position, the position and status of friendly and hostile forces, and so on (see also Chapter 7). Situation assessment is the process of achieving that state of knowledge. A *mental model* is the representation in short- and long-term memory of information obtained from the environment. *Multitasking* models the process of managing multiple, concurrent tasks (see also Chapter 4). *Learning* models the process of altering knowledge, factual or procedural (see also Chapter 5). *Decision making* models the process of generating and selecting alternatives (see also Chapter 6).

Motor behavior, broadly speaking, models the functions performed by the neuromuscular system to carry out the physical actions selected by the above-mentioned processes. Planning, decision making, and other "invisible" cognitive

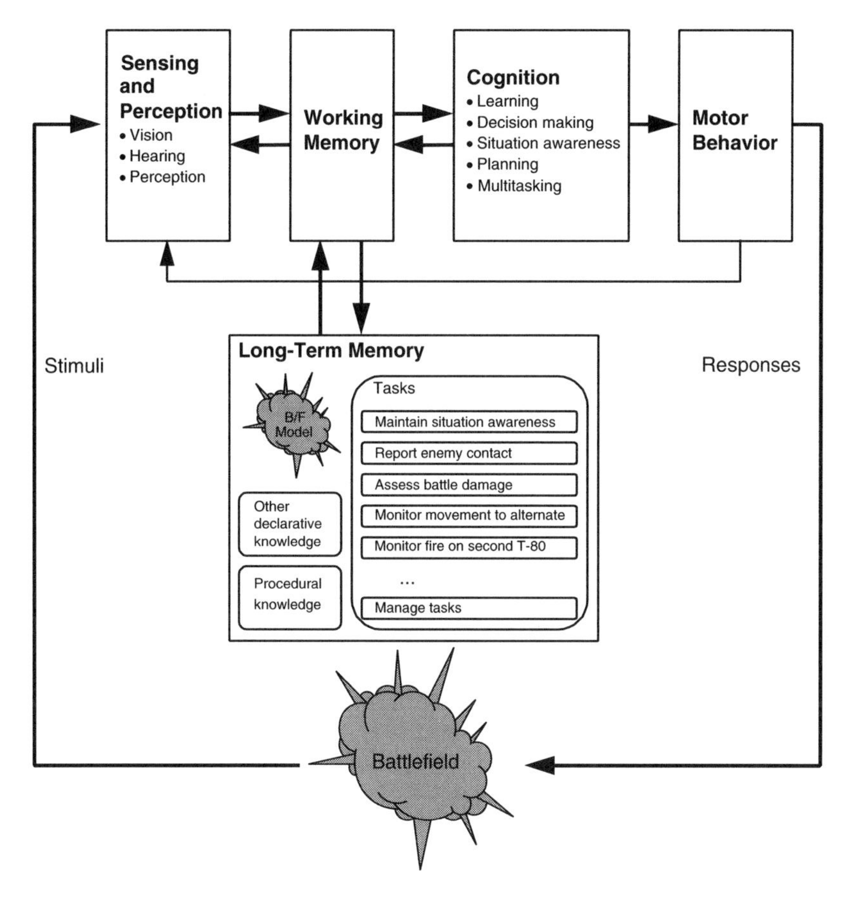

FIGURE 3.1 Modified stage model. NOTE: Tasks shown are derived from the vignette presented in Chapter 2.

behaviors are ultimately manifested in observable behaviors that must be simulated with varying degrees of realism, depending on actual applications; aspects of this realism include response delays, speed/accuracy tradeoffs, and anthropometric considerations, to name but a few.

Each of the architectures reviewed incorporates these components to some extent. What is common to the architectures is not only their inclusion of submodels of human behavior (sensing, perception, cognition, and so on), but also the integration of the submodels into a large and coherent framework. It would be possible to bring a set of specific submodels of human behavior together in an ad hoc manner, with little thought to how they interact and the emergent properties that result. But such a model would not be an integrative architecture. On the contrary, each of these integrative architectures in some way reflects or instantiates a unified theory of human behavior (at least in the minds of its developers) in which related submodels interact through common representations of intermediate information processing results, and in which consistent conceptual representations and similar tools are used throughout.

REVIEW OF INTEGRATIVE ARCHITECTURES

This section reviews 11 integrative architectures (presented in alphabetical order):

- Adaptive control of thought (ACT-R)
- COGnition as a NEtwork of Tasks (COGNET)
- Executive-process interactive control (EPIC)
- Human operator simulator (HOS)
- Micro Saint
- Man machine integrated design and analysis system (MIDAS)
- MIDAS redesign
- Neural networks
- Operator model architecture (OMAR)
- Situation awareness model for pilot-in-the-loop evaluation (SAMPLE)
- Soar

The following aspects of each architecture are addressed:

- Its purpose and use
- Its general underlying assumptions
- Its architecture and functionality
- Its operation
- Features of its current implementation
- Its support environment
- The extent to which it has been validated
- The panel's assessment of its applicability for military simulations

It should be noted that the discussion of these models is based on documentation available to the panel at the time of writing. Most of the architectures are still in development and are likely to change—perhaps in very fundamental ways. The discussion here is intended to serve as a starting point for understanding the structure, function, and potential usefulness of the architectures. The organizations responsible for their development should be contacted for more detailed and timely information.

Adaptive Control of Thought (ACT-R)[1]

ACT-R is a "hybrid" cognitive architecture that aspires to provide an integrated

[1]This section draws heavily from Anderson (1993).

account of many aspects of human cognition (see also the later section on hybrid architectures). It is a successor to previous ACT production-system theories (Anderson, 1976, 1983), with emphasis on activation-based processing as the mechanism for relating a production system to a declarative memory (see also Chapter 5).

Purpose and Use

ACT-R as originally developed (Anderson, 1993) was a model of higher-level cognition. That model has been applied to modeling domains such as Tower of Hanoi, mathematical problem solving in the classroom, navigation in a computer maze, computer programming, human memory, learning, and other tasks. Recently, a theory of vision and motor movement was added to the basic cognitive capability (Byrne and Anderson, 1998), so theoretically sound interaction with an external environment can now be implemented.

Assumptions

In general, ACT-R adheres to the assumptions inherent in the modified stage model (Figure 3.1), with the minor exception that all processors, including the motor processors, communicate through the contents of working memory (not directly from cognition).

ACT-R assumes that there are two types of knowledge—declarative and procedural—and that these are architecturally distinct. Declarative knowledge is represented in terms of chunks (Miller, 1956; Servan-Schreiber, 1991), which are schema-like structures consisting of an *isa* pointer specifying their category and some number of additional pointers encoding their contents. Procedural knowledge is represented in production rules. ACT-R's pattern-matching facility allows partial matches between the conditions of productions and chunks in declarative memory (Anderson et al., 1996).

Both declarative and procedural knowledge exist permanently in long-term memory. Working memory is that portion of declarative knowledge that is currently active. Thus, the limitation on working memory capacity in ACT-R concerns *access* to declarative knowledge, not the *capacity* of declarative knowledge.

ACT-R assumes several learning mechanisms. New declarative chunks can be learned from the outside world or as the result of problem solving. Associations between declarative memory elements can be tuned through experience. New productions can be learned through analogy to old procedural knowledge. Production strengths change through experience. The visual attention model embedded within ACT-R assumes a synthesis of the spotlight metaphor of Posner (1980), the feature-synthesis model of Treisman (Treisman and Sato, 1990), and the attentional model of Wolfe (1994).

Details of the ACT-R architecture have been strongly guided by the rational analysis of Anderson (1990). As a consequence of that rational analysis, ACT-R is a production system tuned to perform adaptively given the statistical structure of the environment.

Architecture and Functionality

ACT-R's declarative and procedural knowledge work together as follows. All production rules in ACT-R have the basic character of responding to some goal (encoded as a special type of declarative memory element), retrieving information from declarative memory, and possibly taking some action or setting a subgoal. In ACT-R, cognition proceeds forward step by step through the firing of such production rules.

All production instantiations are matched against declarative memory in parallel, but the ACT-R architecture requires only one of these candidate productions to fire on every cycle. To choose between competing productions, ACT-R has a conflict-resolution system that explicitly tries to minimize computational cost while still firing the production rule most likely to lead to the best result. The time needed to match a production depends on an intricate relationship among the strength of the production, the complexity of its conditions, and the level of activation of the matching declarative memory elements. In addition to taking different times to match, productions also differ in their contribution to the success of the overall task. This value is a relationship among the probability that the production will lead to the goal state, the cost of achieving the goal by this means, and the value of the goal itself. The architecture chooses a single production by *satisficing*: when the expected cost of continuing the match process exceeds the expected value of the next retrieved production, the instantiation process is halted, and the production that matches the highest value is chosen. To implement this conflict-resolution scheme, ACT-R models require many numerical parameters. Although most early ACT-R models set these parameters by matching to the data they were trying to explain, more recent models have been able to use stable initial estimates (Anderson and Lebiere, forthcoming).

As mentioned above, ACT-R includes several learning mechanisms. Declarative knowledge structures can be created as the encoding of external events (e.g., reading from a screen) or created in the action side of a production. The base-level activation of a declarative knowledge element can also be learned automatically. Associative learning can automatically adjust the strength of association between declarative memory elements. ACT-R learns new procedural knowledge (productions) through inductive inferences from existing procedural knowledge and worked examples. Finally, production rules are tuned through

---

[2]In some ways, the ACT-R implementation is intended to be a tool for exploring architectural assumptions. One method to support such exploration is the ability to turn each learning mechanism on or off independently. ACT-R is implemented in Lisp as an invitation to change the architecture.

learning strengths and updating of the estimates of success-probability and cost parameters. All of these learning mechanisms can be turned on and off depending on the needs of the model.[2]

Operation

Most existing ACT-R models stand alone; all of the action is cognitive, while perception and motor behavior are finessed. However, some models have been built that interact with an external world implemented in Macintosh Common Lisp or HyperCard™. To our knowledge, no ACT-R models currently interact with systems not implemented by the modelers themselves (e.g., with a commercially available system such as Flight Simulator™ or a military system).

Initial knowledge, both declarative and procedural, is hand-coded by the human modeler, along with initial numerical parameters for the strength of productions, the cost and probability of success of productions with respect to a goal, the base-level activation of declarative knowledge structures, and the like. (Default values for these parameters are also available, but not assumed to work for every new task modeled.) The output of an ACT-R model is the trace of productions that fire, the way they changed working memory, and the details of what declarative knowledge was used by those productions. If learning is turned on, additional outputs include final parameter settings, new declarative memory elements, and new productions; what the model learns is highly inspectable.

Current Implementation

The currently supported versions of ACT-R are ACT-R 3.0 and ACT-R 4.0. ACT-R 3.0 is an efficient reimplementation of the system distributed with *Rules of the Mind* (Anderson, 1993), while ACT-R 4.0 implements a successor theory described in *Atomic Components of Thought* (Anderson and Lebiere, 1998). Since both systems are written in Common Lisp, they are easily extensible and can run without modification on any Common Lisp implementation for Macintosh, UNIX, and DOS/Windows platforms. ACT-R models can run up to 100 times as fast as real time on current desktop computers, depending on the complexity of the task.

Support Environment

In addition to the fully functional and portable implementation of the ACT-R system, a number of tools are available. There is a graphical environment for the development of ACT-R models, including a structured editor; inspecting, tracing, and debugging tools; and built-in tutoring support for beginners. A perceptual/motor layer extending ACT-R's theory of cognition to perception and action is also available. This system, called ACT-R/PM, consists of a number of modules for visual and auditory perception, motor action, and speech production, which can be added in modular fashion to the basic ACT-R system. Both ACT-R and ACT-R/PM are currently available only for the Macintosh, but there are plans to port them to the Windows platform or to some platform-independent format, such as CLIM.

The basic system and the additional tools are fully documented, and manuals are available both in MS Word and over the World Wide Web. The Web manuals are integrated using a concept-based system called Interbook, with a tutorial guiding beginners through the theory and practice of ACT-R modeling in 10 lessons. The Web-based tutorial is used every year for teaching ACT-R modeling to students and researchers as part of classes at Carnegie Mellon University (CMU) and other universities, as well as summer schools at CMU and in Europe. The yearly summer school at CMU is coupled with a workshop in which ACT-R researchers can present their work and discuss future developments. The ACT-R community also uses an electronic mailing list to announce new software releases and papers and discuss related issues. Finally, the ACT-R Web site (http://act.psy.cmu.edu) acts as a centralized source of information, allowing users to download the software, access the Web-based tutorial and manuals, consult papers, search the mailing list archive, exchange models, and even run ACT-R models over the Web. The latter capacity is provided by the ACT-R-on-the-Web server, which can run any number of independent ACT-R models in parallel, allowing even beginners to run ACT-R models over the Web without downloading or installing ACT-R.

Validation

ACT-R has been evaluated extensively as a cognitive architecture against human behavior and learning in a wide variety of tasks. Although a comprehensive list of ACT-R models and their validation is beyond the scope of this chapter, the most complete sources of validation data and references to archival publications are Anderson's series of books on the successive versions of ACT (Anderson, 1983, 1990, 1993; Anderson and Lebiere, 1998). The latter reference also contains a detailed comparison of four cognitive architectures: ACT-R, executive-process interactive control (EPIC), Soar, and CAPS (a less well-known neural-cognitive architecture not reviewed here).

Applicability for Military Simulations

The vast majority of ACT-R models have been for relatively small problem-solving or memory tasks. However, there is nothing in principle that prevents ACT-R from being applicable to military simulations. There may be some problems associated with scaling up the current implementation to extremely large tasks that require extensive knowledge (only because the architecture has not been pushed in this manner), but any efficiency problems could presumably be solved with optimized matchers and other algorithmic or software engineering techniques such as those that have been applied to Soar. The intent of some currently funded military research contracts (through the Office of Naval Research) is to use ACT-R to model the tasks of the tactical action officer in submarines (Gray et al., 1997) and of radar operators on Aegis-like military ships (Marshall, 1995), but the work is too preliminary at this time to be reported in the archival literature.

Although ACT-R's applicability to military simulations is as yet underdeveloped, ACT-R is highly applicable to military training in cognitive skill. ACT-R-based intelligent tutors (called *cognitive tutors*) are being used to teach high-school-level mathematics and computer programming in many schools around the world, including the DoD schools for the children of military personnel in Germany. These cognitive tutors reliably increase the math SAT scores of students by one standard deviation. Any cognitive skill the military currently teaches, such as the operation of a dedicated tactical workstation, could be built into a cognitive tutor for delivery anywhere (e.g., on-board training for Navy personnel).

COGnition as a NEtwork of Tasks (COGNET)

COGNET is a framework for creating and exercising models of human operators engaged in primarily cognitive (as opposed to psychomotor) tasks (Zachary et al., 1992; Zachary et al., 1996).

Purpose and Use

COGNET's primary use is for developing user models for intelligent interfaces. It has also been used to model surrogate operators and opponents in submarine warfare simulators.

Assumptions

COGNET allows the creation of models of cognitive behavior and is not designed for modeling psychomotor behavior. The most important assumption behind COGNET is that humans perform multiple tasks in parallel. These tasks compete for the human's attention, but ultimately combine to solve an overall information processing problem. COGNET is based on a theory of weak task concurrence, in which there are at any one time several tasks in various states of completion, though only one of these tasks is executing. That is, COGNET assumes serial processing with rapid attention switching, which gives the overall appearance of true parallelism.

The basis for the management of multiple, competing tasks in COGNET is a pandemonium metaphor of cognitive processes composed of "shrieking demons," proposed by Selfridge (1959). In this metaphor, a task competing for attention is a demon whose shrieks vary in loudness depending on the problem context. The louder a demon shrieks, the more likely it is to get attention. At any given time, the demon shrieking loudest is the focus of attention and is permitted to execute.

Architecture and Functionality

The COGNET architecture consists of a problem context, a perception process, tasks, a trigger evaluation process, an attention focus manager, a task execution process, and an action effector. It is a layered system in which an outer shell serves as the interface between the COGNET model and other components of the larger simulation. There is no explicit environment representation in COGNET. Rather, COGNET interfaces with an external environment representation through its shell.

The *problem context* is a multipanel blackboard that serves as a common problem representation and a means of communication and coordination among tasks (see below). It provides problem context information to tasks. Its panels represent different parts of the information processing problem and may be divided into areas corresponding to different levels of abstraction of these problem parts.

Perception is modeled in COGNET using a *perception process* consisting of perceptual demons. These software modules recognize perceptual events in the simulated environment and post information about them (e.g., messages and hypotheses) on the blackboard.

Tasks in COGNET are independent problem-solving agents. Each task has a set of trigger conditions. When those conditions are satisfied, the task is activated and eligible to execute. An activated task competes for attention based on the priority of its associated goal (i.e., the loudness of its shrieks in the "shrieking demon" metaphor). Its priority, in turn, is based on specific blackboard content.

Task behaviors are defined by a procedural representation called COGNET executable language. COGNET executable language is a text-based language, but there is also a graphical form (graphical COGNET representation). COGNET executable language (and by extension, graphical COGNET representation) contains a set of primitives supporting information processing behavior. A goal is defined by a name and a set of conditions specifying the requirements for the goal to become active (relevant) and be satisfied. Goals may be composed of subgoals.

Four types of COGNET operators support the information processing and control behaviors of COGNET models. System environment operators are used to activate a workstation function, select an object in the environment (e.g., on a display), enter information into an input device, and communicate. Directed perceptual operators obtain information from displays, controls, and other sources. Cognitive operators post and unpost and transform objects on the blackboard.

Control operators suspend a task until the specified condition exists and subrogate (turn over control) to other tasks.

COGNET executive language also includes conditionals and selection rules, which are used to express branching and iteration at the level of the lowest goals. Conditionals include IF and REPEAT constructs. Selection rules (IF condition THEN action) can be deterministic (the action is performed unconditionally when the condition is true) or probabilistic (the action is performed with some probability when the condition is true).

The COGNET *trigger evaluation process* monitors the blackboard to determine which tasks have their trigger conditions satisfied. It activates triggered tasks so they can compete for attention.

The *attention focus manager* monitors task priorities and controls the focus of attention by controlling task state. A task may be executing (that is, its COGNET executive language procedure is being performed) or not executing. A task that has begun executing but is preempted by a higher-priority task is said to be interrupted. Sufficient information about an interrupted task is stored so that the task can be resumed when priority conditions permit. Using this method, the attention focus manager starts, interrupts, and resumes tasks on the basis of priorities. As mentioned above, COGNET assumes serial behavior with rapid attention switching, so only the highest-priority task runs at any time. Tasks are executed by the *task execution process*, which is controlled by the attention focus manager. The *action effector* changes the environment.

Operation

The operation of a COGNET model can be described by the following general example, adapted from Zachary et al. (1992):

1. Perceptual demon recognizes event, posts information on blackboard.
2. New blackboard contents satisfy triggering condition for high-level goal of task B, and that task is activated and gains the focus of attention.
3. B subrogates (explicitly turns control over) to task A for more localized or complementary analysis.
4. A reads information from blackboard, makes an inference, and posts this new information on the blackboard.
5. New information satisfies triggering condition for task D, but since it lacks sufficient priority, cannot take control.
6. Instead, new information satisfies triggering condition for higher-priority task C, which takes over.
7. C posts new information, then suspends itself to wait for its actions to take effect.
8. Task D takes control and begins posting information to blackboard.
9. etc.

Current Implementation

COGNET is written in C++ and runs in Windows 95, Windows 97, and UNIX on IBM PCs and Silicon Graphics and Sun workstations. A commercial version is in development. COGNET models run substantially faster than real time.

Support Environment

The support environment for COGNET is called GINA (for Generator of INterface Agents). Its purpose is the creation and modification of COGNET models, the implementation and debugging of executable software based on those models, and the application of executable models to interface agents. It consists of model debugging and testing tools, COGNET GCR editing tools, and a translator for switching back and forth between the GCR representation used by the modeler and the COGNET executive language version of the procedural representation necessary to run the model.

Once a model has been developed and debugged within the GINA development environment, it can be separated and run as a stand-alone computational unit, as, for example, a component in distributed simulation or an embedded model within some larger system or simulation.

Validation

A specific COGNET model in an antisubmarine warfare simulation received limited validation against both the simulated information processing problems used to develop it and four other problems created to test it. The investigators collected data on task instance predictions (that is, COGNET predictions that a real human operator would perform a certain task) and task prediction lead (that is, the amount of time the predicted task instance preceded the actual task performance). When tested against the data from the problems used to develop the model, COGNET correctly predicted 90 percent of the task instances (that is, what tasks were actually performed) with a mean task instance prediction lead time of 4 minutes, 6 seconds. When tested against the new set of information processing problems, it correctly predicted 94 percent of the tasks with a mean task prediction lead time of 5 minutes, 38 seconds. The reader should keep in mind that antisubmarine warfare is much less fast-paced than, say, tank warfare, and situations typically evolve over periods of hours rather than minutes or seconds.

Applicability for Military Simulations

COGNET models have been used as surrogate adversaries in a submarine warfare trainer and as surrogate watchstanders in a simulated Aegis combat information center. Although the performance of the models is not described in the documentation available to the panel at the time of writing, the fact that these applications were even attempted is an indication of the potential applicability of COGNET in military simulations. Clearly COGNET is a plausible framework for representing cognitive multitasking behavior. A development environment for creating such models would be of benefit to developers of military simulations. Although validation of COGNET has been limited, the results so far are promising.

COGNET is limited in psychological validity, though its developers are taking steps to remedy that problem. The degree of robustness of its decision-making and problem-solving capabilities is not clear. Its perceptual models appear to be limited, and it is severely constrained in its ability to model motor behavior, though these limitations may not be significant for many applications. Nevertheless, given its ability to provide a good framework for representing cognitive, multitasking behavior, COGNET merits further scrutiny as a potential tool for representing human behavior in military simulations. Moreover, COGNET is being enhanced to include micromodels from the human operator simulator (see below).

Executive-Process Interactive Control (EPIC)[3]

EPIC is a symbolic cognitive architecture particularly suited to the modeling of human multiple-task performance. It includes peripheral sensory-motor processors surrounding a production rule cognitive processor (Meyer and Kieras, 1997a, 1997b).

Purpose and Use

The goal of the EPIC project is to develop a comprehensive computational theory of multiple-task performance that (1) is based on current theory and results in the domains of cognitive psychology and human performance; (2) will support rigorous characterization and quantitative prediction of mental workload and performance, especially in multiple-task situations; and (3) is useful in the practical design of systems, training, and personnel selection.

Assumptions

In general, EPIC adheres to the assumptions inherent in the modified stage model (Figure 3.1). It uses a parsimonious production system (Bovair et al., 1990) as the cognitive processor. In addition, there are separate perceptual pro-

[3]This section borrows heavily from Kieras and Meyer (1997).

cessors with distinct processing time characteristics, and separate motor processors for vocal, manual, and oculomotor (eye) movements.

EPIC assumes that all capacity limitations are a result of limited structural resources, rather than a limited cognitive processor. Thus the parsimonious production system can fire any number of rules simultaneously, but since the peripheral sense organs and effectors are structurally limited, the overall system is sharply limited in capacity. For example, the eyes can fixate on only one place at a time, and the two hands are assumed to be bottlenecked through a single processor.

Like the ACT-class architectures (see the earlier discussion of ACT-R) EPIC assumes a declarative/procedural knowledge distinction that is represented in the form of separate permanent memories. Task procedures and control strategies (procedural knowledge) are represented in production rules. Declarative knowledge is represented in a preloaded database of long-term memory elements that cannot be deleted or modified. At this time, EPIC does not completely specify the properties of working memory because clarification of the types of working memory systems used in multiple-task performance is one of the project's research goals. Currently, working memory is assumed to contain all the temporary information tested and manipulated by the cognitive processor's production rules, including task goals, sequencing information, and representations of sensory inputs (Kieras et al., 1998).

Unlike many other information processing architectures, EPIC does not assume an inherent central-processing bottleneck. Rather, EPIC explains performance decrements in multiple-task situations in terms of the strategic effects of the task instructions and perceptual-motor constraints. Executive processes—those that regulate the priority of multiple tasks—are represented explicitly as additional production rules. For instance, if task instructions say a stimulus-response task should have priority over a continuous tracking task, a hand-crafted production will explicitly encode that priority and execute the hand movements of the former task before those of the latter. These are critical theoretical distinctions between EPIC and other architectures (although they may not make a practical difference in the modeling of military tasks; see Lallement and John, 1998).

At this time, there is no learning in the EPIC cognitive processor—it is currently a system for modeling task performance.

Architecture and Functionality

EPIC's cognitive, perceptual, and motor processors work together as follows. EPIC's perceptual processors are "pipelines" in that an input produces an output at a certain time later, independent of what particular time the input arrives. A single stimulus input to a perceptual processor can produce multiple outputs in working memory at different times. The first output is a representation that a perceptual event has been detected; this representation is assumed to be fixed and fairly short at about 50 milliseconds. This is followed later by a representation that describes the recognized event. The timing of this second representation is dependent on properties of the stimulus. For example, recognizing letters on a screen in a typical experiment might take on the order of 150 milliseconds after the detection time. At present, these parametric recognition times are estimated from the empirical data being modeled. The cognitive processor accepts input only every 50 milliseconds, which is consistent with data such as those of Kristofferson (1967), and is constantly running, not synchronized with outside events. The cognitive processor accepts input only at the beginning of each cycle and produces output at the end of the cycle. In each cognitive processor cycle, any number of rules can fire and execute their actions; this parallelism is a fundamental feature of the parsimonious production system. Thus, in contrast with some other information processing architectures, the EPIC cognitive processor is not constrained to do only one thing at a time. Rather, multiple processing threads can be represented simply as sets of rules that happen to fire simultaneously.

EPIC assumes that its motor processors (voice, hands, oculomotor) operate independently, but, as noted above, the hands are bottlenecked through a single manual processor. Producing motor movement involves a series of steps. First, a symbolic name for the movement to be produced is recoded by the motor processor into a set of movement features. Given this set of features, the motor processor is instructed to initiate the movement. The external device (e.g., keyboard, joystick) then detects the movement after some additional mechanical delay. Under certain circumstances, the features can be pregenerated and the motor processor instructed to execute them at a later time. In addition, the motor processor can generate only one set of features at a time, but this preparation can be done in parallel with the physical execution of a previously commanded movement.

Operation

The inputs to an EPIC model are as follows:

- A production rule representation of the procedures for performing the task
- The physical characteristics of objects in the environment, such as their color or location
- A set of specific instances of the task situation, such as the specific stimulus events and their timing
- Values of certain time parameters in the EPIC architecture

The output of an EPIC model is the predicted times and sequences of actions in the selected task instances.

Once an EPIC model has been constructed, predictions of task performance are generated through simulation of the human interacting with the task environment in simulated real time, in which the processors run independently and in parallel. Each model includes a process that represents the task environment and generates stimuli and collects the responses and their simulated times over a large number of trials. To represent human variability, the processor time parameters can be varied stochastically about their mean values with a regime that produces a coefficient of variation for simple reaction time of about 20 percent, a typical empirical value.

Current Implementation

EPIC is currently available for Macintosh and various UNIX platforms. It is written in Common Lisp and therefore easily portable to other platforms. A graphical environment is provided on the Macintosh platform via Macintosh Common Lisp. An interface between EPIC's peripherals and Soar's cognitive processor is available on the UNIX platforms.

Support Environment

EPIC is a relatively new architecture that has not yet been the object of the development time and effort devoted to other architectures, such as ACT-R, Soar, and Micro Saint. Therefore, its support environment is not specific to EPIC, but depends on a conventional Lisp programming environment. The EPIC user can rely on several sources of documentation. In addition to publications describing the architecture and research results, Kieras and Meyer (1996) describe the various components of EPIC and their interaction. EPIC's source code, written in Lisp, is heavily annotated and relatively accessible.

Documentation and source code relevant to EPIC can be accessed at <ftp://ftp.eecs.umich.edu/people/kieras/>.

Validation

As mentioned earlier, EPIC is a relatively new integrative architecture, so it has generated fewer models than older architectures such as ACT-R and Soar. However, there has been rigorous evaluation of the EPIC models against human data. Currently, models that match human data both qualitatively and quantitatively exist for the psychological refractory period (PRP) (Meyer and Kieras, 1997a, 1997b), a dual tracking/stimulus-response task (Kieras and Meyer, 1997), a tracking/decision-making task (Kieras and Meyer, 1997), verbal working-memory tasks (Kieras et al., 1998), computer interface menu search (Hornoff and Kieras, 1997), and a telephone operator call-completion task (Kieras et al., 1997).

Applicability for Military Simulations

EPIC has not yet been used for any realistic military tasks; researchers have focused more on rigorous validation of psychological tasks and limited computer-interaction tasks. There is nothing in principle that prevents EPIC from being applied for large-scale military tasks, but its current software implementation would probably not scale up to visual scenes as complex as those in military command and control work areas without an investment in software engineering and reimplementation.

Human Operator Simulator (HOS)[4]

The human operator simulator (HOS) simulates a single human operator in a human-machine system, such as an aircraft or a tank (Glenn et al., 1992). It models perception, cognition, and motor response by generating task timelines and task accuracy data.

Purpose and Use

HOS was developed to support the design of human-machine systems by allowing high levels of system performance to be achieved through explicit consideration of human capabilities and limitations. HOS is used to evaluate proposed designs prior to the construction of mockups and prototypes, which can be very time-consuming and expensive. It serves this function by generating task timelines and task accuracy predictions.

The HOS user employs a text-based editor to define environment objects, such as displays and controls. The user then analyzes operator tasks to produce a detailed task/subtask hierarchy, which is the basis for the model, and defines task procedures in a special procedural language. HOS models must be parameterized to account for moderator variables and other factors that affect behavior (see Chapter 9). The user then runs the model on representative scenarios to derive timelines and predicted task accuracy. These results are analyzed to determine whether performance is satisfactory. If not, the results may yield insight into necessary changes to the design of the human-machine interface.

Assumptions

HOS is based on several assumptions that have implications for its validity and its applicability to military simulations. First, HOS assumes that the human has a single channel of attention and time-shares tasks serially through rapid attention switching. Second, the operator's activities are assumed to be highly

[4]This description of HOS is based on documentation provided for HOS V.

proceduralized and predictable. Although decision making can be modeled, it is limited. Finally, it is assumed that the human does not make errors.

Architecture and Functionality

The HOS system consists of editors, libraries, data files, and other components that support the development and use of HOS models specialized to suit particular analysis activities. Those components most closely related to the human operator are described below.

The environment of the HOS operator is object-oriented and consists of simulation objects. Simulation objects are grouped into classes, such as the class of graphical user interface (GUI). At run time, instances of these classes are created, such as a specific GUI. These instances have attributes, such as the size, color, and position of a symbol on the GUI.

The HOS procedural language is used to define procedures the operator can perform. A HOS procedure consists of a sequence of verb-object steps, such as the following:

- Look for an object
- Perceive an attribute of an object
- Decide what to do with respect to an object
- Reach for a control object
- Alter the state of a control object
- Other (e.g., calculate, comment)

The HOS procedural language also includes the following control constructs:

- Block
- Conditional
- Loop
- Permit/prevent interruptions by other procedures
- Branch
- Invoke another procedure

At the heart of HOS functionality are its micro-models, software modules that model motor, perceptual, and cognitive processes. Micro-models calculate the time required to complete an action as a function of the current state of the environment, the current state of the operator, and default or user-specified model parameters. HOS comes equipped with default micro-models, which are configurable by the user and may be replaced by customized models.

Default micro-models include those for motor processes, which calculate the time to complete a motion based on the distance or angle to move, the desired speed, the desired accuracy, and other factors. Specific motor processes modeled include hand movements, hand grasp, hand control operations, foot movement, head movement, eye movement, trunk movement, walking, and speaking.

Micro-models for perceptual processes calculate the time required to perceive and interpret based on user-defined detection and perception probabilities. The default micro-models in this category include perception of visual scene features, reading of text, nonattentive visual perception of scene features, listening to speech, and nonattentive listening.

Micro-models for cognitive processes determine the time required to calculate, based on the type of calculation to be made, or to decide, based on the information content of the decision problem (e.g., the number of options from which to choose). Short-term memory is explicitly modeled; this model calculates retrieval probabilities based on latency.

Physical body attributes, such as position and orientation, are modeled by the HOS physical body model. The desired body object positions are determined by micro-models, and the actual positions are then updated. Body parts modeled include the eyes, right/left hand, right/left foot, seat reference position, right/left hip, right/left shoulder, trunk, head, right/left leg, and right/left arm.

In HOS, channels represent limited processing resources. Each channel can perform one action at a time and is modeled as either busy or idle. HOS channels include vision, hearing/speech, right/left hand, right/left foot, and the cognitive central processor.

HOS selection models associate micro-models with procedural language verbs. They invoke micro-models with appropriate parameters as task procedures are processed.

The HOS attention model maintains a list of active tasks. As each step in a procedure is completed, the model determines what task to work on, based on priority, and decides when to interrupt an ongoing task to work on another. Each task has a base priority assigned by the user. Its momentary priority is based on its base priority, the amount of time it has been attended to, the amount of time it has been idle since initiated, and parameters set by the user. Idle time increases momentary priority to ensure that tasks with lower base priority are eventually completed. Interrupted tasks are resumed from the point of interruption.

The attention model also determines when multiple task activities can be performed in parallel and responds to unanticipated external stimuli. The latter function is performed by a general orienting procedure, which is always active and does not require resources. When a stimulus is present, it invokes an appropriate micro-model to determine stochastically whether the operator responds. If so, the appropriate procedure becomes active.

Operation

HOS operation can be described by the following pseudocode procedure:

1. *Put* mission-level task on active task list
2. *While* active task list not empty and termination condition not present *do*
   2.1. *Get* next step from procedure of highest-priority task
   2.2. *Apply* selection model to interpret verb-object instruction
   2.3. *Invoke* micro-models and initiate channel processing, when able
   2.4. *Compute* and record task times and accuracies
   2.5. *Compute* new momentary task priorities

HOS in its pure form computes task performance metrics (times and accuracies), not actual behaviors that could be used to provide inputs to, say, a tank model. However, it could be adapted to provide such outputs and is in fact being combined with COGNET (discussed above) for that very purpose.

Current Implementation

The current version of HOS, though not fully implemented, is written in C to run on an IBM PC or compatible. Earlier versions were written in FORTRAN and ran much faster than real time on mainframe computers. Major functional elements of HOS (converted to C++) are being incorporated into a new version of COGNET (see above) that runs on a Silicon Graphics Indy.

Support Environment

The HOS operating environment includes editors for creating and modifying objects, and procedures and tools for running HOS models and analyzing the results.

Validation

Many of the individual micro-models are based on well-validated theories, but it is not clear that the overall model has been validated.

Applicability for Military Simulations

HOS has a simple, integrative, flexible architecture that would be useful for its adaptation to a wide variety of military simulation applications. HOS also contains a reasonable set of micro-models for many human-machine system applications, such as modeling a tank commander or a pilot.

On the other hand, HOS has a number of limitations. It is currently a standalone system, not an embeddable module. Also, the current status of its micro-models may prevent its early use for modeling dismounted infantry. For example, many of the micro-models are highly simplified in their default forms and require parameterization by the user. The attention model, though reasonable, appears to be rather ad hoc. The architecture may also cause brittleness and poor applicability in some applications that differ from those for which it was designed. Moreover, HOS computes times, not behavior, though it should be augmentable. Knowledge representation is very simple in HOS and would not likely support sophisticated perception and decision-making representations. Finally, the user must define interruption points to allow multitasking.

The integration of HOS with COGNET should overcome many of the limitations of each.

Micro Saint

Micro Saint is a modern, commercial version of systems analysis of integrated networks of tasks (SAINT), a discrete-event network simulation language long used in the analysis of complex human-machine systems. Micro Saint is not so much a model of human behavior as a simulation language and a collection of simulation tools that can be used to create human behavior models to meet user needs. Yet many Micro Saint models have been developed for military simulations (e.g., Fineberg et al., 1996; LaVine et al., 1993, 1996), and the discussion here, which is based on the descriptions of Laughery and Corker (1997), is in that context. It is worth noting that such Micro Saint models are used as part of or in conjunction with other analysis tools, such as WinCrew, IMPRINT, HOS V (see above), task analytic work load (TAWL), and systems operator loading evaluation (SOLE).

Purpose and Use

Micro Saint is used to construct models for predicting human behavior in complex systems. These models, like the HOS models discussed above, yield estimates of times to complete tasks and task accuracies; they also generate estimates of human operator workload and task load (i.e., the number of tasks an operator has to perform, over time). These are not outputs that could be applied as inputs to other simulation elements, though in theory the models could be adapted to provide that capability.

Assumptions

Micro Saint modeling rests on the assumption that human behavior can be modeled as a set of interrelated tasks. That is, a Micro Saint model has at its heart a task network. To the user of these models, what is important is task completion time and accuracy, which are modeled stochastically using probability distributions whose parameters are selected by the user. It is assumed that the operator workload (which may affect human performance) imposed by individual tasks can be aggregated to arrive at composite workload measures. Further assumptions are that system behavior can be modeled as a task (or function) network and that the environment can be modeled as a sequence of external events.

Architecture and Functionality

The architecture of Micro Saint as a tool or framework is less important than the architecture of specific Micro Saint models. Here we consider the architecture and functionality of a generic model.

The initial inputs to a Micro Saint model typically include estimates (in the form of parameterized probability distributions) for task durations and accuracies. They also include specifications for the levels of workload imposed by tasks. During the simulation, additional inputs are applied as external events that occur in the simulated environment.

Both human operators and the systems with which they interact are modeled by task networks. We focus here on human task networks, with the understanding that system task network concepts form a logical subset. The nodes of a task network are tasks. Human operator tasks fall into the following categories: visual, numerical, cognitive, fine motor (both discrete and continuous), gross motor, and communications (reading, writing, and speaking). The arcs of the network are task relationships, primarily relationships of sequence. Information is passed among tasks by means of shared variables.

Each task has a set of task characteristics and has a name as an identifier. The user must specify the type of probability distribution used to model the task's duration and provide parameters for that distribution. A task's release condition is the condition(s) that must be met before the task can start. Each task can have some effect on the overall system once it starts; this is called its beginning effect. Its ending effect is how the system will change as a result of task completion. Task branching logic defines the decision on which path to take (i.e., which task to initiate) once the task has been completed. For this purpose, the user must specify the decision logic in a C-like programming language. This logic can be probabilistic (branching is randomized, which is useful for modeling error), tactical (a branch goes to the task with the highest calculated value), or multiple (several subsequent tasks are initiated simultaneously). Task duration and accuracy can be altered further by means of performance-shaping functions used to model the effects of various factors on task performance. These factors can include personnel characteristics, level of training, and environmental stressors (see Chapter 9). In practice, some performance-shaping functions are derived empirically, while some are derived from subjective estimates of subject matter experts.

The outputs of a Micro Saint model include mission performance data (task times and accuracies) and workload data.

Operation

Because Micro Saint models have historically been used in constructive (as opposed to virtual) simulations, they execute in fast time (as opposed to real time). The simulation is initialized with task definitions (including time, and accuracy parameters). Internal and external initial events are scheduled; as events are processed, tasks are initiated, beginning effects are computed, accuracy data are computed, workloads are computed, and task termination events are scheduled. As task termination events are processed, the system is updated to reflect task completions (i.e., task ending effects).

Current Implementation

Micro Saint is a commercial product that runs on IBM PCs and compatible computers running Windows. It requires 8 megabytes of random access memory and 8 megabytes of disk space. DOS, OS 2, Macintosh, and UNIX versions are available as well.

Micro Saint is also available in a version integrated with HOS (described above). The Micro Saint-HOS simulator is part of the integrated performance modeling environment (IPME), a network simulation package for building models that simulate human and system performance. The IPME models consist of a workspace design that represents the operator's work environment; a network simulation, which is a Micro Saint task network, and micro-models. These micro-models (which come from HOS) calculate times for very detailed activities such as walking, speaking, and pushing buttons. They provide an interface between the network and the workspace (i.e., the environment), and they offer a much finer level of modeling resolution than is typically found in most Micro Saint networks. The integrated performance package runs on UNIX platforms.

Support Environment

The Micro Saint environment includes editors for constructing task networks, developing task descriptions, and defining task branching decision logic; an expandable function library; data collection and display modules; and an animation viewer used to visualize simulated behavior.

Validation

Since Micro Saint is a tool, it is Micro Saint models that must be validated, not the tool itself. One such model was a workload model showing that a planned three-man helicopter crew design was unworkable; later experimentation with human subjects confirmed that prediction. HARDMAN III, with Micro Saint as the simulation engine, has gone through the Army verification, validation, and accreditation process—one of the few human behavior representations to do so.

Applicability for Military Simulations

Micro Saint has been widely employed in constructive simulations used for analysis to help increase military system effectiveness and reduce operations and support costs through consideration of personnel early in the system design process. Examples include models to determine the effects of nuclear, biological, and chemical agents on crews in M1 tanks, M2 and M3 fighting vehicles, M109 155-millimeter self-propelled howitzers, and AH64 attack helicopters. Micro Saint models have also been developed for the analysis of command and control message traffic, M1A1 tank maintenance, and DDG 51 (Navy destroyer) harbor entry operations.

Although Micro Saint has not yet been used directly for training and other real-time (virtual) simulations, Micro Saint models have been used to derive tables for human performance decrements in time and accuracy. Those tables have been used in real-time ModSAF simulations.

Micro Saint models of human behavior are capable of computing task times and task accuracies, the latter providing a means for explicitly modeling human error—a feature lacking in many models. The models can also be readily configured to compute operator workload as a basis for modeling multitasking. Used in conjunction with HOS V micro-models, Micro Saint has the capability to represent rather detailed human behaviors, at least for simple tasks. As noted, it is a commercial product, which offers the advantage of vendor support, and its software support environment provides tools for rapid construction and testing of models.

On the other hand, since Micro Saint is a tool, not a model, the user is responsible for providing the behavioral modeling details. Also, in the absence of HOS V micro-models, Micro Saint is most suited for higher levels of abstraction (lower model resolution). Being a tool and lacking model substance, Micro Saint also lacks psychological validity, which the user must therefore be responsible for providing. Knowledge representation is rudimentary, and, other than a basic branching capability, there is no built-in inferencing mechanism with which to develop detailed models of complex human cognitive processes; such features must be built from scratch.

Nevertheless, Micro Saint has already shown merit through at least limited validation and accreditation and has further potential as a good tool for building models of human behavior in constructive simulations. Being a commercial product, it is general purpose and ready for use on a wide variety of computer platforms. It also has been used and has good potential for producing human performance tables and other modules that could be used in virtual simulations. With the micro-model extensions of HOS V and a mechanism for converting fast-time outputs to real-time events, it is possible that Micro Saint models could even be used directly in virtual simulations.

Man Machine Integrated Design and Analysis System (MIDAS)

MIDAS is a system for simulating one or more human operators in a simulated world of terrain, vehicles, and other systems (Laughery and Corker, 1997; Banda et al., 1991; see also Chapters 7 and 8).

Purpose and Use

The primary purpose of MIDAS is to evaluate proposed human-machine system designs and to serve as a testbed for behavioral models.

Assumptions

MIDAS assumes that the human operator can perform multiple, concurrent tasks, subject to available perceptual, cognitive, and motor resources.

Architecture and Functionality

The overall architecture of MIDAS comprises a user interface, an anthropometric model of the human operator, symbolic operator models, and a world model. The user interface consists of an input side (an interactive GUI, a cockpit design editor, an equipment editor, a vehicle route editor, and an activity editor) and an output side (display animation software, run-time data graphical displays, summary data graphical displays, and 3D graphical displays).

MIDAS is an object-oriented system consisting of objects (grouped by classes). Objects perform processing by sending messages to each other. More specifically, MIDAS consists of multiple, concurrent, independent agents.

There are two types of physical component agents in MIDAS: equipment agents are the displays and controls with which the human operator interacts; physical world agents include terrain and aeronautical equipment (such as helicopters). Physical component agents are represented as finite-state machines, or they can be time-script-driven or stimulus-response-script-driven. Their behaviors are represented using Lisp methods and associated functions.

The human operator agents are the human performance representations in MIDAS—cognitive, perceptual, and motor. The MIDAS physical agent is Jack™, an animated mannequin (Badler et al., 1993). MIDAS uses Jack to address workstation geometry issues, such as the placement of displays and controls. Jack models the operator's hands, eyes, and feet, though in the MIDAS version, Jack cannot walk.

The visual perception agent computes eye movements, what is imaged on the retina, peripheral and foveal fields of view, what is in and out of focus relative to the fixation plane, preattentional phenomena (such as color and flashing), detected peripheral stimuli (such as color), and detailed information perception. Conditions for the latter are that the image be foveal, in focus, and consciously attended to for at least 200 milliseconds.

The MIDAS updatable world representation is the operator's situation model, which contains information about the external world. Prior to a MIDAS simulation, the updatable world representation is preloaded with mission, procedure, and equipment information. During the simulation, the updatable world representation is constantly updated by the visual perception agent; it can deviate from ground truth because of limitations in perception and attention. Knowledge representation in the updatable world representation is in the form of a semantic net. Information in the updatable world representation is subject to decay and is operated on by daemons and rules (see below).

The MIDAS operator performs activities to modify its environment. The representation for an operator activity consists of the following attributes: the preconditions necessary to begin the action; the satisfaction conditions, which define when an action is complete; spawning specifications, which are constraints on how the activity can be decomposed; decomposition methods to produce child activities (i.e., the simpler activities of which more complex activities are composed); interruption specifications, which define how an activity can be interrupted; activity loads, which are the visual, auditory, cognitive, and psychomotor resources required to perform the activity; the duration of the activity; and the fixed priority of the activity. Activities can be forgotten when interrupted.

MIDAS includes three types of internal reasoning activities. Daemons watch the updatable world representation for significant changes and perform designated operations when such changes are detected. IF-THEN rules provide flexible problem-solving capabilities. Decision activities select from among alternatives. Six generalized decision algorithms are available: weighted additive, equal weighted additive, lexicographic, elimination by aspect, satisficing conjunctive, and majority of confirming decisions.

There are also various types of primitive activities (e.g., motor activities, such as "reach," and perceptual activities, such as "fixate"). Therefore, the overall mission goal is represented as a hierarchy of tasks; any nonprimitive task is considered a goal, while the leaf nodes of the mission task hierarchy are the primitive activities.

MIDAS models multiple, concurrent activities or tasks. To handle task contention, it uses a scheduling algorithm called the Z-Scheduler (Shankar, 1991). The inputs to the Z-Scheduler are the tasks or activities to be scheduled; the available visual, auditory, cognitive, and motor resources; the constraints on tasks and resources (that is, the amount of each resource available and the times those amounts are available); and the primary goal (e.g., to minimize time of completion or balance resource loading). The Z-Scheduler uses a blackboard architecture. It first builds partial dependency graphs representing prerequisite and other task-to-task relations. Next it selects a task and commits resource time slices to that task. The task is selected on the basis of its criticality (the ratio of its estimated execution time to the size of the time window available to perform it). The Z-Scheduler interacts with the task loading model (see below). The output of the Z-Scheduler is a task schedule. There is also the option of using a much simpler scheduler, which selects activities for execution solely on the basis of priorities and does not use a time horizon.

The MIDAS task loading model provides information on task resource requirements to the Z-Scheduler. It assumes fixed amounts of operator resources (visual, auditory, cognitive, and motor), consistent with multiple resource theory (Wickens, 1992). Each task or activity requires a certain amount of each resource. The task loading model keeps track of the available resources and passes that information to the Z-Scheduler.

Operation

In a MIDAS simulation, declarative and procedural information about the mission and equipment is held in the updatable world representation. Information from the external world is filtered by perception, and the updatable world representation is updated. Mission goals are decomposed into lower-level activities or tasks, and these activities are scheduled. As the activities are performed, information is passed to Jack, whose actions affect cockpit equipment. The external world is updated, and the process continues.

Current Implementation

MIDAS is written in the Lisp, C, and C++ programming languages and runs on Silicon Graphics, Inc., workstations. It consists of approximately 350,000 lines of code and requires one or more workstation to run on. It is 30 to 40 times slower than real time, but can be simplified so it can run at nearly real time.

MIDAS Redesign

At the time of this writing, a new version of MIDAS was in development (National Aeronautics and Space Administration, 1996). The following brief summary is based on the information currently available.

The MIDAS redesign is still object-oriented, but is reportedly moving away from a strict agent architecture. The way memory is modeled is better grounded in psychological theory. MIDAS' long-term memory consists of a semantic net; its working memory has limited node capacity and models information decay.

The modeling of central processing (that is, working memory processes) is different from that of the original MIDAS. An event manager interprets the significance of incoming information on the basis of the current context (the goals being pursued and tasks being performed). The context manager updates declarative information in the current context, a frame containing what is known about the current situation. MIDAS maintains a task agenda—a list of goals to be accomplished—and the agenda manager (similar to the Z-Scheduler in the previous MIDAS) determines what is to be done next on the basis of the task agenda and goal priorities. The procedure selector selects a procedure from long-term memory for each high-level goal to be accomplished. The procedure interpreter executes tasks or activities by consulting the current context to determine the best method. Motor behavior is executed directly. A high-level goal is decomposed into a set of subgoals (which are passed to the agenda manager). A decision is passed to the problem solver (see below). The procedure interpreter also sends commands to the attention manager to allocate needed attentional resources. The problem solver, which embodies the inferencing capabilities of the system, does the reasoning required to reach decisions.

The redesigned MIDAS can model multiple operators better than the original version. Like the original, it performs activities in parallel, resources permitting, and higher-priority tasks interrupt lower-priority tasks when sufficient resources are not available. It models motor error processes such as slips and it explicitly models speech output.

Support Environment

The MIDAS support environment has editors and browsers for creating and changing system and equipment specifications, and operator procedures and tools for viewing and analyzing simulation results. Currently much specialized knowledge is required to use these tools to create models, but it is worth noting that a major thrust of the MIDAS redesign is to develop a more self-evident GUI that will allow nonprogrammers and users other than the MIDAS development staff to create new simulation experiments using MIDAS. In addition, this version will eventually include libraries of models for several of the more important domains of MIDAS application (rotorcraft and fixed-wing commercial aircraft). The intent is to make the new version of MIDAS a more generally useful tool for human factors analysis in industry and a more robust testbed for human performance research.

Validation

The original version of MIDAS has been validated in at least one experiment involving human subjects. MIDAS was programmed to model the flight crew of a Boeing 757 aircraft as they responded to descent clearances from air traffic control: the task was to decide whether or not to accept the clearance and if so, when to start the descent. The model was exercised for a variety of scenarios. The experimenters then collected simulator data with four two-pilot crews. The behavior of the model was comparable to that of the human pilots ($0.23 \leq p \leq 0.33$) (Laughery and Corker, 1997).

Applicability for Military Simulations

In both versions, MIDAS is an integrative, versatile model with much (perhaps excess) detail. Its submodels are often based on current psychological and psychomotor theory and data. Its task loading model is consistent with multiple resource theory. MIDAS explicitly models communication, especially in the new version. Much modeling attention has been given to situation awareness with respect to the updatable world representation. There has been some validation of MIDAS.

On the other hand, MIDAS has some limitations. Many MIDAS behaviors, such as operator errors, are not emergent features of the model, but must be explicitly programmed. The Z-Scheduler is of dubious psychological validity. The scale-up of the original MIDAS to multioperator systems would appear to be quite difficult, though this problem is being addressed in the redesign effort. MIDAS is also too big and too slow for most military simulation applications. In addition, it is very labor-intensive, and it contains many details and features not needed in military simulations.

Nevertheless, MIDAS has a great deal of potential for use in military simulations. The MIDAS architecture (either the original version or the redesign) would provide a good base for a human behavior representation. Components of MIDAS could be used selectively and simplified to provide the level of detail and performance required. Furthermore, MIDAS would be a good testbed for behavioral representation research.

Neural Networks

In the past decade, great progress has been made in the development of general cognitive systems called artificial neural networks (Grossberg, 1988), connectionistic networks (Feldman and Ballard, 1982), or parallel distributed processing systems (Rumelhart et al., 1986b; McClelland and Rumelhart, 1986). The approach has been used to model a wide range of cognitive processes and is in widespread use in cognitive science. Yet neural network modeling appears to be quite different from the other architectures reviewed in this section in that it is more of a computational approach than an integrative human behavior architecture. In fact, the other architectures reviewed here could be implemented with neural networks: for example, Touretzky and Hinton (1988) implemented a simple production system and an experimental version of Soar was implemented in neural nets (Cho et al., 1991).

Although the neural network approach is different in kind from the other architectures reviewed, its importance warrants its inclusion here. There are many different approaches to using neural nets, including adaptive resonance theory (Grossberg, 1976; Carpenter and Grossberg, 1990), Hopfield nets (Hopfield, 1982), Boltzman machines (Hinton and Sejnowski, 1986), back propagation networks (Rumelhart et al., 1986a), and many others (see Arbib, 1995, or Haykin, 1994, for comprehensive overviews). The discussion in this section attempts to generalize across them. (Neural networks are also discussed in the section on hybrid architectures, below, because they are likely to play a vital role in overcoming some weaknesses of rule-based architectures.)

Purpose and Use

Unlike the rule-based systems discussed in this chapter, artifical neural networks are motivated by principles of neuroscience, and they have been developed to model a broad range of cognitive processes. Recurrent unsupervised learning models have been used to discover features and to self-organize clusters of stimuli. Error correction feed-forward learning models have been used extensively for pattern recognition and for learning nonlinear mappings and prediction problems. Reinforcement learning models have been used to learn to control dynamic systems. Recurrent models are useful for learning sequential behaviors, such as sequences of motor movements and problem solving.

Assumptions

The two major assumptions behind neural networks for human behavior modeling are that human behavior in general can be well represented by self-organizing networks of very primitive neuronal units and that all complex human behaviors of interest can be learned by neural networks through appropriate training. Through extensive study and use of these systems, neural nets have come to be better understood as a form of statistical inference (Mackey, 1997; White, 1989).

Architecture and Functionality

A neural network is organized into several layers of abstract neural units, or nodes, beginning with an input layer that interfaces with the stimulus input from the environment and ending with an output layer that provides the output response interface with the environment. Between the input and output layers are several hidden layers, each containing a large number of nonlinear neural units that perform the essential computations. There are generally connections between units within layers, as well as between layers.

Neural network systems use distributed representations of information. In a fully distributed system, the connections to each neural unit participate (to various degrees) in the storage of all inputs that are experienced, and each input that is experienced activates all neural units (to varying degrees). Neural networks are also parallel processing systems in the sense that activation spreads and flows through all of the nodes simultaneously over time. Neural networks are content addressable in that an input configuration activates an output retrieval response by resonating to the input, without a sequential search of individual memory locations.

Neural networks are robust in their ability to respond to new inputs, and they are flexible in their ability to produce reasonable outputs when provided noisy inputs. These characteristics follow from the ability of neural networks to generalize and extrapolate from new noisy inputs on the basis of similarity to past experiences.

Universal approximation theorems have been proven to show that these networks have sufficient computational power to approximate a very large class of nonlinear functions. Convergence theorems have been provided for these learning algorithms to prove that the learning algorithm will eventually converge on a maximum. The capability of deriving general mathematical properties from neural networks using general dynamic system theory is one of the advantages of neural networks as compared with rule-based systems.

Operation

The general form of the architecture of a neural network, such as whether it is feed-forward or recurrent, is designed manually. The number of neural units and the location of the units can grow or decay according to pruning algorithms that are designed to construct parsimonious solutions during training. Initial knowledge is based on prior training of the connection weights from past examples. The system is then trained with extensive data sets, and learning is based on algorithms for updating the weights to maximize performance. The system is then tested on its ability to perform on data not in the training sets. The process iterates until a satisfactory level of performance on new data is attained. Models capable of low-level visual pattern recognition, reasoning and inference, and motor movement have been developed.

The neural units may be interconnected within a layer (e.g., lateral inhibition) as well as connected across layers, and activation may pass forward (projections) or backward (feedback connections). Each unit in the network accumulates activation from a number of other units, computes a possibly nonlinear transformation of the cumulative activation, and passes this output on to many other units, with either excitatory or inhibitory connections. When a stimulus is presented, activation originates from the inputs, cycles through the hidden layers, and produces activation at the outputs.

The activity pattern across the units at a particular point in time defines the state of the dynamic system at that time. The state of activation evolves over time until it reaches an equilibrium. This final state of activation represents information that is retrieved from the memory of the network by the input stimulus. The persistence of activation produced by stimulus is interpreted as the short-term memory of the system.

Long-term storage of knowledge is represented by the strengths of the connections from one neural unit to another, which are called connection weights. The initial weights represent prior knowledge before training begins. During training, the weights are updated by a learning process that often uses a gradient ascent algorithm designed to search for a set of weights that will maximize a performance function.

Most neural network systems are stand-alone systems that do not have a real-time connection to an information source or the outside world, but some systems do interact with the real world. For instance, the NavLab project uses neural networks for its road-following navigation system and has built 10 robot vehicles (three vans, two HMMWVs, three sedans, and two full-sized city buses), one of which traveled 98.2 percent of the trip from Washington, D.C. to San Diego, California, without human intervention, a distance of 2,850 miles (Jochem et al., 1995). However, even if performance is in real time, the training prior to performance is usually computationally intensive and done off-line.

Current Implementation

Packages for designing and simulating neural network behavior are available from many sources. Some textbooks come packaged with software, for example, McClelland and Rumelhart,1988, and its successor PDP++ from the Center for the Neural Basis of Cognition (Carnegie Mellon University). Commercial packages abound (and can be found on the Internet by searching for "neural networks"), but they are often not extensible. One useful Windows and menu-based package is available from the commonly used engineering programming language MATLAB. However, it is relatively easy to develop models using any mathematical programming language that includes matrix operators (such as MATHEMATICA).

Support Environment

There are numerous excellent textbooks that can help in designing and programming neural networks (see, e.g., Anderson, 1997; Golden, 1996; Haykin, 1994; Levin, 1991; McClelland and Rumelhart, 1988). Programming documentation for software is available from MATLAB, PDP++, and commercial packages. Many of the newer packages have GUI interfaces and associated editing and debugging tools. Courses in neural networks are commonly taught in cognitive science, or engineering, or computer science departments at major universities.

Validation

Neural networks applied to particular tasks have been extensively and rigorously tested in both cognitive psychology experiments and engineering applications. A comprehensive review of this large literature is beyond the scope of this chapter, but some examples include word perception and individual word reading (McClelland and Rumelhart, 1981; Seidenberg and McClelland, 1989; Hinton and Shallice, 1991; Plaut et al., 1996), many memory tasks (e.g., Knapp and Anderson, 1984; McClelland and Rumelhart, 1985), control of automatic processes (Cohen et al., 1990), problem-solving and reasoning (Rumelhart et al., 1986c; Rumelhart, 1989) and motor control (e.g., Burnod et al., 1992; Kawato et al., 1990).[5] The ART architecture, in particular, is supported by applications of the theory to psychophysical and neurobiological data about vision, visual object recognition, auditory streaming, variable-rate speech perception, cognitive information processing, working memory, temporal planning, and cognitive-emotional interactions, among others (e.g., Bradski et al., 1994; Carpenter and Grossberg, 1991; Grossberg, 1995; Grossberg et al., 1997; Grossberg and Merrill, 1996).

Application for Military Simulations

Most neural network applications have been designed to model a very small part of the cognitive system, often sensory/motor processes. We know of no work that models human behavior at the level of reasoning about military strategy and tactics or performance of the tasks to carry out those decisions. Researchers differ in their opinions as to whether it is possible for neural networks to support a general cognitive architecture with high-level reasoning skills involving structured information: see Sharkey and Sharkey (1995) and Touretzky (1995) for opposing views. Given this open research issue, systems composed of sets of neural networks might be feasible for some tasks, or hybrid systems of neural networks and symbolic architectures (discussed further in the last section of this chapter) might be most suitable for military applications. For instance, the NavLab project on autonomous driving (though not a model of human driving behavior) combined neural networks for reactive tasks, such as road-following and obstacle avoidance, with symbolic knowledge and processes to plan and

[5]But see Pinker and Prince (1988) and Coltheart et al. (1993) for some criticisms of how neural net models are fit to human data, and the ensuing debate: MacWhinney and Lienbach (1991) and Plaut et al. (1996).

execute those plans (Pomerleau et al., 1991). An additional issue is that for high-knowledge domains, e.g., such as the responsibilities of a platoon leader, the off-line training required for a neural net system would be very time consuming and computationally expensive.

Operator Model Architecture

The operator model architecture (OMAR) models human operators in complex systems, such as command and control systems, aircraft, and air traffic control systems (Deutsch et al., 1993, 1997; Deutsch and Adams, 1995; MacMillan et al., 1997).

Purpose and Use

The primary purpose of OMAR is the evaluation of operator procedures in complex systems. It is also intended to be used to evaluate system design (in particular the design of displays and controls) and as a testbed for developing human behavior models.

Assumptions

OMAR is based on certain assumptions about attributes of human behavior. The first is that much human behavior is proactive, that is, goal directed. Humans perform multiple tasks concurrently, so OMAR includes explicit representations of goals. Multiple concurrent tasks contend for limited sensory, cognitive, and motor resources, so OMAR models include task instances. Parallel behavior is limited to cases in which some of the concurrent tasks are overlearned and almost automatic, requiring no conscious thought; therefore, in OMAR the execution of such "automatic" procedures is done in parallel with a cognitive task. Some behaviors are rule based, and humans display differing levels of skill in performing activities. A final and very significant assumption is that operators work in teams, so OMAR has provisions for modeling multiple communicating operators.

Architecture and Functionality

The major architectural components of OMAR are a set of knowledge representation languages, a knowledge acquisition environment (including editors and browsers), the target system model, scenario representations, one or more human operator models, compilers, an event-based simulator, online animation tools, and post-run analysis tools.

OMAR can be described as a set of interrelated layers. The bottom layer is the core simulation layer. It is built in the Common Lisp object system. The simple frame language, used to define simulation objects, adds essential features to the Common Lisp object system. The SCORE language, which is an extension of Common Lisp, adds robust parallel syntax and semantics to support the parallel processing required by the model. Also included is a forward-chaining rule-based language. The core layer simulator is a discrete-event simulator, which uses time-sorted queues of future events called closures. It allows individual threads of execution (parallel execution) and implements asynchronous event monitoring (closure-condition associations).

The perceptor/effector model layer is a place-holder for detailed perceptor (sensory) and effector (motor) models. Default models are provided, and specialized perceptor/effector models have also been developed for event types of particular workplace environments (e.g., commercial aircraft flight decks and air traffic control workstations) and for in-person, telephone, and party-line radio communication to support team-based activities.

At the cognitive level, OMAR consists of agents, which are entities capable of executing goals, plans, and tasks. Generally, a human operator model consists of a single agent, but operator model components (e.g., perception) can be modeled as agents as well. Goals, plans, tasks, and procedures represent proactive components of the human operator model. A goal is an explicit statement of what the agent is to do, specifying the activation circumstances and the procedure for computing the goal's priority (also called activation level). Goals are decomposed into and/or trees. A plan is the set of leaves of a goal tree. A task is the instantiation of a plan; it encapsulates and activates a network of procedures to perform the task and describes the resources needed by the procedures. The goal achievement simulator attempts to execute subgoals and plans that make up goal trees. OMAR tries to catch the interplay between proactive and reactive behaviors: goals and plans set up proactive capabilities, forming a dynamically structured network of procedures responsive to sensory input that "guides" the reactive response to events through pattern matching to appropriate goal-directed behaviors.

In OMAR, tasks compete for perceptor, cognitive, and effector resources. There may be contention within classes of tasks (e.g., two tasks competing for the eyes) or between classes of tasks (e.g., a task to speak vs. a task to listen). Task contention takes into account the dynamic priorities (activation levels) of competing tasks, the class membership of competing tasks, how close tasks are to completion, and the cost of interrupting an ongoing task (inertia). Deliberative (or thoughtful) task scheduling is modeled as well.

Operation

As OMAR runs, stimuli simulated in the external world are processed by perceptor models. Task priorities or activation levels are dynamically computed on the basis of existing conditions. The task with the highest activation level runs until it is finished or is preempted by another task that has acquired sufficient activation credits. The execution of a task is performed by a network of procedures, approximating a data flow system.

The OMAR operator generally runs serially, with some parallelism for tasks with high levels of automaticity. Effector models simulate behavioral responses that change the simulated environment, and new goals are then created.

Current Implementation

The current implementation of OMAR is programmed in Common Lisp and extensions, as described above, and runs on Sun and Silicon Graphics workstations. At the time of this writing, it was being ported to IBM PCs, and to increase portability, all interface code was being converted to Java. For many applications, the current implementation runs faster than real time.

Support Environment

OMAR's developers have provided a toolkit to facilitate building models of human performance. It consists of a concept editor for creating and modifying concepts (objects and relations that comprise networks), a procedure browser for viewing procedures and procedure networks, a simulator for running the model, and tools for analyzing model performance both during and after a run. A complete manual is available (BBN, 1997).

Validation

Many OMAR submodels are well grounded in psychological theory, but there has been little or no validation of the model as a whole. However, OMAR has been used in mixed human-in-the-loop and model simulations, and the OMAR models seem to interact well with the real human operators.

Applicability for Military Simulations

As noted above, OMAR is an integrative architecture that is well grounded in psychological theory. It appears to be flexible and versatile. A powerful toolkit is available for creating and exercising models. OMAR provides much detail and, important for the simulation of unit behavior, can be used to model teams of operators. Its limitations are that it is a very complex model and may be difficult for users not well trained in behavioral modeling. It is also likely to be labor intensive. Nevertheless, OMAR offers one of the best tradeoffs between capabilities and limitations of all the integrative architectures reviewed here.

Situation Awareness Model for Pilot-in-the-Loop Evaluation (SAMPLE)

SAMPLE is the most recent version of a line of models that began with the procedure oriented crew (PROCRU) model. SAMPLE and its antecedents have been used to represent individual operators, as well as crews of complex human-machine systems (Baron et al., 1980; Zacharias et al., 1981, 1994, 1996). The following discussion takes a general perspective, and not all of the characteristics described apply to every variant.

Purpose and Use

SAMPLE and its antecedents have been used in the examination of crew procedures. Some variations are used to model single operators. Variations of SAMPLE have been used in analysis of the approach to landing procedures of commercial transport aircraft (PROCRU), analysis of antiaircraft artillery system procedures (AAACRU), evaluation of nuclear power plant control automation/ decision aiding (crew/system integration model [CSIM]), and air combat situation awareness analysis. Most recently, SAMPLE has been used to evaluate air traffic alerting systems in a free flight environment.

Assumptions

Assumptions basic to SAMPLE are that the behavior of the crew (in some cases an individual operator) is guided by highly structured, standard procedures and driven by detected events and assessed situations. Some variants assume a multitasking environment, though this aspect is less emphasized in the later model. In all cases, however, the crew is concerned primarily with performing situation assessment, discrete procedure execution, continuous control, and communication.

Architecture and Functionality

The SAMPLE architecture consists of a system model and one or more human operator models.

The system model includes system dynamics (e.g., ownship, the plant, or a target), which are modeled by partial differential equations of motion (e.g., point mass equations for vehicle trajectory). The system dynamics can be modeled at any level of complexity desired.

A human operator model exists for each crew member. It consists of sensory and effector channels and several processors.

The sensory channels model visual and auditory sensing. Both are based on an optimal control model with no perceptual delay. They model limitations due to observation noise and stimulus energy with respect to defined thresholds. All auditory information is assumed to be heard correctly. The operator cannot process all sources of information simultaneously, so the model must decide what to attend to. The attention allocator uses suboptimal attention allocation algorithms for this purpose.

The information processor monitors the system and environment based on inputs from the rest of the simulation. It consists of two time-varying Kalman filters: the continuous state estimator produces estimates and other statistics of continuous system state; the discrete event detector generates occurrence probabilities of operationally relevant event cues.

The situation assessor uses event cue probabilities to generate occurrence probabilities of possible situations facing the operator. Earlier versions of SAMPLE used a fixed, predefined set of possible situations facing the pilot or operator; more recent versions allow for flexible situational definitions by building up a situation from a number of situational attributes (e.g., an overall situation may be defined by the component threat situation, friendly situation, fuel situation, armament situation, and so on). The situation assessor uses belief nets to model the process of situation assessment. Belief net nodes represent situations, events, and event cues detected by the discrete event detector. Belief net arcs (or arrows) represent associative and inferential dependencies between nodes. Situation awareness is modeled as an inferential diagnostic process: event cues are detected, events are deduced, and situations are inferred using Bayesian logic. (See Chapter 7 for a more detailed discussion of this material.)

Procedures form the core of the SAMPLE human operator model. The modeled operator faces a number of procedures or tasks that can be performed, either continuously or in response to events. All crew actions (except decisions about what procedures to perform) are determined by procedures. Procedures may comprise subprocedures; they are interruptible; and they are assumed to be performed essentially correctly, though a wrong procedure can be selected, and errors can result from improper decisions based on lack of information quantity or quality.

There are several classes of procedures, depending on the domain application. In early PROCRU efforts, there are six classes: system control procedures (e.g., aircraft or antiaircraft system control), system monitoring procedures (for system status and events), verbal request and callout procedures, subsystem monitoring and control procedures (e.g., servicing an altitude alert subsystem), verbal acknowledgment procedures, and miscellaneous procedures (for modeling convenience). Procedures are represented by both procedural code and rules. In more recent applications of SAMPLE, procedural classes have been extended to cover a number of tactical air combat procedures.

The execution of a procedure results in a sequence of actions that affect the modeled system and/or its environment. Procedures also compute the time required to complete each associated action. Completion times may be deterministic or sampled from a probability distribution. As mentioned above, procedure execution is assumed to be error free.

The procedure selector is the component that determines which procedure the operator model will execute next. Given an assessed situation, it generates selected procedures, determining which procedure to execute next as a function of expected gain. A procedure's expected gain is a function of a constant relative "value" of the procedure, the situational relevance of the procedure at the current instant, and time. Expected gain increases with time until the procedure is executed or is declared as missed.

Procedure actions are implemented by the procedure effector, which generates attention requests, control actions, and communication actions. Attention requests are internal "actions" that guide the attentional focus of the sensory channels toward specific information sources. Control actions (which, for example, control the airplane) may be continuous or discrete. Communication actions transfer information to some extrinsic node in the simulation. The information may be about the system state, it may be a command, or it may be about some event.

SAMPLE is implemented as an object-oriented software agent. Each of the key system components (sensory channel, information processor, situation assessor, decision maker, and procedure effector) is a software object that communicates with its neighbors via message passing. The topology of the model is defined by specifying the functionality of each component (which may have its own hierarchy of subcomponents linked via message passing) and the connectivity among the components. Multiple agents communicate with each other by means of the same message-passing methodology, which is based on the command and control simulation interface language (CCSIL) framework used in distributed interactive simulation.

In SAMPLE, no restrictions are placed on the algorithmic composition of any of the object classes, provided they conform to a well-defined class interface. The software interface is structured to facilitate integration and run-time linking of third-party software libraries (e.g., expert system languages such as NASA's CLIPS, numerical algorithms, database access, and any other custom software tools). The topology is defined using a set of ASCII data files that specify the contents and connectivity of each system block.

Operation

SAMPLE operates in the following manner. Continuous system control procedures are always enabled. Events in the system or environment are detected by the discrete event detector, and discrete procedures are enabled by event detections. Enabled procedures compete for the operator's attention on the basis of expected gain, and the individual operator can execute only one procedure at a time. Procedure execution results in actions that affect the simulated system and its environment.

The model generates several outputs, most of them time series. These include system state trajectory information, each crew member's estimate of the system and environment state, each crew member's attention allocation, system control inputs, a list of significant events, assessed situations, a procedural time line denoting the procedures being executed by each crew member, a message time line indicating communication traffic and auditory signals, and a milestone time line showing important milestones and events as defined by the modeler.

Current Implementation

Recent variants of SAMPLE are written in C++. They run on an IBM PC under the Windows 95 operating system.

Support Environment

SAMPLE has a limited programming environment, but work is under way on developing a graphical editor for model specifications.

Validation

SAMPLE and its descendants have been exercised in a wide variety of analysis projects, so their performance has come under the scrutiny of individuals familiar with human operator performance. However, there has been no formal validation.

Applicability for Military Simulations

The SAMPLE architecture provides a general framework for constructing models of operators of complex systems, particularly in cases in which the operators are engaged in information processing and control tasks. SAMPLE draws heavily on modern control theory, which has enjoyed considerable success in the modeling of human control behavior. The belief net core of the situation assessor of later variants appears to have considerable potential for representing situation awareness. However, procedure development for SAMPLE models would appear to be quite labor-intensive since there seems to be no high-level procedure representation language.

Soar

Soar is a symbolic cognitive architecture that implements goal-oriented behavior as a search through a problem space and learns the results of its problem solving (Newell, 1990; see also Chapters 5 and 8). Complete discussions of Soar describe a hierarchy from an abstract knowledge level down to a hardware- or wetware-dependent technology level (Polk and Rosenbloom, 1994). However, the two middle levels—the problem-space level and the architecture level—are of primary interest when one is comparing Soar with other cognitive architectures for human behavior representation; these levels are the focus of the following discussion.

Purpose and Use

As a unified theory of cognition, Soar is used to model all the capabilities of a generally intelligent agent. It has the capability to work on the full range of tasks expected of an intelligent agent, from highly routine to extremely difficult, open-ended problems. It interacts with the outside world, either simulated or actual. It represents and uses many different kinds of knowledge, such as that thought of as procedural, declarative, or episodic. Finally, it learns all of these types of knowledge as it performs tasks, learning from instruction, from success and failure, through exploration, and so on.

Soar's development has always involved two factions: (1) artificial intelligence researchers and system builders who wish to use all of Soar's capabilities to solve problems as efficiently as possible, and (2) cognitive psychology researchers who wish to understand human behavior by modeling it in the Soar architecture. The artificial intelligence contingent has driven the development of Soar to very large-scale models that perform rationally, run efficiently in real time, and interact with a real environment (e.g., robots or intelligent forces in a simulated theater of war). The psychological contingent has driven Soar to cognitive plausibility in many areas, such as multitasking performance under time pressure, problem solving, learning, and errors. Although any one Soar program may emphasize the goals of only one of these contingents, both sets of goals could in principle be served by a single Soar model.

Assumptions

Soar assumes that human behavior can be modeled as the operation of a cognitive processor and a set of perceptual and motor processors, all acting in parallel. In general, Soar is another example of a modified stage model (Figure 3.1), with the minor exception that all processors, including the motor processors, communicate through the contents of working memory (not directly from cognition as shown in Figure 3.1). Working memory is represented symbolically and holds the current interpretation of the state of knowledge, which can be considered a Soar model's situation awareness as defined in Chapter 7. However, Soar assumes that working memory cannot be understood independently of learning and long-term memory (Young and Lewis, forthcoming); these memories and processes work together to generate situation awareness.

Soar's research approach has involved adding a few more assumptions, applying the philosophy that the number of distinct architectural mechanisms should be minimized. For example, in Soar there is a single framework for all tasks and subtasks (problem spaces); a single representation of permanent knowledge (productions), so that there is no procedural/declarative distinction; a single representation of temporary knowledge (objects with attributes and values); a single mechanism for generating goals (automatic subgoaling); and a single elementary learning mechanism (chunking). To date, Soar has been able to model a broad range of human behavior without the need for more complex architecturally supported mechanisms.

Finally, Soar makes quantitative predictions about human behavior, as well as qualitative predictions about which knowledge, associations, and operators will come into play during performing and learning. In particular, Soar assumes that each elementary cognitive operator takes approximately 50 milliseconds to complete. This parameter allows predictions of the time to perform, the time to learn, and interesting interactions with a changing environment (e.g., errors occur because of time pressure).

Architecture and Functionality

At the problem-space level, Soar can be described as a set of interacting problem spaces, where each problem space contains a set of operators that are applied to states to produce new states. A state is a set of attribute-value pairs that encodes the current knowledge of the situation. A task, or goal, in a problem space is modeled by the specification of an initial state and one or more desired states. (Note that the desired states do not have to be completely specified; that is, open-ended problems can be represented by desired states that simply contain more knowledge than the initial state, without the particular type or content of that knowledge being specified.) When sufficient knowledge is available in the problem space for a single operator to be selected and applied to the current state, the behavior of a Soar model is strongly directed and smooth, as in skilled human behavior. When knowledge is lacking, either search in additional problem spaces may be necessary to locate the knowledge or decisions must be made without the knowledge, leaving open the probability of errors and thus error-recovery activities. This more complex branching and backtracking performance models human problem-solving behavior.

The architecture is itself a hierarchy. At the lowest level, Soar consists of perceptual and motor modules that provide the means of perceiving and acting upon an external world. At the next level, associations in the form of symbolic productions match the contents of working memory (comprising the inputs of perception, the output parameters for the motor modules, and purely internal structure) to retrieve additional information from long-term memory. In contrast with most classical production systems, Soar's associations match and fire in parallel,[6] are limited in their action repertoire to the generation of preferences for

---

[6]Though implemented in productions, Soar's associations are more similar to ACT-R's activation processes, since they are at a fine grain, happen in parallel, and take no action other than to bring

the activation of working-memory structure, and automatically retract these preferences when their conditions no longer match. Associations repeatedly fire and retract until no further associations are eligible to do so; Soar's decision-level process then weighs all the active preferences and chooses a new operator or state. Whenever the activations are not sufficient to allow a unique choice, the architecture responds by setting a subtask to resolve this impasse, and the entire process recurs. If the recursive processing adds new preferences that are active in the original task, Soar's architectural learning mechanism (called chunking) creates a new association between those working-memory structures in the original task that led, through a chain of associations in the subtask, to the new preferences in the original task. Thus chunking effectively transfers knowledge from the subtask space to the original task space. It straightforwardly produces speedup, but can also inductively acquire new knowledge (Rosenbloom et al., 1991).

Operation

Soar has been used in many different ways: as a stand-alone model in which time has no real meaning; in simulated real time, with each decision cycle being defined as taking 50 milliseconds and able to run in synchronicity with other systems constrained to the same timescale; or in real time, interacting with a totally separate, independently acting system. In all models, initial knowledge is encoded by the human modeler in production rules. In many models, the initial state of the model is coded manually, but other models start with no knowledge of the problem state and must acquire that knowledge from the external environment. In some models, the termination state (desired state) is coded manually; in others, the model simply runs until the human modeler stops it. External events can be simulated wholly within Soar with a prototyping facility such as Tcl/Tk, or the model can be linked to a totally independent external system. When run, all Soar models can output a trace of what operators were selected and applied and what knowledge was used to perform the task. In addition, some Soar models literally change the external environment, interacting with real-world systems (e.g., controlling robots, flying a flight simulator, participating in a simulated theater of war). If learning is turned on, the new productions produced by chunking are also a system output; the human modeler can directly examine what the model learned.

Current Implementation

The currently supported version of Soar is Soar 7, a C-based implementation

more knowledge into working memory. Soar's operators are similar to the production rules in ACT-R and EPIC, and, like ACT-R but unlike EPIC, Soar can perform one operator at a time. However, these operators may be arbitrarily complex.

available for UNIX and Macintosh platforms. Several releases of Soar 6 are available for UNIX, Macintosh, and DOS platforms. Soar 7 replaces the user interface of Soar 6 with Tcl.

Both Soar 6 and Soar 7 have been used with external environments, for example, to control robots and to participate in simulated theaters of war. Depending on the complexity of the task, they can run faster than real time.

Support Environment

Soar is a highly developed programming environment with full-fledged user and reference manuals and program editing, tracing, and debugging facilities. Soar users are an active group, and many programming tools have been developed and shared. For instance, there are emacs-based development environments for UNIX users and a Tcl/Tk environment for those who prefer a GUI. There are also a Macintosh-specific environment specifically targeted at supporting the development of psychological models and a tool for evaluating the match between the behavior of a Soar model and verbal protocol data collected from humans.

Tutorials on Soar are held periodically in both the United States and Europe, and tutorial material is available on the World Wide Web. Workshops are held approximately yearly in the United States and Europe at which Soar users share their models and results. There are also active mailing lists for soliciting assistance and sharing results, and a frequently asked questions (FAQ) site to help users get started.

In addition, there are user libraries of general capabilities and models that can be adopted into new models. For instance, MONGSU is a UNIX socket package enabling external systems to interact more easily with Soar. The SimTime system simulates the passage of real time as a function of the cognitive behavior of a Soar model so independent external events can be simulated without linking up to an external environment. NL-Soar is a general theory of natural-language comprehension and generation that can be incorporated into new models. (Note, however, that according to the Soar theory, natural-language use is highly integrated into the performance of a task, so using this code in the service of modeling military systems such as command and control communication will not be as simple as adding new vocabulary and calling a subroutine.) All of this material can be accessed from the Soar World Wide Web site: <http://www.isi.edu/soar/soar.html>.

Validation

Soar has been evaluated extensively as a cognitive architecture against human behavior in a wide variety of tasks. Although a comprehensive list of Soar models and their validation is beyond the scope of this chapter, some examples of these models are natural-language comprehension (Lewis, 1996, 1997a, 1997b), syllogistic reasoning (Polk and Newell, 1995), concept acquisition (Miller and Laird, 1996), use of a help system (Peck and John, 1992; Ritter and Larkin, 1994), learning and use of episodic information (Altmann and John, forthcoming), and various human-computer interaction tasks (Howes and Young, 1996, 1997; Rieman et al., 1996). In addition, other work that compares Soar models with human behavior include models of covert visual search (Weismeyer, 1992), abductive reasoning (Krems and Johnson, 1995), a simplified air traffic controller task (Bass et al., 1995), the job of the NASA test director (Nelson et al., 1994a, 1994b), job-shop scheduling (Nerb et al., 1993), and driving a car (Aasman, 1995).

Recent work has joined the Soar cognitive architecture to EPIC's perceptual and motor processors (Chong and Laird, 1997). The resulting perceptual-cognitive-motor model has been validated against human dual-task data and shown to be as accurate as EPIC alone, despite the differences between the cognitive processing of EPIC and SOAR. This approach of incorporating the mechanisms of other architectures and models and "inheriting" their validation against human data promises to result in rapid progress as parallel developments by other architectures emerge.

Applicability for Military Simulations

Soar has been used extensively for military problems. Most notably, the agents of Soar-intelligent forces (Soar-IFOR) have been participants in simulated theaters of war (STOW-E in 1995, STOW-97 in October 1997). These agents have simulated the behavior of fixed-wing and rotary-wing pilots on combat and reconnaissance missions, competing favorably with human pilots. They have acted as very large expert systems, using Soar's architecture to encode knowledge prior to performance and then acting on it, or solving problems using that knowledge. They use this knowledge to act as individuals and as groups of cooperating individuals. However, these systems have not yet run with learning capabilities.

Other research Soar systems have demonstrated more far-reaching capabilities and would require some additional research before becoming robust, real-time systems for military use. An example is the system noted above that integrated natural-language capabilities, visual search, and task knowledge to model the decision process of the NASA test director in simulated real time (slower than actual real time) (Nelson et al., 1994a, 1994b). Another system used ubiquitous learning to store and recall situations and decisions during operation for post-mission debriefing (Johnson, 1994). Although some work has been done on modeling frequency learning in laboratory tasks (Miller and Laird, 1996), it is doubtful that Soar's purely symbolic learning mechanism will extend far enough to capture human-like adaptivity to the frequency of occurrences in real-world environments. In addition, as with most of the other architectures reviewed here, no actual modeling has been done to account for the effects of moderator variables, although this is an active area of interest for the Soar research community (see Chapter 9).

In summary, the Soar architecture has been shown to be sufficiently powerful to represent believable individual human behavior in military situations and sufficiently robust to run very large-scale models connected to real-world systems in real time (without learning). However, this is not yet true for all aspects of learning or moderator variables.

COMPARISON OF ARCHITECTURES

Table 3.1 presents a summary of the integrative architectures reviewed in this chapter, comparing them across several dimensions that should help the reader assess their relative potential in meeting the needs of intended applications. The following sections describe each of these dimensions in turn, attempting to point out features that distinguish some architectures from others.

The reader should keep two things in mind when comparing the architectures. First, our assessments are based on our interpretations of the materials available to us at the time of writing, and in most cases we have very limited experience in working directly with the architectures. Second, from direct conversations with developers and users of the architectures, it is clear that there were very promising developments taking place at the time of this writing for every architecture reviewed. Each therefore warrants additional scrutiny based on the reader's needs.

The following dimensions are included in Table 3.1 and discussed below: original purpose; submodels (sensing and perception, working/short-term memory, long-term memory, motor); knowledge representation (declarative, procedural); higher-level cognitive functions (learning, situation assessment, planning, decision making); multitasking; multiple human modeling; implementation (platform, language); support environment; and validation.

Original Purpose

The purpose for which the architecture was originally created may influence its usefulness for certain applications. Architectures such as HOS and OMAR were created primarily for evaluating equipment and procedure designs as part of the process of human-machine system development. Though they may have many attractive elements, certain features, such as their execution speed and the type of output they produce, may make them less desirable for modeling human problem solving, which is the express purpose of other architectures, such as ACT-R and Soar.

Submodels

Sensing and Perception

Some of the architectures, such as EPIC and MIDAS, have fairly detailed submodels of human sensory and perceptual processes. Such submodels make these architectures particularly applicable in situations where failures in stimulus detection or discrimination are likely to have a significant effect on overall human performance. Other architectures, such as Micro Saint, model such processes only in terms of times and accuracy probabilities, and still others, such as COGNET, have only very abstract representations of such processes, though they may be perfectly acceptable for applications in which detection and discrimination are not limiting factors. Even in cases in which detection and discrimination are limiting factors, most of the architectures offer modelers the capability to incorporate more sophisticated sensing and perception modules.

Working/Short-Term Memory

As discussed in Chapter 5, working or short-term memory plays a major role in human information processing. Applications in which attention and decision making are important considerations are likely to benefit from veridical representations of working memory. The most sophisticated working memory submodels among the architectures reviewed by the panel are those of ACT-R and the newer version of MIDAS, both of which treat working memory as an area of activation of long-term memory. Other architectures, such as Micro Saint, do not model working memory explicitly, and are therefore incapable of directly capturing performance degradation due to working memory capacity and duration limitations.

Long-Term Memory

All the architectures provide some means of long-term information storage. Some, however, such as HOS and Micro Saint, do not explicitly model such a store. Rather, such information is implicitly stored in procedures and task networks, respectively.

Motor

The architectures reflect a wide range of motor submodels. The motor submodels of HOS and Micro Saint provide time and accuracy outputs. EPIC, ACT-R, and Soar have effectors that change the environment directly. MIDAS and OMAR drive the Jack animated mannequin, though the computational costs of that submodel may be too great for most simulations. OMAR has also devel-

TABLE 3.1 Integrative Architectures

		Submodels
Architecture	Original purpose	Sensing and Perception
ACT-R	Model problem solving and learning	Perceptual processors: visual, auditory, tactile
COGNET	Develop user models in intelligent interfaces, surrogate users and adversaries	Abstract—perceptual daemons, with provision of user-defined models
EPIC	Develop and test theories of multiple task performance	Perceptual processors: visual, auditory, tactile
HOS	Generate timelines for HMS analysis and evaluation	Visual and auditory sensing and perception micro-models
Micro Saint based network tools	Evaluate systems and procedures	Detection/identification probabilities, times, and accuracies
MIDAS	Evaluate interfaces and procedures	Visual perception agent
MIDAS Redesign	Evaluate interfaces and procedures	Visual and auditory perception
Neural network based tools	Multiple constraint, satisfaction in memory, language, thought, pattern recognition	Visual and auditory perception
OMAR	Evaluate procedures and interfaces	Default perceptor models
SAMPLE	Evaluate crew procedures, equipment	Visual, auditory channels (optimal control model-based)
Soar	Model problem solving and learning	Perceptual processors: visual, auditory

Working/ Short-Term Memory	Long-Term Memory	Motor	Outputs
Activation-based part of long-term memory	Network of schema-like structures plus productions	Motor processors: manual, vocal, oculomotor	Behaviors
Extended working memory through multipanel blackboard	Multipanel blackboard	Abstract, with provision for user-defined models	Behaviors
Unlimited capacity and duration	Propositions and productions	Motor processors: manual, vocal, oculomotor	Behaviors
Micro-model with limited capacity and decay	Not explicitly modeled	Eye, hand, trunk, foot, etc., micro-models	Metrics
Not explicitly modeled	Not explicitly modeled	Times and accuracies plus micro-models	Metrics
Not explicitly modeled	Semantic net, updatable world representation	Jack-animated mannequin	Behaviors
Subset of nodes in long-term memory, limited capacity and decay	Frames	Jack-animated mannequin	Behaviors
Activation-based limited capacity	Connection weights	Sensor/motor integration, occularmotor	Behaviors
Emergent property—not explicitly modeled	Not explicitly modeled—network of frames	Default effector models	Behaviors
Not explicitly modeled	Procedural knowledge base	Time-delayed procedural actions	Behaviors
Unlimited capacity, duration tied to goal slack	Productions	Motor processors: manual, vocal, oculo-motor	Behaviors

continued on pages 100-105

TABLE 3.1 Integrative Architectures (continued)

	Knowledge Representation	
Architecture	Declarative	Procedural
ACT-R	Schema-like structures	Productions
COGNET	Blackboard elements, semantic network	Goal hierarchies, COGNET task description language
EPIC	Propositions	Productions
HOS	Object attributes stored in short-term memory	HOS procedural language
Micro Saint based network tools	Global variables	Micro Saint language, task networks
MIDAS	Objects	Functions, daemons, rules, decision procedures
MIDAS Redesign	Frames	Plan frames, rules, functions
Neural network based tools	Active: pattern activation Latent: connection weights	Sequences of activation patterns over time
OMAR	Augmented frames (built on CLOS)	SCORE language (built on Common Lisp)
SAMPLE	Objects	Production rules
Soar	Productions	Productions

Higher-Level CognitiveFunctions			
Learning	Planning	Decision Making	Situation Assessment
Weight adjustment, production strength adjustments, new productions, new schemas	Creates new plans	Knowledge-based, Bayesian	Overt and inferred
No learning	Instantiates general plans	Knowledge-based	Overt and inferred
No learning	Instantiates general plans	Knowledge-based	Overt and inferred
No learning	No planning	None	Overt
No learning	No planning	Knowledge-based	Overt
No learning	Instantiates general plans	Knowledge-based, Bayesian	Overt
No learning	Instantiates general plans	Knowledge-based, Bayesian	Overt
Weight updating by gradient ascent on objective function	Motor planning	Competitive systems of activation	Primarily inferred
No learning	Instantiates general plans	Knowledge-based, Bayesian	Overt
No learning	Instantiates general plans	Knowledge-based, Bayesian	Overt
Learning by chunking (flexible—see text)	Creates new plans	Knowledge-based	Overt and inferred

continued

TABLE 3.1 Integrative Architectures (continued)

	Multitasking	
Architecture	Serial/Parallel	Resource Representation
ACT-R	Serial with switching (no interruptions)	Amount of declarative memory activation
COGNET	Serial with switching and interruptions	Limited focus on attention, parallel motor/perceptual processors
EPIC	Parallel	Limited perceptual and motor processors, unlimited cognitive processor
HOS	Serial with switching, plus parallel movement possible	Speech, hand, foot, and cognitive channels
Micro Saint based network tools	Parallel with switching to serial resources limited	Visual, auditory, cognitive, psychomotor workload
MIDAS	Resource-limited parallel	Visual, auditory, cognitive, motor resources
MIDAS Redesign	Resource-limited parallel	Visual, auditory, cognitive, motor resources
Neural network based tools	Contention scheduling via competition among components	Units and connections allocated to a task
OMAR	Serial with some parallelism for automatic tasks	Perceptor, cognitive, effector resources
SAMPLE	Serial with interruptions	Sensory, cognition, and action channels
Soar	Serial with switching and interruptions	Serial cognitive processor, limited perceptual and motor resources

Goal/Task Management	Multiple Human Modeling	Implementation Platform
None (one goal at a time)	Potential through multiple ACT models	Mac, PC (limited)
Pandemonium model—priorities based on task context	Through multiple COGNET models	PC, SGI Sun
Priority-based task deferment	No	Lisp platforms
Priority-based attention switching	No	PC
Simple dynamic prioritization	Yes	PC, Mac, Unix platforms
Z-Scheduler uses time, resource constraints	Limited	SGI
Agenda Manager uses time and resource constraints, goal priorities	Yes	
Competitive systems of activation	Potential through multiple networks	MAC, PC, Unix platforms
Tasks compute own priorities	Yes	Sun, SGI, PC (future)
Priority based on base value + "situational relevance"	Yes	PC, Unix
Preference-based attention allocation	Yes	Mac, PC, Unix platforms

continued

TABLE 3.1 Integrative Architectures (Continued)

Architecture	Language	Support Environment
ACT-R	Lisp	Editors, debuggers
COGNET	C++	Graphical editors, syntax/semantic checkers, debuggers, structural SPI to external processes
EPIC	Lisp	Lisp programming environment tools
HOS	FORTRAN	Editors (mostly text-based)
Micro Saint based network tools	C, C++	Editors, debuggers
MIDAS	Lisp, C, C++	Graphical editors, graphical data displays
MIDAS Redesign	C++	Similar to original MIDAS
Neural network based tools	C, NETLAB	Commercial products, GUIS
OMAR	Lisp	Compilers, editors, browsers, online animation tools, post-run analysis tools
SAMPLE	C++	Editors
Soar	C	Editors, debuggers

Validation	Comments
Extensive at many levels	ACT-R models focus on single, specific information processing tasks; has not yet been scaled up to complex multitasking situations or high-knowledge domains
Full model	Used in low-fidelity submarine training simulations and high-fidelity AEGIS CIC training simulations to provide surrogate operators and adversaries
Extensive for reaction time	EPIC models focus on simple, dual-task situations; has not yet been scaled up to complex multitasking situations or high-knowledge domains
Components (micro-models)	Currently capable of scripted behaviors only
Some micro-models; at least one implementation	Used extensively in military simulations
Full model	Currently capable of scripted behaviors only
None	In development
Extensive for component tasks	Focus on sensory/motor integration, have not yet been scaled up to complex multitasking situations or high-knowledge domains
Components	
Control tasks (OCM)	Has been used in small-scale military simulations
Extensive at multiple levels	Has been used in military simulations (e.g., synthetic theater of war-Europe [STOW-E], STOW-97)

oped KATE, a highly-simplified stick figure mannequin that can execute OMAR tasks. Many of the architectures can be extended by building custom motor modules.

Most of the architectures generate outputs in the form of behaviors of the human being represented (e.g., motions, gestures, utterances) that can or could be used as inputs to other simulation modules, such as tank or aircraft models. HOS (in its original version) and Micro Saint, however, generate performance metrics, such as task start and end times and task accuracy data. The former are more applicable for virtual simulations, which need to generate visible images of human and system behaviors, while the latter are more applicable to constructive simulations. In principle, however, it should be possible to modify the two systems to produce such behaviors; in fact, as discussed earlier, modified HOS micro-models have recently been added to COGNET for that very purpose.

Knowledge Representation

Declarative

Representation of declarative knowledge ranges from simple variables to complex frames and schemas. Those applications in which complex factual knowledge structures must be represented explicitly would be better served by the more sophisticated techniques of such architectures as ACT-R and OMAR.

Procedural

All the architectures, except neural net architectures, have a means of representing procedural knowledge beyond that provided by the languages in which they are written. The production rule languages of ACT-R and Soar appear to be the most flexible and powerful, though the related complexity of such languages may not be warranted for those applications in which behavior is highly procedural (perhaps even to the point of being scripted).

Higher-Level Cognitive Functions

Learning

Only three architectures offer learning: ACT-R, neural net architectures, and Soar. Of these, ACT-R provides the most flexibility.

Situation Assessment

In most of the architectures, situation assessment is overt, in the sense that the programmer/modeler must explicitly code all its aspects. Some of the architectures, however, have the capability of inferring details of the situation from other information. This capability obviously reduces the programmer's burden, provided the mechanism is suitable for the application.

Planning

HOS and Micro Saint do not provide a planning function (though presumably the programmer can implement it). Several of the architectures (e.g., MIDAS) have the capability of instantiating general plans based on the specifics of a given situation. ACT-R and Soar models can create new plans. This capability is obviously valuable for some applications; however, for those applications not requiring planning (i.e., where procedures can be initiated in a prespecified way), the development and computational overhead may be undesirable.

Decision Making

HOS does not provide decision-making capability, except to account for the time required for the operator to make a decision given certain characteristics of the decision problem. Most of the other architectures provide for knowledge-based decision making, with a few offering Bayesian techniques. Some form of decision-making capability seems essential for most military simulations, and architectures such as ACT-R, OMAR, and Soar seem most capable in this regard.

Multitasking

All the architectures permit multitasking in some sense, though ACT-R enforces strict serial execution of tasks with no interruptions allowed. The degree of multitasking varies from that of HOS, which allows only a single cognitive task (with parallel, ballistic body motion possible) to EPIC, which permits any number of concurrent tasks provided resource requirements are not exceeded. From the standpoint of multitasking flexibility, then, architectures such as EPIC and MIDAS appear to offer the greatest potential. However, there is disagreement over the psychological validity of such systems, and architectures such as OMAR that limit the operator to only one task requiring conscious thought may be more realistic. It is clear, however, that interruptability is an important feature; all architectures, except ACT-R, have this feature.

Multiple Human Modeling

In principle, multiple copies of any of the architectures could be integrated to allow modeling of multiple operators. However, only about half of the architectures explicitly provide this capability. It is worth noting that this capability was a major driving force in the development of OMAR and the MIDAS redesign.

Implementation

Platform

Most of the architectures have been implemented on one or only a few computer platforms. COGNET and OMAR are more flexible, and Micro Saint and Soar are available for almost any platform a user could require.

Language

For those users needing to modify the architectures to fit the needs of their applications, the language in which the architectures are written is of some concern. To many military simulation modelers who have done most of their work in FORTRAN and C, architectures such as ACT-R, EPIC, and OMAR (written in Lisp) may pose some (small) problems.

Support Environment

The software provided to assist the modeler in creating models in the various architectures ranges widely. At one extreme is EPIC, for which there exists very little such software; at the other extreme is OMAR, whose toolkit appears well thought out and tailored to the user not intimately familiar with OMAR's inner workings. To their credit, though, ACT-R and Soar both have a very large base of users, many of whom have developed additional tools to facilitate the use of the two systems.

Validation

Some of the architectures are too new for any substantive validation to have been done. In other cases (e.g., HOS), certain submodels have been extensively validated. Some of the architectures have been validated in their entirety ("full model" validation in Table 3.1), some of these extensively. At least one (Micro Saint) has instances that have received military accreditation. Unfortunately, most of the validation of "full models" has been based on subjective assessments of subject matter experts, not real human performance data. This last consideration dictates considerable caution on the part of users.

HYBRID ARCHITECTURES: A POSSIBLE RESEARCH PATH

As discussed above, the various architectures have different strengths and weaknesses with respect to their relevance for military simulations. To model all the details of the complex behavior involved in the tank platoon's hasty defense described in the vignette of Chapter 2, an architecture would have to encompass all the phenomena addressed in all the following chapters: attention and multitasking, memory and learning, decision making, situation assessment, and planning, all subject to behavioral moderators. No existing architectures approach this all-encompassing capability. This does not make these architectures useless—many have been demonstrated to be quite useful despite their limitations—but it does suggest areas of improvement for each.

One research and development path is obvious: start with one of the existing architectures, and expand it in the areas in which it is weak. For example, an emerging ACT architecture, ACT-R/PM, adds EPIC-like perception and motor capabilities to the ACT-R architecture (Byrne and Anderson, 1998). Each of the reviewed architectures could be expanded in this way, as needed for task performance. This approach represents a time-honored path with almost guaranteed incremental payoff, but may eventually encounter boundaries as the architectures reach their limits of expressibility.

Another research and development path might prove more fruitful: combine the strengths of two or more architectures to produce a hybrid that better encompasses human phenomena. A simple example, the combination of Soar's cognitive processor with EPIC's perceptual and motor processors, has already been mentioned. In contrast with ACT-R/PM's reimplementation of EPIC-like processors in the ACT architecture itself, neither Soar nor EPIC was rewritten, but communicate through a shared working memory. More fundamental combinations of architectures are the subject of an ongoing basic research program at the Office of Naval Research (Hybrid Architectures as Models of Human Learning), which supported the infancy of several hybrid architectures. To address the effects of environmental frequency in Soar, that architecture was combined with Echo (Thagard, 1989), a statistical technique for belief updating. Neural nets were augmented with a symbolic explanation-based learning system (Dietterich and Flann, 1997) to address the learning of long procedures when the effects of actions are widely separated from the actions, credit and blame are difficult to assign, and the right combination of moves should count more toward learning than the myriad of failures along the way. The CLARION architecture (Sun, 1995) integrates reactive routines, generic rules, learning, and decision making to develop versatile agents that learn in situated contexts and generalize resulting knowledge to different environments. Gordan (1995) extends Marshall's schema theory of human learning, supplementing its high-level planner with a low-level stimulus-response capability. Marshall (1995) is using a neural net for the identification component of schema theory and ACT-R for the elaboration, planning, and execution components. Cohen and Thompson (1995) use a localist/connectionist model to support rapid recognitional domain reasoning, a distributed connectionist architecture for the learning of metacognitive behavior, and a symbolic subsystem for the control of inferencing that mediates between the domain and metacognitive systems.

These preliminary architectural explorations are promising examples of an area ripe for research. Through extensive use of the reviewed architectures, researchers and developers in the field now know enough about their relative strengths and weaknesses to judiciously explore plausible combinations for achieving the greatest impact of human behavior models.

CONCLUSIONS AND GOALS

None of the architectures described in this chapter is precisely what is needed for military simulations. Collectively, however, they offer a foundation on which to build models that will be truly useful and practical for military simulations. Once thorough task analyses of the combatant's activities have been conducted to determine exactly what behaviors are to be modeled, these architectures offer frameworks and a wide variety of detailed models of specific perceptual, cognitive, and motor processes that can be used to represent these behaviors. Elements of the various architectures could be combined to yield representations with psychological validity, model fidelity, and computational efficiency.

The problem with these architectures is not what they do not model. Among the architectures reviewed here, there is hardly an interesting and potentially useful phenomenon that is not considered in some way. The problem is that the architectures are not validated. Many of these architectures have submodels that are well validated and have reached a level of maturity suitable for application in military simulations. However, few of the architectures have been validated overall, and their emergent properties are not well understood. More experience with them is needed. A careful, measured expansion of their application in military simulations may just be the proper path to take.

Short-Term Goals

- Make validation of integrative architectures a priority. Continued use of any integrative architecture or tool should be predicated on its validation. Early effort will be required to define performance measures and standards for use in the validation, as well as scenarios to be used as test cases. Data from exercises and human-in-the-loop simulations can be collected in preparation for validation of models to be developed, and human behavior models currently used in military simulations can be tested against these data.
- As a first cut at upgrading these architectures, gradually incorporate the concepts, theories, and tools presented in subsequent chapters of this report into existing simulations to incrementally improve existing models. One approach to this end would be to use a simplified stage model to augment current human

behavior representations. Such an augmented representation should incorporate, at a minimum, a module representing human perceptual processes, a module representing motor processes, a means for representing multiple active tasks (e.g., planning, decision making, communicating, moving), and a mechanism for selecting a subset of active tasks for execution at any given time. The perceptual module could be based on simple detection and identification probability data derived from existing psychological data. Similarly, the motor process module could be based on simple movement models that yield time and accuracy estimates, or on time and accuracy probabilities derived from human performance data. All current human behavior representations have some means of representing tasks, and most of these could be modified to permit the existence of more than one task at a time. The mechanism for selecting among competing tasks could be based on a simple, static priority system with task priorities derived from task analyses. Such an approach would yield simple architectures with greater face validity than that of current architectures. Although it is unlikely that such architectures would be entirely satisfactory, they would give modelers more experience in developing and validating human behavior representations.

Intermediate-Term Goals

- Continue validation into the intermediate term as more sophisticated integrative architectures are developed. This generation of architectures can be expected to draw on the integrative architectures reviewed in this chapter, as well as newer architectures that will emerge after this report is published.
- Continue the development of hybrid architectures, such as those described in this chapter, combining the best elements of existing and emerging integrative architectures. For such hybridization to proceed in a timely manner, it will also be necessary to conduct research and development activities to modularize existing architectures and yield interchangeable components.
- Apply these architectures in sustained and intensive development of human behavior representations that incorporate specific military tasks in selected domains, such as tank warfare.
- Compare the different modeling approaches by developing alternative architectures for a domain and comparing them against data from field exercises and human-in-the-loop simulations.

Long-Term Goals

- Continue architecture validation.
- Continue to refine new architectures created in the intermediate term.
- In addition to continued efforts to improve the quality of existing modeling approaches, explore entirely new approaches that will result in architectures as yet unconceived.

4

Attention and Multitasking

INTRODUCTION

Divided attention and multitasking—doing several things at once—are ubiquitous in combat operations. An infantryman may have to decide on a general course of action, plan his path of movement, run, and fire his weapon simultaneously. When engaging multiple targets, a tank crew must continuously navigate and control the vehicle, search for targets, aim and fire the gun, and assess battle damage. A pilot must simultaneously control his aircraft, plan maneuvers, navigate, communicate with his wingman, control sensors, aim and fire weapons, and monitor and manage other aircraft systems. A commander responsible for several units must divide his attention among the units as he attempts to accomplish multiple, concurrent, perhaps conflicting goals. In each of these settings, several decisions and actions may have to be evaluated and then executed in overlapping time frames.

In most situations lasting more than a few seconds, the individual or the team ideally should or actually does review current goals to consider and prioritize them; assess progress made toward accomplishing each goal; and then allocate immediate attention to tasks in accordance with scheduling priorities, importance, urgency, probabilities, training, and anticipated ability to accomplish certain tasks or processes in parallel, with specified loss due to sharing. This management-like activity should occur continuously and generally represents an attempt to allocate cognitive resources efficiently. In some cases, the decision maker may choose to deal with competing tasks by devoting attention to each in turn. In most cases, however, a realistic representation of the performance of competing tasks or processes will require some degree of overlap or sharing. In this chapter, models of such situations are termed *multitasking models*. Both theories and models of attention and multitasking behavior are reviewed. Conclusions and goals emerging from this review are presented in the final section. First, however, some essential details related to attention and multitasking are added to the vignette presented in Chapter 2, and some key concepts and terms are defined.

Hasty Defense Vignette: Additional Details

To frame the discussion and provide examples of attention and multitasking concepts, it is necessary to add some detail to the hasty defense vignette described in Chapter 2. These details include specific tasks the platoon leader is responsible for performing.

Suppose that after the initial engagement, the tank platoon has moved to battle position 1 (BP1). All tanks have moved into initial hide positions, and all tank commanders have identified fire and alternative hide positions. All the tank commanders and gunners (including the platoon leader and his gunner) are scanning for additional enemy forces. The scenario unfolds according to the event sequence found in Exhibit 4.1. At this point in the scenario, the platoon leader is attempting to perform the following tasks:

- Maintain general situation awareness, and initiate appropriate tasks
- Report enemy contact to A Company commander
- Assess battle damage (to first T-80)
- Monitor movement to alternate position
- Monitor fire on second T-80—interrupted by third T-80
- Assess damage to own tank
- Direct turret slew toward target (third T-80)
- Communicate with platoon
- Reset radio
- Monitor firing (on T-80)

Clearly, the platoon leader cannot perform all these tasks simultaneously. Furthermore—and significant to the theme of this chapter—the way he allocates his attention to these tasks will have a significant effect on the outcome of the battle.

Key Concepts and Terms

Relation to Learning

The relationship between learning and attention and multitasking has intrigued researchers and theorists from the earliest days of experimental psychology. For example, Bryan and Harter (1899) studied improvements in the sending and receiving of telegraphy. They proposed that naive performers needed to

Exhibit 4.1 Tank Platoon Scenario Event Sequence

1. Platoon leader and platoon leader's gunner detect T-80 on trail, advancing toward BP1.
2. Platoon leader commands driver to move to firing position, begins monitoring movement.
3. Platoon leader initiates communication to alert platoon.
4. Platoon leader initiates report to A Company commander.
5. Tank reaches firing position.
6. Platoon leader commands engagement of T-80, begins monitoring firing.
7. Gunner begins laying gun.
8. Gunner acquires target and fires.
9. Platoon leader commands driver to move to predesignated hide position, begins monitoring movement.
10. Blast occurs in vicinity of target. Platoon leader begins battle damage assessment.
11. Gunner detects and identifies second T-80 behind first T-80, alerts platoon leader.
12. Platoon leader confirms gunner's identification, commands fire on second T-80, begins monitoring firing.
13. Own tank hit by round from undetected third T-80 on left flank (minor damage to own tank, no injuries to crew). Platoon leader begins assessing damage to own tank.
14. Platoon leader detects and identifies third T-80.
15. Platoon leader begins slewing turret to third T-80.
16. Platoon leader initiates communication to alert platoon, but finds radio failed.
17. Platoon leader finds radio reset switch, initiates corrective actions.
18. Platoon leader designates third T-80 to gunner, commands fire.

allocate scarce attentional resources to the tasks involved, but that training allowed automatization of processes, or automatism (discussed further below), freeing attention for increasingly higher-level cognitive activities. Perhaps the most easily observable and largest effects in the fields of performance and cognitive behavior are those that occur during the often quite extended periods of deliberate practice, known as the development of expertise (and skill) (see Ericsson and Smith, 1991). Part of this gain in skill is known to depend on the storage in memory of a vast amount of relevant knowledge and behavioral procedures that can be accessed and executed with relatively low demands on attention (e.g., see Chase and Simon, 1973).

Other researchers have studied the degree to which training allows performers to accomplish two largely unrelated simultaneous tasks. For example, Downey and Anderson (1915) showed that extensive training would allow performers to "read chapters while writing memorized verses" with little cost in terms of a performance decrement or errors in the written passages. Schneider and Shiffrin (1977) and Shiffrin and Schneider (1977) carried these ideas further, tested them empirically in a number of visual and memory search studies, and proposed a general theory of attentive and automatic processing. The idea was that certain processes that are trained consistently may be learned as automatic units, reducing demands for attentional resources. Logan (1988) also explored such issues by showing how recourse to learned procedures allowed performers to bypass the need to accomplish tasks by algorithmic means, that is, by a slower, sequential series of smaller steps. In both of these instances, the authors identified specific processes that were learned and thereby reduced processing demands, but these processes were specific to the tasks under study. A valid criticism would note that the general theory simply describes the shift from procedures requiring scarce resources to ones that bypass such demands, without providing a method for predicting what sort of learning might take place in other situations.

Models of learning are not the subject of this chapter. It is assumed that the participant is at a given level of skill development, one at which the need for multitasking is critical to carrying out the task. At a fixed stage of learning, or level of skill, the issue is not what changes with training, but what set of available cognitive resources is allocated to accomplish the tasks at hand. The general term used to describe such allocation is *selective attention*. Both some brief historical notes on selective attention and a discussion of current models are presented later in the chapter.

Relation to Working Memory

Although the allocation of limited attentional resources is often described as selective attention, these processes are difficult to separate from the general operations that control the cognitive processing system. Such control processes are usually thought to reside in a temporarily active memory system, and are also referred to as *working memory* (see Chapter 5). It would probably not be theoretically possible to draw lines between working memory, selective attention, and multitasking. However, it has been traditional to talk about selective attention with respect to tasks that involve perception and motor performance and are usually fairly simple; to talk about multitasking with respect to several tasks or processes that are complex, relatively independent, and distinct; and to talk about working memory with respect to tasks involving coding, learning, and retrieval from memory. Although these terms are used in this chapter (and in Chapter 5), no theoretically important distinction is implied by the choice of a particular term.

Tasks and Processes

Sometimes an individual (or group) must try to accomplish two or more externally defined tasks that are largely unrelated and independent. An example involving the platoon leader in the above vignette would be directing the slew of the turret while trying to reset the radio. The demands on the performer in such cases are perhaps most aptly described as multitasking. In other cases, multiple internal processes must be used to accomplish a single externally defined task. For example, a driver may need to reach a designated location, nominally a single task, but to do so must scan for obstacles and enemies, plan a route, manipulate the controls, listen to and act on orders from the tank commander, and so forth. The language of selective attention is more often used in such cases. There is really no hard and fast line between uses of the terms, and multitasking is used in this chapter to refer to both situations; that is, tasks are taken to refer to both external and internal activities.

Automatism

Automatism offers a means of accomplishing multitasking with less sharing and less sequential application of attentional resources. It is usually developed through extended training. For example, driving a tank requires that multiple concurrent tasks be accomplished; a novice must usually focus on just one task at a time, such as steering. After extended training, a skilled tank crewmember may be carrying out 10 or more tasks in a generally concurrent fashion, sharing attention among them, and even have enough attention left over to carry on a simultaneous conversation on an unrelated topic. Automatism is closely related to the development of skill and expertise. For a few experimental paradigms, the processes of automatism have been worked out. Examples include chunking in memory search (Schneider and Shiffrin, 1977), attention attraction in visual search (Shiffrin and Schneider, 1977), memorization in alphabetic arithmetic (Logan and Klapp, 1991), learning of responses in Stroop situations (MacLeod and Dunbar, 1988), and unitization in the perception of novel characters (Shiffrin and Lightfoot, 1997). The processes by which automatism develops generally are not yet well understood, although some existing models, such as adaptive control of thought (ACT) and Soar, incorporate automatization (see Chapter 3 for further discussion). Thus simulation modeling for a given task would require specific implementation of the components of automatism appropriate for that task.

ATTENTION

Informally, attention may be thought of as the focus of conscious thought, though this is an inadequate definition. Somewhat more formally, attention may be thought of as the means by which scarce or limited processing resources are allocated to accomplish multitasking. There are several broad reviews of attention as it relates potentially to human behavior modeling (e.g., Parasuraman and Davies, 1984; Shiffrin, 1988). The following brief summary is based heavily on Wickens (1992:74-115).

Selective Attention

Selective attention is a process through which the human selectively allocates processing resources to some things over others. The things involved could include internal processes of all kinds, but the term was used more restrictively in the early years of the field to refer to perceptual processing of sensory stimuli. Thus the term originally referred to decisions to attend to some stimuli, or to some aspects or attributes of stimuli, in preference to others (Kahneman, 1973:3). Examples of selective attention drawn from our vignette would include visual sampling (in which the platoon leader, early in the vignette, selectively attends to his integrated display), visual target search (in which he scans his vision blocks or the independent thermal viewer for enemy armor), and auditory selective attention (in which he monitors his radio for transmissions related to him and his platoon).

Selective attention is by definition limited. A human can attend substantially or fully to a relatively small number of stimuli and/or stimulus attributes at one time. Not only is attention limited, but the selection process may be inaccurate and inappropriate for the tasks at hand. For example, the platoon leader could become momentarily preoccupied with the radio problem when the more immediately important task is to slew the turret to the third T-80 threat.

Focused Attention

Focused attention is a process in which the human rejects some processes in favor of others; in perceptual domains the term usually denotes the rejection of irrelevant stimuli (Schneider et al., 1984:9). For example, the platoon leader uses focused attention to ignore vegetation and cultural features as he searches for enemy forces. He also uses it to filter out radio transmissions that are not relevant to him and his platoon.

The inability to reject irrelevant stimuli and/or information or, more generally, irrelevant processing, marks a failure of focused attention. Focusing sometimes fails because attention is attracted by a singular or intrusive event in the environment (e.g., an irrelevant stimulus that attracts attention, as in Shiffrin and Schneider [1977], Yantis [1993], and Theeuwes [1994], or a target that is noticed and causes a subsequent target to be missed, as in Shapiro and Raymond [1944]). For example, the platoon leader may be distracted by his gunner's complaints about heat and humidity inside the tank and miss an important radio transmission.

Divided Attention

Divided attention describes a situation in which the human attempts to carry on many processes simultaneously, distributing resources among them. In the perceptual domain, such situations usually involve processing more than one stimulus at a time. For example, the platoon leader may have to try to scan his independent thermal viewer for enemy forces and watch his integrated display simultaneously. He may have to listen to a status report from his gunner while listening to a potentially relevant radio transmission from the A Company commander to the platoon on his right flank. Humans are clearly limited in their ability to divide attention in this manner. For example, the platoon leader may miss an enemy tank while watching his integrated display, or miss relevant information from the radio while listening to the gunner.

Automatization of component processes is the typical means by which people increase their ability to carry out simultaneous tasks. Relatively little is known about interactions among tasks, but when tasks differ substantially, there is often a cost associated with switching between them, even after some practice (e.g., Allport et al., 1994).

Theories and Models of Selective Attention

Theories and models of selective attention are still in an early formative stage (as are the models of working memory of which they are a subset). Historically, Broadbent (e.g., 1957) introduced his *filter* theory for application to the processing of sensory information. According to Broadbent, primitive sensory features are processed in parallel, preattentively, without capacity limitations. Slightly later in the processing stream, a filter or blockage is reached, and further processing requires selective allocation of attention to some feature or dimension of the input information (such as one of the two ears); information having some other feature or arriving on some other dimension, termed a channel, is blocked from further processing (e.g., the information arriving on the other ear will not be processed further). Subsequent research demonstrated that certain information on unattended channels does get processed deeply enough for its meaning to have an impact (e.g., one's own name presented to an unattended ear). This finding led Treisman (e.g., 1969) to modify Broadbent's theory and propose that the processing of information on unattended channels is attenuated rather than blocked. An alternative theoretical approach was suggested by Deutsch and Deutsch (1963). They posited that all incoming information is processed to deep levels, but that the attentional capacity limitations are those of memory (e.g., selective forgetting of processed information from short-term memory). A more fully developed version of this concept, the theory of automatic and attentive processing (then called automatic and controlled processing) was presented by Shiffrin and Schneider (1977) and summarized and updated by Shiffrin (1988).

All the above approaches share the assumption that the difficulty for the processing system is limited processing capacity. Early theories explicitly or implicitly assumed that capacity represents a common pool of allocatable resources. Later researchers proposed that capacity is better conceived as a group of overlapping pools, so that increasing the difficulty of a task may require sharing of resources and attention across similar domains (such as between the two ears), but not across dissimilar domains (such as between the ears and eyes). Examples of this view include Wickens (1984) and Navon and Gopher (1979).

It must be emphasized that these issues continue to undergo intense empirical testing and theoretical development today, and a general or simple resolution has not been achieved. For certain tasks, the capacity limitations are almost certainly quite central and related to deep processes such as decision making and/or forgetting in short-term memory (e.g., Shiffrin and Gardner, 1972; Palmer, 1994); for others, however, the blockage may be at a more peripheral locus.

Because the field of attention is so complex and relatively new and comprises largely empirical studies, theory development is still at an early age. Some models are little more than metaphors; an example is the "spotlight" theory, in which attention is spatially compact and moves continuously across the visual field. Detailed and well-developed computer simulation and mathematical models have been devised for particular tasks with some success, but they are based on extensive collection of data within those tasks and tailored to those domains (e.g., Schneider and Shiffrin, 1977; Wolfe, 1994; Sperling and Weichselgartner, 1995; Meyer and Kieras, 1997a, 1997b). Theory development has not yet proceeded to the point where the theories can generalize well across tasks or allow extrapolation to new domains in which extensive empirical research has not occurred. In this regard, the work of Meyer and Kieras (1997a, 1997b) on executive-process interactive control (EPIC) offers a promising approach. (EPIC is discussed in detail in Chapter 3.)

If one broadens the notion of attending to stimuli to encompass attending to processes and tasks, theories and models of attention can be expanded from a focus on perceptual processing to become theories and models of multitasking—the topic of the next section.

MULTITASKING

It is difficult to imagine a situation in which the modeling of multitasking in the general sense would not be needed for military simulations. Whether in the guise of a model of working memory or of selective attention or multitasking, this modeling will have similar conceptual underpinnings. Applications are sometimes needed when there are two or more externally defined and somewhat independent tasks to be accomplished, and sometimes when one (complex) task requires a variety of internal processes that need more resources than are available concurrently. Of course, some military simulations do incorporate limited multitasking, if only to permit the interruption of one task for another, but the extent to which such capability is based on psychological theory is not always clear. When overload occurs, there are several potential outcomes including (1) continuing to do everything but less well, (2) reducing the number of things being attended to, (3) putting tasks in a queue, and (4) dropping everything and walking away. It is worth noting that there may be situations in which one would not want to model a real human with real limitations; rather, one might find it useful to assume that the human has unlimited parallel processing capabilities (e.g., if a model of a "superopponent" were desired).

Theories and Models of Multitasking

There are a number of excellent reviews of multitasking and related topics, including several by Wickens (1984, 1989, 1992) and Adams et al. (1991). The discussion below covers engineering and psychological theories and models of multitasking. Then, to summarize that information in a form more likely to be usable by military simulation modelers, a composite theory of multitasking is presented. Note that the models described here are generally not computational models and are therefore not directly applicable to military simulations. But these models and the theories they interpret could serve as a valuable base on which to construct computational models. Relevant computational models of multitasking are described in more detail in the context of integrative models, which are discussed in Chapter 3.

Engineering Theories and Models of Multitasking

Engineering theories of human behavior are generally concerned with describing gross human behaviors, not the cognitive or psychomotor mechanisms that underlie them. This is particularly true of engineering theories and models of multitasking. Pattipati and Kleinman (1991) present a summary of such models; the summary here is based on but also expands on their account.

As mentioned above, multitasking theories and models can be viewed as an extension of theories and models of attention. An example of this point is *multitasking theories and models based on queuing theory*. Queuing theory is a branch of operations research that addresses systems capable of being described in terms of one or more servers and a population of customers queuing (lining up) for service. Queuing theory was first applied to the domain of human operator modeling by Carbonell and colleagues, who used it to model the visual scanning behavior of a pilot or other operator obtaining information from several different displays (Carbonell, 1966; Carbonell et al., 1968). Their theory was that the operator's visual attention could be described as a server and the instruments to be read as customers queuing for service. They were then able to use queuing theory formulas to generate estimates of sampling frequencies and other parameters.

The notion of attending to multiple displays was expanded to the broader issue of attending to multiple tasks by a number of researchers (e.g., Walden and Rouse, 1978; Chu and Rouse, 1979; Greenstein and Rouse, 1982). Their general approach was to describe human high-level attention as a server with a given service time probability distribution having specified parameters. They described tasks competing for the operator's attention as customers queuing up for service according to one or more arrival time distributions with specified parameters. This approach allowed them to model the operator's attention allocation policy in terms of queuing discipline, such as first-in-first-out or last-in-first-out, possibly involving balking (the decision of a customer to leave the queue if waiting time becomes too great). To describe multitasking as a queuing process enabled these researchers to use queuing theory formulas to develop general information about the multitasking behavior of the operator. For example, the ability to compute mean waiting time gives insight into the time required before a task must be attended to and helps in deriving mean task execution time.

The value of such estimates of overall multitasking behavior in the present context may be limited to constructive military simulations, in which the moment-to-moment activity of the modeled human is relatively unimportant to the user. However, these queuing-based theories and models of multitasking behavior provide one basis for the higher-resolution, discrete-event computational models described briefly below and in more detail in Chapter 3 of this report. For a more thorough review of queuing theory models, see Liu (1996).

Engineering research has also contributed *multitasking theories and models based on control and estimation theory*. These have their roots in optimal control theory and optimal control models of the human operator (e.g., Kleinman et al., 1970, 1971). The application of optimal control theory to human behavior is based on the assumption that the performance of an experienced human operator in controlling a continuous-state system (such as flying an aircraft or controlling a chemical plant) approaches that of a nonhuman optimal control system. Optimal control models of human performance have been shown to predict accurately the performance of real humans who are well practiced at a control task.

In an attempt to extend this success to the modeling of human multitasking behavior, a number of researchers have applied optimal control theory to that domain (e.g., Tulga and Sheridan, 1980; Pattipati and Kleinman, 1991). An optimal control theory of human multitasking behavior has the following elements. A task is represented as a dynamic subsystem of the controlled system (the plant, in optimal control terminology). Thus the plant represents not just an airplane or a tank, but an airplane or a tank augmented by the tasks the operator is trying to perform. The plant is acted upon by disturbances beyond the operator's control. The task state is the state of the plant (possibly including its environment) with respect to the task. The decision state is the time required for each task, the time available for the task, and so on. A display element delays and adds noise to true task states. A human limitations and monitoring element yields perceived task states. A Kalman filter/predictor yields estimates of true task states. The decision state submodel calculates decision variables. The attractiveness measures submodel yields attractiveness measures for each task, and the stochastic choice model computes probabilities of working on each task, which in turn affect the plant. These elements are synthesized through optimal control theory in relation with a carefully specified performance index selected to minimize the system dynamic error.

Like queuing theory models of multitasking, control and estimation theory models can yield estimates of overall human performance, especially for well-trained individuals or groups. Since they are applicable to hardware and software controllers that must operate in real time (or more rapidly), they are also potentially capable of generating representations of moment-to-moment human behavior. As a result, such models may be useful in both constructive and virtual military simulations. One must keep in mind, however, that models assuming optimum control are almost certainly best applied to humans who have already developed considerable expertise and levels of skill.

Psychological Theories and Models of Multitasking

In general, psychological theories and models of multitasking are distinguished from engineering theories and models in that the former are more concerned with understanding and representing the mechanisms underlying behaviors.

Resource Theories and Models Theories of perceptual attention treat the visual or auditory system as a limited resource to be allocated among two or more competing stimuli or information channels. This view has been naturally extended to the concept of multitasking, in which more complex resources must be allocated to tasks (e.g., Navon and Gopher, 1979). Resource theories are typically based on data obtained in dual-task experiments, in which subjects perform two concurrent tasks (such as tracking a target on a cathode ray tube screen while doing mental arithmetic) while performance on each task is measured.

In *single resource theory* (Wickens, 1992:366-374), cognitive mechanisms, including those used for memory and decision making, are viewed as a single, undifferentiated resource pool. Task performance is dependent on the amount of resources allocated to the task, and is sometimes defined more formally by means of a performance resource function, which gives the performance of a task as a function of the amount of resources allocated to that task. The performance resource function can be used to characterize a task with respect to whether it is resource limited (i.e., not enough resources to perform it perfectly) or data limited (i.e., limited by the quantity and/or quality of information available to perform it).

When two tasks compete concurrently for such resources and there are insufficient resources available to perform both tasks perfectly, a tradeoff occurs: one task is performed better at the expense of poorer performance on the other. This relationship is sometimes specified more formally by means of a performance operating characteristic. The performance operating characteristic is a function that describes the performance on two concurrent tasks as a function of resource allocation policy (the amount of resources allocated to each task). Here, too, resource-limited vs. data-limited performance is characterized by the shape of the curve.

The single resource theory is a significant step toward understanding multitasking, but has limitations. For example, single resource theory cannot easily account for dual-task data indicating that interference between two tasks could not be predicted from their difficulty, only from their structure. Another failure of the single resource theory is its inability to explain why, in some cases, two demanding tasks can be time-shared perfectly.

Such limitations led to the development of two related alternative theories. One of these is the *theory of automatic and controlled processing*, discussed earlier (e.g., Shiffrin and Schneider, 1977; Shiffrin, 1988). In this theory, differential ability to carry on two or more simultaneous tasks is due to differential development of automatic processes and procedures that allow attentional limitations to be bypassed. The other approach is that of *multiple resource theory* (Wickens, 1992:375-382). In multiple resource theory, resources are differentiated according to information processing stages (encoding and central processing or responding), perceptual modality (auditory or visual), and processing codes (spatial or verbal). Different tasks require different amounts of different resources. For example, for the platoon leader to detect and interpret a symbol on his commander's integrated display would require more encoding resources (processing stage category), while for him to acknowledge a radio transmission from the A Company commander would require more responding resources.

With this refinement, the concept of resources can be used to explain some of the dual-task data beyond the capabilities of the single resource theory. For example, near-perfect time sharing of the two tasks described in the previous paragraph can be explained by their need for different resources. Presumably, the platoon leader can allocate sufficient encoding resources to display interpretation and at the same time allocate adequate responding resources to perform the acknowledgment task.

The multiple resource theory has been used in an attempt to formalize the notion of *mental workload* (Wickens, 1992:389-402). Here mental workload is defined as the resource demand. Thus poor performance in situations deemed to impose "high workload" is explained in terms of excess demands for specific resources. The multiple resource theory and derivative theories of mental workload have been at least partially validated in realistically complex domains and are already in use in applications designed to evaluate human-machine interfaces and operator procedures (e.g., W/INDEX, North and Riley, 1989).

It is important to realize that both the theory of automatic and controlled processing and multiple resource theory are really general frameworks and for the most part do not provide specific models for new tasks and task environments. However, they can be the basis for such models. One example is a queuing model that integrates aspects of single-channel queuing with multiple-resource-based parallel processing (Liu, 1997).

Strategic Workload Management Theories and Models The multiple resource theory and workload theories and models generally do not address explicitly the issue of how the human allocates resources. Recognizing this, Hart (1989) observed that pilots and operators of other complex systems seem to schedule task performance and augment and reduce task load so as to maintain a "comfortable" level of workload. Raby and Wickens (1994) validated this theory in a study examining how pilots manage activities while flying simulated landing approaches. Moray et al. (1991) found that while Hart's theory may be true, humans are suboptimal in scheduling tasks, especially when time pressure is great. A recent theory relying heavily on scheduling priorities and timing to explain capacity limitations and multitasking is that of Meyer and Kieras (1997a, 1997b; see the discussion of EPIC in Chapter 3).

Theories and Models of Task Interruptions An issue closely related to managing activities is interruptions. In a simulator study of airline pilot multitasking behavior, Latorella (1996a, 1996b) found that task modality, level of goals in the mission goal hierarchy, task interrelationships, and level of environmental stress affect the way humans handle interrupting tasks and the ongoing tasks that are interrupted. Damos (forthcoming) is currently trying to identify how airline pilots prioritize tasks.

Theories and Models of Task Management There have been a number of efforts to define the process by which operators of complex systems (especially pilots of modern aircraft) manage tasks. For example, Funk and colleagues developed a preliminary normative theory of cockpit task management (Funk, 1991; Chou et al., 1996; Funk and McCoy, 1996). According to this theory, managing a set of cockpit tasks involves the following activities:

- Assessing the current situation
- Activating new tasks in response to recent events
- Assessing task status to determine whether each task is being performed satisfactorily
- Terminating tasks with achieved or unachievable goals
- Assessing task resource requirements (both human and machine)
- Prioritizing active tasks
- Allocating resources to tasks in order of priority (initiating, interrupting, and resuming them, as necessary)
- Updating the task set

Rogers (1996) used structured interviews to refine and expand the concept of task management, and Schutte and Trujillo (1996) studied task management in non-normal flight situations. The conclusions to be drawn from these studies are that task management is ubiquitous and significant and that it plays an important role in aviation safety. These conclusions almost certainly generalize to other complex systems. Detailed models of these complex activities are still under development.

Connectionist and Neurally Based Models Neurally motivated computer simulations yield *connectionist theories and models of multitasking*. For example, Detweiler and Schneider (1991) describe a connectionist model in which separate, radiating columns or modules of nodes and connections represent separate channels or operator "resources" (visual, auditory, motor, and speech systems). In their model, all columns are connected to a single inner loop so that pathways can be established between modules. Multiple pathways imply the capacity for parallel performance. The nature and extent of the connections dictate the nature and extent of multiple task performance. In Detweiler and Schneider's model, the development of connections can be used to model the acquisition of multitasking skill.

Prospects for a Composite Theory of Multitasking

A composite, comprehensive account of multitasking is essentially equivalent to a comprehensive model of human cognition. That is, almost any task of reasonable complexity, especially one likely to be incorporated in real-world military simulations, will involve resource allocation, motor performance, strategic use of working memory, scheduling, retrieval from long-term memory, decision making, and all other components of a general model of cognition and performance. No one would pretend that anyone has yet come close to producing such a model. The closest approximations available are applications of varying degrees of specificity that are tailored to particular task environments (e.g., detailed process models of sequential choice reaction time in the laboratory or less detailed models of pilot performance). A number of such approaches have been described in this section. Although continued progress can be expected in the development of large-scale models, it is unlikely that anything like a comprehensive model will be available within a time horizon of immediate interest to the armed forces. Thus for the near future, applications with the greatest utility will be based on models developed for and tailored to specific task environments.

INTEGRATING CONCEPTUAL FRAMEWORKS

Theories and models of attention and multitasking cover a wide and complex range of human behavior, and there have been several attempts to integrate them into a single, coherent framework. One example is Adams et al. (1991). Their objective was to summarize what is known about attention and multitasking, including task management. Their approach was to review existing psychological literature, to extend and extrapolate research findings to realistically complex domains, and to present a framework for understanding multitasking and task management. The results of their efforts are summarized in the following paragraphs.

Task management involves task prioritization. Task prioritization depends on situation awareness, which in turn depends on perception. Perception is schema based; that is, input information is interpreted in the context of structured expectations about situations and events. This implies that information used to update situation models must be anticipated and prepared for. Long-term memory is a connectionist, associative structure, and the attentional focus corresponds to areas of long-term memory activation. Multitasking in turn depends on attentional shifts, which are cognitively difficult and take measurable time. Human behavior is goal driven, and goals help determine how and where attention will be shifted. A goal hierarchy comprising goals and subgoals is the basis for task ordering or prioritization when simultaneous performance of all tasks is impossible. Tasks correspond to knowledge structures in long-term memory (one structure per task, though certain structural elements are shared across tasks). Since information processing is resource limited, the human can allocate conscious mental effort to only one task while queuing others. This is the motivation for task prioritization.

There is a tendency to process only that incoming information which is relevant to the task currently being attended to. If incoming information is not relevant to that task, the human must interrupt it to determine which queued task (if any) the information concerns. Such information tends to be "elusive" and subject to neglect, since there is no schema-based expectation or preparation for it. However, noticing and processing a stimulus or event implies interrupting the ongoing task. Humans resist interruptions and can even become irritable when they occur. Tasks associated with lower-level (more specific) goals are more resistant to interruption. But interruptions do occur; fortunately, memory for interrupted tasks is highly persistent.

Task management further involves task scheduling. The ability to schedule depends on the individual's understanding of temporal constraints on goals and tasks. Subjects in task-scheduling studies are capable of responding appropriately to task priority, but scheduling performance may break down under time pressure and other stressors. Task management itself is an information processing function, and it is most crucial when information processing load is at its highest, for example, when there are more tasks to manage. Therefore, task management is a significant element of human behavior.

The conceptual framework of Adams et al. (1991) formed part of the basis for the operator model architecture (OMAR) model of human performance (see Chapter 3). Another framework for attention and multitasking that is beginning to yield interesting results from validation study is EPIC (Meyer and Kieras, 1997a, 1997b) (also described in Chapter 3).

This review of theories and models of attention and multitasking has focused on the engineering and psychological literature, which often proposes explicit mechanisms of attention allocation and task management. There is also some promising computer science work in which attention and multitasking are emergent phenomena that result not necessarily from explicit representations, but from properties of information processing mechanisms. Selfridge's pandemonium theory (Selfridge, 1959) is an early example.

CONCLUSIONS AND GOALS

Modeling of multitasking is clearly relevant to military simulations and to human performance generally. The field has reached a point at which there is real potential to produce useful simulations of multitasking behaviors. However, doing so will not be easy or quick. Currently, the relevant theories and models are not well developed or validated, and the computational models are somewhat ad hoc. But current theories and models do provide a starting point from which acceptable computational models can be built. We offer the following goals for further development.

Short-Term Goals

- Conduct studies to identify the factors that influence the allocation of attention to tasks (e.g., importance, urgency, salience of stimuli), using domain experts as subjects.
- When possible, use existing military models that support multitasking. Otherwise, augment existing serial models to support the execution of concurrent tasks.
- Delineate carefully the different approaches of sharing tasks versus switching between tasks.
- Because most research has focused on attentional effects on perception, devote resources to incorporating in the models the effects of attentional allocation on memory management, decision making, and translation of cognitive activities into action and motor control.

Intermediate-Term Goals

- Develop models of attentional capacity, expanding on the concepts of sharing and switching.
- Identify the factors that influence the allocation of attention to tasks (e.g., importance, urgency, salience of stimuli).
- Validate the behavior of the models by comparing the tasks they accomplish by the model and their performance with similar data from human domain experts.
- Begin the process of collecting data on cognitive resource sharing in military domain situations.

• Explore alternative model representations, including rule-based, exemplar-based, and neural net representations.

• Investigate the effects of various factors (e.g., salience and uniqueness of stimuli) on attention allocation strategies.

Long-Term Goals

• Develop models of multitasking behavior in realistically complex military domains, incorporating a wide range of cognitive and motor processes that are affected by attention and resource allocation.

• Validate such models against real-world data, especially from the military domain.

• Expand the applicability of models of attentional allocation to other cognitive modules mentioned in this report, from memory and learning (Chapter 5) through planning and decision making (Chapters 8 and 6, respectively), to group behavior (Chapter 10) and information warfare (Chapter 11).

5

Memory and Learning

Memory and learning have been the chief concern of experimental psychology since that field's inception and have been at the center of cognitive psychology and cognitive science. Thus there is an enormous literature on the subject, both empirical and theoretical, ranging from animal behavior to high-level human cognition, from behavioral studies to neuroscience, from applied research to basic mechanisms, and from very short-term memories (a few hundred milliseconds) to the development of expertise and skill (perhaps 10 years). A great portion of this research is directly or potentially relevant to the representation of human behavior in military simulations, but even a survey of the field would require several long books. Thus this chapter provides an extremely selective survey of those elements that seem particularly relevant to simulations of interest to defense forces. After briefly reviewing the basic structures associated with memory and learning, the chapter examines modeling of the various types of memory and then of human learning. The final section presents conclusions and goals in the area of memory and learning.

BASIC STRUCTURES

There is general agreement that human memory is best modeled in terms of a basic division between active short-term memory and a passive storehouse called long-term memory.[1] Short-term memories include sensory memories, such

[1]The following discussion is based on the work of early psychologists such as James (1890) and the more detailed treatment of the subject by Atkinson and Shiffrin (1968).

as the visual icons originally studied by Sperling (1960), which operate fairly automatically and decay autonomously within several hundred milliseconds, and a variety of more central memories, including so-called "working memories," that are used to carry out all the operations of cognition (e.g., Baddeley, 1986, 1990). Short-term memories allow control processes to be carried out because they hold information for temporary periods of, say, up to a minute or more. The long-term memory system can be divided by function or other properties into separate systems, but this is a matter of current research. The operations associated with memory include those of rehearsal and coding used to store new information in long-term memory; operations that govern the activities of the cognitive system, such as decisions; and operations used to retrieve information from long-term memory. Almost all simulations incorporate such structural features.

Different aspects of memory and learning are important for simulations aimed at different purposes. For example, a simulation of a tank gunner might rely heavily on short-term vision, short-term working memory, and whatever task procedures have been learned previously and stored in long-term memory. Retrieval from visual short-term memory may be important, for example, when the positions of recent targets and terrain features must be retained during a series of maneuvers that move these targets and features out of direct vision. For a gunner in a short firefight, learning during the task may be relatively unimportant since the gunner will rely primarily on previously trained procedures and strategies or decisions made in working memory. In addition, the process of accessing learned procedures in long-term memory may not need to be modeled, only assumed (since this process may be well learned and automatic). On the other hand, strategic decision making by a battle commander will involve working memory and the processes of long-term retrieval. Sensory memories and the retention role of short-term memories may be relatively unimportant. Finally, in the course of a war lasting several years, learning will probably be critical for models of personnel at all levels of command.

It should be obvious that the process of learning itself is critical whenever training and the development of skill and expertise are at issue. The nature of storage, the best means of retrieval, ways to utilize error correction, and best modes of practice are all essential components of a simulation of the learner. It is well known that the largest effects on performance observed in the literature are associated with the development of expertise (see also Chapter 4). The development of skills can proceed quite quickly in short periods of time, but typically continues for long periods. Given motivation and the right kinds of practice, development of the highest levels of skill in a task may require 3 to 4 hours a day of rigorous practice for a period of 10 years (e.g., Ericcson et al., 1993; Ericcson, 1996). Although the nature of the learning process in such situations is not yet understood in detail, there is much empirical evidence (starting with Chase and Simon, 1973) demonstrating that a critical component of the learning of a skill is the storage of massive amounts of information about particular situations and procedures, information that apparently becomes retrievable in parallel with increasing ease.

Several researchers (Van Lehn, 1996; Logicon, 1989; Brecke and Young, 1990) have constructed frameworks showing the relationship between the development of expertise and changes in the content and structure of knowledge stored by the individual as he or she moves from novice to expert. These researchers argue that experts have simpler but more effective and more richly related knowledge about a domain or situation than do novices, who may know more facts but fewer useful relationships among them. Thus, a key to the evolution of expertise is the evolution of representation: as expertise increases, knowledge is represented in a more abstract way, reducing the need to manipulate massive amounts of information.

Modeling the learning process itself, as in the development of expertise, can be important in another way. Even if one is concerned only with the end state of learning and wishes to model the procedures used by a well-trained individual in a given military situation, there are usually many possible models of such procedures. Modeling of the course of learning itself can be used to generate an effective representation for a given level of skill.

Although modeling the course of learning over 10 years of intensive practice may be useful for military simulations only in rare instances, other sorts of learning over shorter time spans will often be important. To carry this discussion further, it is necessary to discuss the different types of memory and their modeling.

MODELING OF THE DIFFERENT TYPES OF MEMORY

Memory can be broken down into three broad categories: (1) episodic, generic, and implicit memory; (2) short-term and working memory; and (3) long-term memory and retrieval.

Episodic, Generic, and Implicit Memory

Episodic memory refers to memory of events that tend to be identified by the context of storage and by personal interaction with the rememberer; successful retrieval of such memories tends to require contextual and personal cues in the memory probe. Examples might include the positions of enemy forces identified in the last several minutes of a firefight and, in general, all attended recent events. Many terms have been used to refer to aspects of general knowledge, some with particular theoretical connotations. In an attempt to remain neutral, we use the term *generic memory* to refer to general knowledge—knowledge not necessarily identified by circumstances of storage or requiring contextual cues for retrieval to occur. Examples might include the sequence of moves needed to aim and fire a familiar weapon, the rules for addition, and most general knowledge of facts and procedures. Finally, *implicit memory* refers to effects on retrieval of general knowledge that result from recent experience, even if the recent experience is not recalled. For example, a gunner who saw a picture of a particular weapon in a magazine the previous evening would be more likely to notice that same weapon under battlefield conditions in the field the following day, even if the magazine event were totally forgotten and unavailable.

Although the existence of implicit memory effects is well established, the importance of including them in military models and simulations is not yet clear. Therefore, the focus here is on episodic and generic storage and retrieval, both of which are essential components of any simulation of human behavior. It is important to note that retrieval in all episodic tasks relies on a combination of generic and episodic access. For example, an externally provided retrieval cue, such as a word, will first be used to access generic, lexical, and semantic knowledge, which will join the surface features in a probe of episodic traces. A soldier prompted by a loud sound to recall the position of a recently seen tank, for example, will automatically be prompted to access long-term memory for associations with such a sound, perhaps recovering knowledge of the sorts of weapon systems that produce such a sound. The issue of whether both types of retrieval occur in episodic retrieval is largely independent of whether both must be included in a simulation; that answer depends on the particular application.

Short-Term and Working Memory

Short-term memory is the term used most often when one is interested more in the retention properties of the short-term system than in the control operations involved. Conversely, *working memory* is the term used most often when one is concerned with the operations carried out in active memory. These are not universal usages, however, and there is no hard and fast line between the two terms. The retention properties of the short-term systems are still an active area of study today, and there is no shortage of mathematical and simulation models of the retention of item and order information (e.g., Schweikert and Boruff, 1986; Cowan, 1993). Models of working memory that address the control of cognition are still in their infancy, and until very recently dealt only with certain specific control processes, such as rehearsal, coding, and memory search. What seems to be clear is that a successful model of short-term memory (and of memory in general) must deal with both the retention and control properties of the system. It seems clear that a model useful for military simulations must incorporate both aspects. Recent models of working memory are increasingly modular, with greater limitations of capacity within rather than between functional modules. Examples are the Soar architecture (e.g., Laird et al., 1987), the adaptive control of thought (ACT)-R model (Anderson, 1993), and the recent executive-process interactive control (EPIC) model of Meyer and Kieras (1997a); these are discussed in detail in Chapter 3.

Having pointed out the need for such a model, we must also note that a complete model of short-term memory (in this general sense, which includes all the control processes used to store, retrieve, and decide) is tantamount to a complete model of cognition. The field is a long way from having such a model. Thus for the foreseeable future, short-term memory simulations will need to be tailored carefully to the requirements and specifications of the particular task being modeled.

Long-Term Memory and Retrieval

For most modelers, *long-term memory* denotes a relatively permanent repository for information. Thus models of long-term memory consist of representation assumptions (e.g., separate storage of exemplars, distributed representations, networks of connected symbolic nodes) and retrieval assumptions. The latter assumptions are actually models of a certain set of control and attention operations in short-term memory, but it is traditional to treat them separately from the other operations of short-term memory.

For episodic storage and retrieval, the best current models assume separate storage of events (e.g., the search of associative memory [SAM] model of Raaijmakers and Shiffrin, 1981; the ACT-R model of Anderson, 1993; the retrieving effectively from memory [REM] model of Shiffrin and Steyvers, 1997; Altmann and John, forthcoming); however, there may be equivalent ways to represent such models within sparsely populated distributed neural nets. Retrieval of all types is universally acknowledged to be driven by similarities between probe cues and memory traces, with contextual cues playing an especially significant role in episodic storage and retrieval.

It seems likely at present that different modes of retrieval are used for different tasks. For example, free recall (recall of as many items as possible from some loosely defined set) is always modeled as a memory search, with sequential sampling from memory (e.g., Raaijmakers and Shiffrin, 1981). Recognition (identification of a test item as having occurred in a specific recent context) is almost always modeled as a process of comparing the text item in parallel with all traces in memory (e.g., Gillund and Shiffrin, 1984). Cued recall (recall of one of a pair of studied items when presented with the other) is still a matter of debate, though response time data favor a sequential search model for episodic tasks (e.g., Nobel, 1996).

Most current models, and those with an elaborated structure, have been developed to predict data from explicit memory tests. Models for retrieval from generic memory are still in their infancy. One problem hindering progress in this arena is the likelihood that a model of retrieval from generic memory may require, in principle, an adequate representation of the structure of human knowledge. Except for some applications in very limited domains, the field is quite far from having such representations, and this state of affairs will probably continue for some years to come.

The situation may not be hopeless, however, because several types of simplified representations sufficient for certain needs are currently being used. First, there are exemplar-based systems (described later in the section on learning). These systems generally assume that knowledge consists of the storage of multiple copies (or strengthened copies) of repeated events. The more sophisticated of these systems place these copies in a high-dimensional metric space, in which smaller distances between exemplars represent increased similarity (e.g., Logan and Bundesen, 1997; Hintzman, 1986; Nosofsky and Palmeri, 1997; Landauer and Dumais, 1997). For some purposes, such as modeling improvements with practice and the production of errors based on similarity, this approach has proved fruitful. Second, neural net-based learning systems develop limited structural representations through the process of modifying connections between nodes. Although the structure thus developed is usually not transparent to the external observer, typically it is limited to connections between a few layers of nodes, and may not have any relation to actual cognitive structure. The developed representations capture much of the statistical structure of the inputs that are presented to the nets during learning and have often proved useful (Plaut et al., 1996). Other models that capture the statistical structure of the environment have also proved extremely promising (such as the modeling of word meanings developed by Landauer and Dumais, 1997). Third, in a number of application domains, experts have attempted to build into their representations those aspects they feel are necessary for adequate performance; examples include rule-based systems such as the Soar system (see the next section) and certain applications within the ACT-R system (see Anderson, 1993). Although none of these approaches could be considered adequate in general, separately or in combination they provide a useful starting point for the representation of knowledge.

A second problem just now surfacing in the field is the nature of retrieval from generic memory: To what degree are generic retrieval processes similar to those used in explicit retrieval (i.e., retrieval from episodic memory)? Even more generally, many operations of cognition require a continuing process of retrieval from long-term (episodic and generic) memory and processing of that information in short-term memory that can last for periods longer than those usually thought to represent short-term retention. These processes are sometimes called "long-term" or "extended" working memory. Some recent neurophysiological evidence may suggest that different neural structures are active during explicit and generic retrieval, but even if true this observation says little about the similarity of the retrieval rules used to access the stored information. The models thus far have largely assumed that retrieval processes involved in explicit and generic retrieval are similar (e.g., Humphreys et al., 1989; Anderson, 1993; Shiffrin and Steyvers, 1997). This assumption is not yet well tested, but probably provides a useful starting point for model development.

With regard to models of human behavior of use to the military, there is little question that appropriate models of explicit and generic retrieval and their interaction (as studied in implicit memory, for example) are all critical. Because accurate and rapid access to general and military knowledge, as opposed to access to recent events, is of great importance in producing effective action and in modeling the cognition of participants in military situations, and because this area of modeling is not yet well developed, the military may eventually need to allocate additional resources to this area of work.

MODELING OF HUMAN LEARNING

Two types of learning are of potential importance for military simulations. The first is, of course, the learning process of military or other personnel. The second, superficially unrelated, is the possibility of learning within and by the simulations themselves. That is, any simulation is at best a gross approximation of the real processes underlying behavior. Given appropriate feedback, a simulation program itself could be made to learn and adapt and improve its representation. There is no necessary reason why the learning process by which a simulation program adapts its representation should match the learning process of people, but experience suggests the processes assumed for people would provide a good starting point. First, many recent models of cognition have made considerable headway by adopting the assumption that people operate in optimal fashion (subject to certain environmental and biological constraints); examples are found in the work of Anderson (1990) and Shiffrin and Steyvers (1997). Second, there is a virtually infinite set of possible learning mechanisms. Evolution may well have explored many of these in settling on the mechanisms now employed by humans. Given the complexity of representing general human behavior in simulations, the potential benefit of allowing simulations themselves to learn and evolve should not be underestimated. The discussion here concentrates, however, on models of human learning.

Learning from experience with the outcomes of previous decisions is a necessary component of the design of an intelligent computer-generated agent. No intelligent system would continue repeating the same mistakes without correction, nor would it continue to solve the same problem from first principles. Human commanders can learn how decision strategies perform across a variety of engagements and different types of opponents. Human soldiers can learn how to become more effective in carrying out their orders on the basis on successive encounters with a weapon system. The capability of learning to anticipate the consequences of an action is essential for the survival of any intelligent organism. Thus whenever the goal is to provide a realistic approximation of human behavior, computer-generated agents must have the capability of learning from experience. However, most simulations do not have such lofty goals.

There are two cases in which it is especially important to build models of learning into a simulation. The first occurs whenever the task being modeled lasts long enough that the operator learns and adapts his or her behavior accordingly. The development of skill during training is an obvious example, but some learning can occur even in relatively short time spans, such as the course of a battle or even a short firefight. If one is interested in modeling learning that occurs with extended practice, one must include a model of the way the storage of episodic memories leads to the storage of lexical/semantic generic memories and generic procedural memories. Models of such a transition are still at an early stage of development (e.g., Shiffrin and Steyvers, 1997)

The second case requiring the incorporation of learning into a simulation is one in which it is assumed that performance, strategies, and decisions are relatively static during the task being modeled, so that the participants are not learning over the course of the task. Nonetheless, the procedures and rules used by a participant in such a situation are frequently quite complicated, and specifying them is often inordinately difficult. Current simulation developers have expended a great deal of effort on extracting knowledge from experts and coding this knowledge into forms suitable for computer programs. These efforts are typically very demanding and not always successful. Thus the need for learning mechanisms in military simulations has been clear from the beginning: learning mechanisms have the potential to provide a way of automatically training a computerized agent to have various levels of knowledge and skill. In fact, machine-learning algorithms and the technology for introducing them into simulations have existed for quite some time (see, e.g., Langley, 1996), but the technology has either been underutilized or omitted in computer-generated agents for military simulations (for example, the Soar-intelligent forces [IFOR] project that does not include learning).

Recently, a technology assessment of command decision modeling (U.S. Army Artificial Intelligence Center, 1996) identified a number of approaches for modeling learning that are promising candidates for improving the performance of military simulations. The technology assessment included rule-based learning models, exemplar- or case-based learning models, and neural network learning models. These three types of learning models, which have also been extensively developed and tested in cognitive psychology, are reviewed in the following subsections. The discussion focuses on models that have undergone extensive experimental testing and achieved considerable success in explaining a broad range of empirical phenomena. Rule-based models are reviewed first because they are the easiest to implement within the current rule-based systems developed for military simulations. The exemplar (case)-based models are discussed next; they have been more successful than rule-based models in explaining a broad range of human learning data. Neural network models are reviewed last; although they hold the most promise for providing powerful learning models, they are also the most difficult to integrate into current military simulations. Research is needed to develop a hybrid system that will integrate the newer neural network technology into the current rule-based technology.

Rule-Based Models

Some of the most advanced systems for computer-generated agents used in military simulations are based on the Soar architecture (Laird et al., 1987; see also Chapter 3). Soar, originally constructed as an attempt to provide a unified theory of human cognition (Newell, 1991), is a production rule system in which all knowledge is represented in the form of condition-action rules. It provides a learning mechanism called chunking, but this option is not activated in the current military simulations. When this mechanism is activated, it works as follows. The production rule system takes input and produces output during each interval defined as a decision cycle. Whenever the decision cycle of the production system reaches an impasse or conflict, a problem-solving process is activated. Eventually, this problem-solving process results in a solution, and Soar overcomes the impasse. Chunking is applied at this point by forming a new rule that encodes the conditions preceding the impasse and encodes the solution as the action for this new rule. Thus the next time the same situation is encountered, the same solution can be provided immediately by the production rule formed by the chunking mechanism, and the problem-solving process can be bypassed. There are also other principles for creating new productions with greater discrimination (additional conditions in the antecedent of the production) or generalization (fewer conditions in the antecedent) and for masking the effects of the older productions.

A few empirical studies have been conducted to evaluate the validity of the chunking process (Miller and Laird, 1996; Newell and Rosenbloom, 1981; Rieman et al., 1996). More extensive experimental testing is needed, however, to determine how closely this process approximates human learning. There are reasons for questioning chunking as the primary model of human learning in military situations. One problem with the current applications of Soar in military simulations thus far is that the system needs to select one operator to execute from the many that are applicable in each decision cycle, and this selection depends on preference values associated with each operator. Currently, these preference values are programmed directly into the Soar-IFOR models, rather than learned from experience. In future military Soar systems, when new operators are learned, preferences for those operators will also need to be learned, and Soar's chunking mechanism is well suited for this purpose. Given the uncertain and dynamic environment of a military simulation, this preference knowledge will need to be continually updated and adaptive to its experience with the environment. Again, chunking can, in principle, continually refine and adjust preference knowledge, but this capability must be shown to work in practice for large military simulations and rapidly changing environments.

A second potential problem is that the conditions forming the antecedent of a production rule must be matched exactly before the production will fire. If the current situation deviates slightly from the conditions of a rule because of noise in the environment or changes in the environment over time, the appropriate rule will not apply. Although present technology makes it possible to add a huge number of productions to simulations to handle generalization and encounters with new situations, this is not the most elegant or efficient solution. Systems with similarity-based generalization and alternative learning approaches are worth considering for this reason. Some artificial intelligence research with regard to learning in continuous environments has demonstrated proof-of-concept models in which Soar learns appropriately (Modi et al., 1990; Rogers, 1996). However, it is uncertain whether these techniques are veridical models of human learning in continuous domains.

An alternative to the Soar model of learning is ACT-R, a rule-based learning model (see Chapter 3). During the past 25 years, Anderson (1993) and colleagues have been developing an alternative architecture of cognition. During this time they also have been systematically collecting empirical support for this theory across a variety of applications, including human memory, category learning, skill learning, and problem solving. There are many similarities between ACT-R and Soar. For example, some versions of ACT-R include a chunking process for learning solutions to problems. But there are also some important differences.

First, ACT-R assumes two different types knowledge—declarative knowledge, representing facts and their semantic relations (not explicit in Soar), and production rules, used for problem solving (as in Soar). Anderson and Lebiere (1998) have accumulated strong empirical evidence from human memory research, including fan effects and interference effects, to support his assumptions regarding declarative memory.

A second important feature of ACT-R concerns the principles for selecting productions. Associated with each production rule is an estimate of its expected gain and cost, and a production rule is selected probabilistically as a function of the difference between the two. The expected gain and cost are estimated by a Bayesian updating or learning model, which provides adaptive estimates for the values of each rule.

A third important feature of ACT-R concerns the principles for activating a production rule. A pattern-matching process is used to match the current situation to the conditions of a rule, and an associative retrieval process is used to retrieve information probabilistically from long-term memory. The retrieval process also includes learning mechanisms that change the associative strengths based on experience.

One drawback of ACT-R is that it is a relatively complex system involving numerous detailed assumptions that are important for its successful operation. Finally, it is necessary to compare the performance of ACT-R with that of some simpler alternatives presented below.

Some of the most advanced military simulation models employ the Soar architecture, but without the use of the chunking learning option. The simplest way to incorporate learning into these simulations would be to activate this option. In the long run, it may prove useful to explore the use of ACT-R in such situations because it has been built systematically on principles that have been shown empirically to approximate human behavior.

Exemplar-Based Models

Exemplar-based learning models are simple yet surprisingly successful in providing extremely good predictions for human behavior in a variety of applications, including skill learning (Logan, 1992), recognition memory (Hintzman, 1988), and category learning (Medin and Shaffer, 1978; Nosofsky and Palmeri, 1997). Furthermore, exemplar-based models have provided a more complete account of learning phenomena in these areas than rule-based learning models. Essentially, learning occurs in an exemplar-based model through storage of a multitude of experiences with past problems; new problems are solved through the retrieval of solutions to similar past problems. However, it is important to select the right assumptions about the retrieval of past examples if exemplar-based learning models are to work well.

Before these assumptions are described, it may be helpful to have a simple example in mind. Suppose a commander is trying to decide whether a dangerous threat is present on the basis of a pattern of cues provided by intelligence information. The cues are fuzzy and conflicting, with some positive evidence signaling danger and other, negative evidence signaling no danger. The cues also differ according to their validity, some being highly diagnostic and others having low diagnosticity. The optimal or rational way to make this inference is to use a Bayesian inference approach (see also Chapter 7). But using this approach would require precise knowledge of the joint probability distributions over the competing hypotheses, which is unrealistic for many military scenarios. Usually the decision maker does not have this information available, and instead must base the decision on experience with previous situations that are similar to the present case.

According to exemplar-based learning models, each experience or episode is coded or represented as an exemplar and stored in long-term memory. After, say, n episodes or training trials, the memory contains n stored exemplars, and new learning is essentially the storage of new exemplars. Thus one of the key assumptions of the exemplar-based models is massive long-term memory storage capacity. A second critical assumption is that each exemplar is represented as a point in a multidimensional feature space, denoted X_j for the *jth* exemplar. Multidimensional scaling procedures are used to locate the exemplars in this multidimensional space, and the distances among the points are used to represent the dissimilarities among the episodes. Each coordinate X_{ij} of this multidimensional space of exemplars represents a feature that is used to describe an episode. Some features receive more weight or attention than others for a particular inference task, and the amount of attention given to each feature is also learned from experience.

When a new inference is required, such as deciding whether a dangerous threat is present based on a pattern of cues from intelligence information, the cue pattern is used to retrieve exemplars from memory. The cue pattern is represented as a point within the same multidimensional space as the exemplars, denoted C. It activates each memory trace in parallel, similar to an echo or resonance effect. The activation a_j of a stored exemplar by the cue pattern is based on the similarity between the cue pattern C and the stored exemplar X_j. The similarity is assumed to be inversely related to the distance between X_j and C:

$$a_j(C) = exp[- \| X_j - C \|^{\alpha}] \quad (4.1)$$

Evidence for one hypothesis, such as the presence of a threat, is computed by summing all of the activated exemplars that share the hypothesis (threatening feature):

$$S_T = \Sigma_{j \varepsilon T} a_j(C) \quad (4.2a)$$

Evidence for the alternative hypothesis, for example the absence of a threat, is computed by summing all of the activated exemplars that share the alternative hypothesis (absence feature):

$$S_A = \Sigma_{j \varepsilon A} a_j(C) \quad (4.2b)$$

The final decision is based on a comparison of the evidence for each hypothesis, and the probability of choosing the threat hypothesis over the absence hypothesis is obtained from the following ratio:

$$P_r[T, A] = S_T / (S_T + S_A) \quad (4.3)$$

Exemplar-based and rule-based models are not necessarily mutually exclusive alternatives. First, it is possible to include exemplar learning processes into the current rule-based architectures by modifying the principles for activation and selection of rules. The gain achieved by including this simple but effective learning process would offset the cost of making such a major modification. Second, hybrid models have been proposed that allow the use of either rules or exemplars, depending on experience within a domain (see, e.g., Logan, 1988, 1992).

Although exemplar-based models have proven to be highly successful in explaining human learning across a wide range of applications, there are some problems with this theoretical approach. One problem is determining how to allocate attention to features, depending on the task. Complex environments involve a very large number of features, and some allocation of attention is required to focus on the critical or most diagnostic features. This allocation of attention needs to be learned from experience for each type of inference task, and current exemplar models have failed to provide such a learning mechanism. Another problem with these models is that they fail to account for sequential effects that occur during training; this failure results in systematic deviations between the predicted and observed learning curves. Some of these problems are overcome by the neural network learning models discussed next.

Neural Networks

The greatest progress in learning theory during the past 10 years has occurred in the field of artificial neural networks, although the early beginnings of these ideas can be traced back to seminal works by the early learning theorists (Hull, 1943; Estes, 1950; Bush and Mostellor, 1955). These models have proven to be highly successful in approximating human learning in many areas, including perception (Grossberg, 1980), probability learning (Anderson et al., 1977), sequence learning (Cleermans and McClelland, 1991), language learning (Plaut et al., 1996), and category learning (Kruschke, 1992). Moreover, several direct comparisons of neural network learning models with exemplar-based learning models have shown that the former models provide a more accurate account of the details of human learning than do exemplar models (Gluck and Bower, 1988; Nosofsky and Kruschke, 1992). Because neural networks provide robust statistical learning in noisy, uncertain, and dynamically changing environments, these models hold great promise for future technological developments in learning theory (see Haykin, 1994, for a comprehensive introduction).

Artificial neural networks are based on abstract principles derived from fundamental facts of neuroscience. The essential element is a *neural unit*, which cumulates activation provided by inputs from other units; when this activation exceeds a certain threshold, the neural unit fires and passes its outputs to other units. *Connection weights* are used to represent the synaptic strengths and inhibitory connections that interconnect the neural units. A neural unit does not necessarily correspond to a single neuron, and it may be more appropriate to think of a neural unit as a group of neurons with a common function.

A large collection of such units is interconnected to form a network. A typical network is organized into several layers of neural units, beginning with an input layer that interfaces with the stimulus input from the environment and ending with an output layer that provides the interface of the output response with the environment. Between the input and output layers are several hidden layers, each containing a large number of neural units.

The neural units may be interconnected within a layer (e.g., lateral inhibition) or across layers, and activation may pass forward (projections) or backward (feedback connections). Each unit cumulates activation from a number of other units, computes a possibly nonlinear transformation of the cumulative activation, and passes this output on to many other units with either excitatory or inhibitory connections. When a stimulus is presented, activation originates from the inputs, cycles through the hidden layers, and produces activation at the outputs.

The activity pattern across the units at a particular point in time defines the *state* of the dynamic system at that time. The state of activation evolves over time until it reaches an equilibrium. This final state of activation represents information that is retrieved from the memory of the network as a result of the input stimulus. The persistence of activation produced by stimulus is interpreted as the *short-term memory* of the system.

Long-term storage of knowledge is represented by the strengths of the connection weights. The initial weights represent knowledge before training begins; during training, the weights are updated by a learning algorithm. The learning algorithm is designed to search the weight space for a set of weights that maximizes a performance function (e.g., maximum likelihood estimation), and the weights are updated in the direction of steepest ascent for the objective function.

Universal approximation theorems have shown that these networks have sufficient computational power to approximate a very large class of nonlinear functions. Convergence theorems have been used to prove that the learning algorithm will eventually converge on a maximum. The capability of deriving general mathematical properties from neural networks using general dynamic system theory is one of the advantages of neural networks as compared with rule-based systems.

Neural networks can be roughly categorized according to (1) the type of feedback provided on each trial (unsupervised learning with no feedback, reinforcement learning using summary evaluations, error correction with target output feedback), and (2) whether the system contains feedback connections (strictly feedforward versus recurrent networks with feedback connections).

Recurrent unsupervised learning models have been used to discover features and self-organize clusters of stimuli (Grossberg, 1980, 1988). Error correction feedforward learning models (also known as backpropagation models) have been used extensively for performing pattern recognition, for learning nonlinear mappings, and for solving prediction problems (see Rumelhart and McClelland, 1986). Reinforcement learning is useful for decision making and for learning to control dynamic systems (Sutton, 1992). Recurrent backpropagation models are useful for learning of sequential behaviors, such as sequences of motor movements (Jordan, 1990), or for problem solving. Most recently, work has been under way on developing modular neural network models, in which each module learns to become an expert for a particular problem domain, and a learning algorithm provides a mechanism for learning how to select each module depending on the current context (Jacobs et al., 1991).

Neural networks offer great potential and power for developing new learning models for use in computer-generated agents. However, these models are also the most difficult to integrate into existing military simulations. Recently, a considerable effort has been made to integrate rule-based systems with neural networks (see the recent volume by Sun and Alexandre, 1997). But this is currently an underdeveloped area, and a substantial amount of further research is needed to develop hybrid systems that will make it possible to begin systematically integrating the two approaches.

Adaptive Resonance Theory: An Illustration of Neural Networks

ART is a sophisticated and broadly applicable neural network architecture that is currently available. It is based on the need for the brain to continue to learn about a rapidly changing world in a stable fashion throughout life. This learning includes top-down expectations, the matching of these expectations to bottom-up data, the focusing of attention on the expected clusters of information, and the development of resonant states between bottom-up and top-down processes as they reach an attentive consensus between what is expected and what is there in the outside world. In ART these resonant states trigger learning of sensory and cognitive representations. ART networks create stable memories in response to arbitrary input sequences with either fast or slow learning. ART systems provide an alternative to classical expert systems from artificial intelligence. These self-organizing systems can stably remember old classifications, rules, and predictions about a previous environment as they learn to classify new environments and build rules and predictions about them.

ART was introduced as a theory of human cognitive information processing (Grossberg, 1976). It led to an evolving series of real-time neural network models that perform unsupervised and supervised category learning and classification, pattern recognition, and prediction: ART 1 (Carpenter and Grossberg, 1987a) for binary input patterns, ART 2 (Carpenter and Grossberg, 1987b) for analog and binary input patterns, and ART 3 (Carpenter and Grossberg, 1990) for parallel search of distributed recognition codes in a multilevel network hierarchy. The ART architecture is supported by applications of the theory to psychophysical and neurobiological data about vision, visual object recognition, auditory streaming, variable-rate speech perception, cognitive information processing, working memory, temporal planning, and cognitive-emotional interactions, among others (e.g., Bradski et al., 1994; Carpenter and Grossberg, 1991; Grossberg, 1995; Grossberg et al., 1997; Grossberg and Merrill, 1996). The basic ART modules have recently been augmented by ARTMAP and Fuzzy ARTMAP models (Carpenter et al., 1992) to carry out supervised learning. Fuzzy ARTMAP (e.g., Gaussian ARTMAP by Williamson, 1996) is a self-organizing radial basis production system; ART-EMAP (Carpenter and Ross, 1995) and VIEWNET (Bradski and Grossberg, 1995) use working memory to accumulate evidence about targets as they move with respect to a sensor through time.

Match-Based and Error-Based Learning Match-based learning achieves ART stability by allowing memories to change only when attended portions of the external world match internal expectations or when something completely new occurs. When the external world fails to match an ART network's expectations or predictions, a search or hypothesis testing process selects a new category to represent a new hypothesis about important features of the present environment. The stability of match-based learning makes ART well suited to problems that require online learning of large and evolving databases. Error-based learning, as in back propagation and vector associative maps, is better suited to problems, such as the learning of sensory-motor maps, that require adaptation to present statistics rather than construction of a knowledge system.

ART Search Figure 5.1 illustrates a typical ART model, while Figure 5.2 illustrates a typical ART search cycle. Level F1 in Figure 5.2 contains a network of nodes, each representing a particular combination of sensory features. Level F2 contains a second network that represents recognition codes, or categories, selectively activated by patterns at F1. The activity of a node in F1 or F2 is called a short-term memory trace, and can be rapidly reset without leaving an enduring trace. In an ART model, adaptive weights in both bottom-up (F1 to F2) paths and top-down (F2 to F1) paths store long-term memory. Auxiliary gain control and

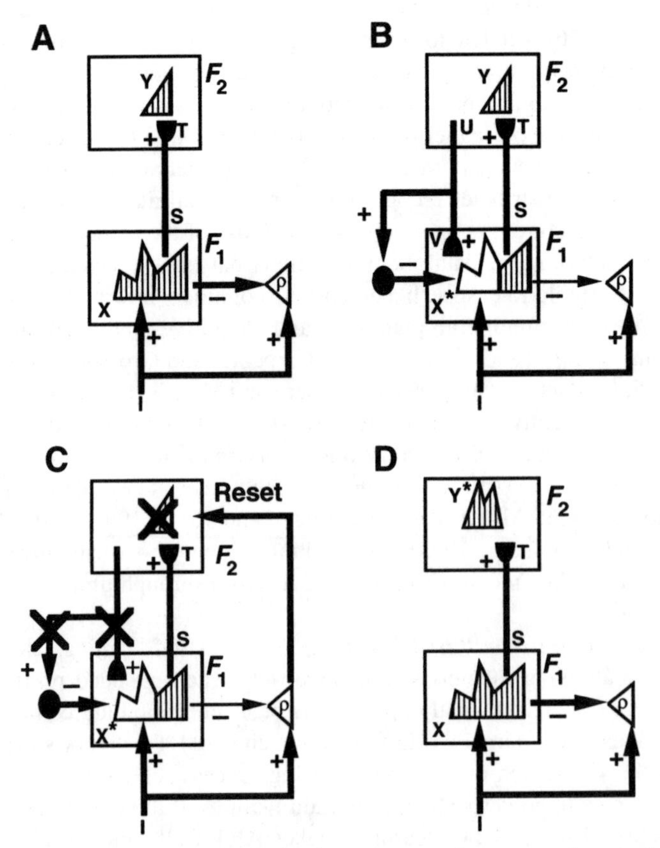

FIGURE 5.1 A typical unsupervised adaptive resonance theory neural network.

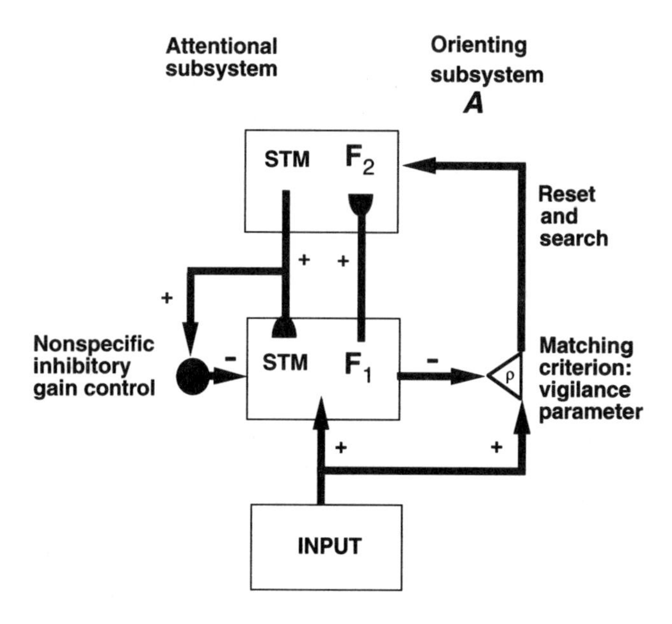

FIGURE 5.2 Adaptive resonance theory (ART) search cycle.

orienting processes regulate category selection, search, and learning, as described below.

An ART input vector I registers itself as a pattern X of activity across level F1 (Figure 5.2A). The F1 output vector S is transmitted through converging and diverging pathways from F1 to F2. Vector S is multiplied by a matrix of adaptive weights, or long-term memory traces, to generate a net input vector T to F2. Competition within F2 contrast-enhances vector T. The resulting F2 activity vector Y includes only one or a few active nodes—the ones that receive maximal filtered input from F1. Competition may be sharply tuned to select only that F2 node which receives the maximal F1 to F2 input (i.e., "winner-take-all"), or several highly activated nodes may be selected. Activation of these nodes defines the category, or symbol, of the input pattern I. Such a category represents all the F1 inputs I that send maximal input to the corresponding F2 node. These rules constitute a self-organizing feature map (Grossberg, 1972, 1976; Kohonen, 1989; von der Malsburg, 1973), also called competitive learning or learned vector quantization.

Activation of an F2 node "makes a hypothesis" about an input I. When Y becomes active, it sends an output vector U top-down through the second adaptive filter. After multiplication by the adaptive weight matrix of the top-down filter, a learned expectation vector V becomes the input to F1 (Figure 3.3B). Activation of V by Y "tests the hypothesis" Y. The network then matches the "expected prototype" V of the category against the active input pattern, or exemplar, I. Nodes in F1 that were activated by I are suppressed if they are not supported by large long-term memory traces in the prototype pattern V. Thus F1 features that are not "expected" by V or not "confirmed" by hypothesis Y are suppressed. The resultant matched pattern X* encodes the features in I that are relevant to hypothesis Y. Pattern X* is the feature set to which the network "pays attention."

Resonance, Attention, and Learning If expectation V is close enough to input I, resonance develops: the attended pattern X* reactivates hypothesis Y, and Y reactivates X*. Resonance persists long enough for learning to occur. ART systems learn prototypes, rather than exemplars, because memories encode the attended feature vector X*, rather than the input I itself.

The ART matching rule enables the system to search selectively for objects or events that are of interest while suppressing information that is not. An active top-down expectation can thereby "prime" an ART system to get ready to experience a sought-after target or event so that when it finally occurs, the system can react to it quickly and vigorously.

To achieve priming, a top-down expectation, by itself, cannot activate cells well enough to generate output signals. When large bottom-up inputs and top-down expectations converge on these cells, they can generate outputs. Bottom-up inputs can also, by themselves, initiate automatic processing of inputs.

Vigilance Control of Generalization A vigilance parameter defines the criterion of an acceptable ART match. Vigilance weighs how close input I must be to prototype V for resonance to occur. Varying vigilance across learning trials lets the system learn recognition categories of widely differing generalization, or morphological variability. Low vigilance leads to broad generalization and abstract prototypes. High vigilance leads to narrow generalization and to prototypes that represent few input exemplars. At very high vigilance, prototype learning reduces to exemplar learning. Thus a single ART system may learn abstract categories of vehicles and airplanes, as well as individual types.

If expectation V and input I are too novel, or unexpected, to satisfy the vigilance criterion, the orienting subsystem (Figure 5.1) triggers hypothesis testing, or memory search. Search allows the network to select a better recognition code or hypothesis to represent input I. This search process allows ART to learn new representations of novel events without risking unselective forgetting of previous knowledge. Search prevents associations from forming between Y and X* if X* is too different from I to satisfy the vigilance criterion. Search resets Y before such an association can form (Figure 5.2C). Search may select a familiar category if its prototype is similar enough to input I to satisfy the vigilance criterion. Learning may then refine the prototype using the information carried by I. If I is too different from any previously learned prototype, the uncommitted

F2 nodes are selected to establish a new category. A network parameter controls the depth of search.

Direct Access to the Globally Best Match Over learning trials, the system establishes stable recognition categories as the inputs become "familiar." All familiar inputs activate the category that provides the globally best match in a one-pass fashion, and search is automatically disengaged. Learning of unfamiliar inputs can continue online, even as familiar inputs achieve direct access, until the network fully uses its memory capacity, which can be chosen arbitrarily large.

Supervised Learning and Prediction Supervised ARTMAP architectures learn maps that associate categories of one input space with another. ARTMAP can organize dissimilar patterns (e.g., multiple views of different types of tanks) into distinct recognition categories that then make the same prediction (e.g., "enemy tank"). ARTMAP systems have an automatic way of matching vigilance to task demands: following a predictive failure during learning, mismatch of a prototype with a real target raises vigilance just enough to trigger memory search for a better category. ARTMAP thereby creates multiple scales of generalization, from fine to coarse, as needed. This "match tracking" process realizes a minimax learning rule that conjointly minimizes predictive error and maximizes generalization using only information that is locally available under incremental learning conditions in a nonstationary environment.

Automatic Rule Extraction ARTMAP learned weights translate into IF-THEN rules at any stage of learning. Suppose, for example, that the input vectors encode different biochemicals and that the output vectors encode different clinical effects of these biochemicals on humans. Different biochemicals may achieve the same clinical effects for different chemical reasons. At any time during learning, an operator can test how to achieve a desired clinical effect by checking which long-term memory traces are large in the pathways from recognition categories to the nodes coding that effect. The prototype of each recognition category characterizes a "rule," or bundle of biochemical features, that predicts the desired clinical effect. A list of these prototype vectors provides a transparent set of rules that all predict the desired outcome. Many such rules may coexist without mutual interference in ARTMAP, in contrast with many other learning models, such as back propagation.

Other Learning Models

Beyond neural networks, there are several other more complex forms of learning that may prove useful. One promising area is research on simulated annealing programs that are used to search very high-dimensional parameter spaces for global minimums or maximums (Kirkpatrick et al., 1983). Simulated annealing programs use a probabilistic search mechanism that helps avoid getting stuck in local minimums or maximums. Another area of note is research on genetic programming algorithms that are used to construct new rules (Holland, 1975). New rules are formed from old ones by mutation and evolutionary principles derived from biology and genetics. Finally, another important approach to learning is provided by what are called decision-tree algorithms—a machine-learning approach to complex concept formation (Quillen, 1986; Utgoff, 1989).

CONCLUSIONS AND GOALS

Models of short-term and working memory are an obvious necessity that have been and should continue to be an active focus of work within the field and in the military simulation community. Models of explicit retrieval (retrieval of recent and personal events) have been well developed, and could be applied in military simulations where needed. Models of retrieval from generic memory (i.e., retrieval of knowledge) are of critical importance to military simulations, but are as yet least well developed in the field. The approaches used thus far in military situations are relatively primitive. It is this area that may require the greatest commitment of resources from defense sources.

Learning is an essential ability for intelligent systems, but current military simulations make little or no use of learning models. There are several steps that can be taken to correct this lack. A simple short-term solution is to continue working within the rule-based systems that provide the basis for most of the simulations developed to date, but to activate the learning mechanisms within these systems. An intermediate solution is to modify the current rule-based systems to incorporate exemplar (or case)-based learning processes, which have enjoyed greater empirical support in the psychological literature. Neural networks may offer the greatest potential and power for the development of robust learning models, but their integration into current simulation models will require much more time and effort. One way to facilitate this integration is to develop hybrid models that make use of rule-based problem-solving processes, exemplar representations, and neural network learning mechanisms.

Short-Term Goals

- Incorporate the latest models of memory and retrieval into existing simulation models.
- Add learning mechanisms to current simulations.
- Begin the process of extending current models to include both generic (knowledge) and explicit (episodic) storage and retrieval.
- Better integrate the operations of working memory and its capacity into models of long-term memory and learning.

Intermediate-Term Goals

- Validate the simulation models against complex data gathered in military domain situations.
- Compare and contrast alternative models with respect to their ability to predict both laboratory and real-world data.
- Within the models, integrate the storage and retrieval of generic knowledge, of episodic knowledge, and of learning.
- Start the process of developing hybrid models of learning and memory, incorporating the best features of rule-based, exemplar-based, evolutionary, and neural net representations.
- Develop process models of those factors treated as moderator variables.

Long-Term Goals

- Develop larger-scale models that are capable of handling explicit, generic, and implicit memory and learning and are applicable to complex real-world situations.
- Validate these models against real-world data.
- Begin to explore the effects of learning, memory, and retrieval processes on group behavior.

6

Decision Making

Traditionally, decision theory is divided into two major topics, one concerned primarily with choice among competing actions and the other with modification of beliefs based on incoming evidence. This chapter focuses on the former topic; the latter is treated in detail in the chapters on situation awareness (Chapter 7) and planning (Chapter 8).

A *decision episode* occurs whenever the flow of action is interrupted by a choice between conflicting alternatives. A *decision* is made when one of the competing alternatives is executed, producing a change in the environment and yielding consequences relevant to the decision maker. For example, in a typical decision episode, a commander under enemy fire needs to decide quickly whether to search, communicate, attack, or withdraw based on his or her current awareness of the situation. Although perception (e.g., target location), attention (e.g., threat monitoring), and motor performance (e.g., shooting accuracy) are all involved in carrying out a decision, the critical event that controls most of the behavior within the episode is the decision to follow a particular course of action (e.g., decide to attack). The centrality of decision making for computer-generated agents in military simulations makes it critical to employ models that closely approximate real human decision-making behavior.

Two variables are fundamental to framing a decision episode: timeframe and aggregation level. Timeframe can vary from the order of seconds and minutes (as in the tactical episode sketched above) to the order of years and even decades (as in decisions concerning major weapon systems and strategic posture). This chapter considers primarily decisions falling toward the shorter end of this range. Second, there are at least two levels of aggregation: *entity*-level decisions represent the decisions of a solitary soldier or an individual commanding a single vehicle or controlling a lone weapon system; *unit*-level decisions represent the decisions made by a collection of entities (e.g., group decisions of a staff or the group behavior of a platoon). This chapter addresses only entity-level decisions; unit-level models are discussed in Chapter 10.

Past simulation models have almost always included assumptions about decision processes, and the complexity with which they incorporate these assumptions ranges from simple decision tables (e.g., the recognize-act framework proposed by Kline, 1997) to complex game theoretic analyses (e.g., Myerson, 1991). However, the users of these simulations have cited at least two serious types of problems with the previous models (see, e.g., Defense Advanced Research Projects Agency, 1996:35-39). First, the decision process is too stereotypical, predictable, rigid, and doctrine limited, so it fails to provide a realistic characterization of the variability, flexibility, and adaptability exhibited by a single entity across many episodes. Variability, flexibility, and adaptability are essential for effective decision making in a military environment. Variability in selection of actions is needed to generate unpredictability in the decision maker's behavior, and also to provide opportunities to learn and explore new alternatives within an uncertain and changing environment. Adaptability and flexibility are needed to reallocate resources and reevaluate plans in light of unanticipated events and experience with similar episodes in the past, rather than adhering rigidly to doctrinal protocols. Second, the decision process in previous models is too uniform, homogenous, and invariable, so it fails to incorporate the role of such factors as stress, fatigue, experience, aggressiveness, impulsiveness, and attitudes toward risk, which vary widely across entities. A third serious problem, noted by the panel, is that these military models fail to take into account known limitations, biases, or judgmental errors.

The remainder of this chapter is organized as follows. First is a brief summary of recent progress in utility theory, which is important for understanding the foundation of more complex models of decision making. The second section reviews by example several new decision models that could ameliorate the stereotypic and rigid character of the current models. The models reviewed are sufficiently formalized to provide computational models that can be used to modify existing computer simulation programs, and are at least moderately supported by empirical research. The next section illustrates one way in which individual differences and moderating states can be incorporated into decision models through the alteration of various parameters of the example decision model. The fourth section reviews, again by example, some human biases or judgmental errors whose incorporation could improve the realism of the simulations. The final section presents conclusions and goals in the area of decision making.

As a final orienting comment, it should be noted that the work reviewed here comes from two rather different traditions within decision research. The first, deriving from ideas of expected value and expected utility, proposes rather general models of the idealized decision maker. He or she might, for example, be postulated to make decisions that maximize expected utility, leaving the relevant utilities and expectations to be specified for each actual choice situation. Such general models seek descriptive realism through modifications of the postulated utility mechanism, depth of search, and so on within the context of a single overarching model. A second tradition of decision research is based more directly on specific phenomena of decision making, often those arguably associated with decision traps or errors, such as sunk cost dependence. Work in this second tradition achieves some generality by building from particular observations of the relevant phenomenon to explore its domain, moderators, scope, and remedies. In rough terms, then, the two traditions can be seen as *top-down* and *bottom-up* approaches to realistic modeling of decision processes. We illustrate in this chapter the potential value of both approaches for military simulations.

SYNOPSIS OF UTILITY THEORY

Utility theory provides a starting point for more complex decision-making models. In their pure, idealistic, and elegant forms, these theories are too sterile, rigid, and static to provide the realism required by decision models for military simulations. However, more complex and realistic models, including examples described later in this chapter, are built on this simpler theoretical foundation. Therefore, it is useful to briefly review the past and point to the future of theoretical work in this area.

Expected Value

Before the eighteenth century, it was generally thought that the optimal way to make decisions involving risk was in accordance with expected value, that is, to choose the action that would maximize the long-run average value (the value of an object is defined by a numerical measurement of the object, such as dollars). This model of optimality was later rejected because many situations were discovered in which almost everyone refused to behave in the so-called optimal way. For example, according to the expected-value model, everyone should refuse to pay for insurance because the premium one pays exceeds the expected loss (otherwise insurance companies could not make a profit). But the vast majority of people are risk averse, and would prefer to pay extra to avoid catastrophic losses.

Expected Utility

After the beginning of the eighteenth century, Bernoulli proposed a modification of expected value, called *expected utility*, that circumvented many of the practical problems encountered by expected-value theory. According to this new theory, if the decision maker accepts its axioms (see below), then to be logically consistent with those axioms he/she should choose the action that maximizes expected utility, not expected value. Utility is a nonlinear transformation of a physical measurement or value. According to the theory, the shape of the utility function is used to represent the decision maker's attitude toward risk. For example, if utility is a concave function of dollars, then the expected utility of a high-risk gamble may be less than the utility of paying an insurance premium that avoids the gamble. The new theory initially failed to gain acceptance because it seemed ad hoc and lacked a rigorous axiomatic justification. Almost two centuries later, however, Von Neumann and Morgenstern (1947) proposed an axiomatic basis for expected-utility theory, which then rapidly gained acceptance as the rational or optimal model for making risky decisions (see, e.g., Clemen, 1996; Keeney and Raiffa, 1976; von Winterfeldt and Edwards, 1986).

The justification for calling expected utility the optimal way to make decisions depends entirely on the acceptance of a small number of behavioral axioms concerning human preference. Recently, these axioms have been called into question, generating another revolution among decision theorists. Empirical evidence against the behavioral axioms of expected-utility theory began to be reported soon after the theory was first propounded by Von Neumann and Morgenstern (Allais, 1953), and this evidence systematically accumulated over time (see, e.g., early work by MacCrimmon and Larsson, 1979, and Kahneman and Tversky, 1979) until these violations became accepted as well-established fact (see, e.g., the review by Weber and Camerer, 1987). Expected-utility theory is thus experiencing a decline paralleling that of expected-value theory.

A simple example will help illustrate the problem. Suppose a commander is faced with the task of transporting 50 seriously wounded soldiers to a field hospital across a battle zone. First, the transport must leave the battlefield, a highly dangerous situation in which there is an even chance of getting hit and losing all of the wounded (but the only other option is to let them all die on the field). If the transport gets through the battlefield, there is a choice between going west on a mountain road to a small medical clinic, where 40 of the 50 are expected to survive, or going north on a highway to a more well-equipped and fully staffed hospital, where all 50 are expected to survive. Both routes take about the same amount of time, not a significant factor in the decision. But the mountain road is well guarded, so that safe passage is guaranteed, whereas the highway is occasionally ambushed, and there is a 10 percent chance of getting hit and losing all of the wounded.

If a battlefield decision is made to go north to the hospital, the implication is that the expected utility of going north on the paved road (given that the battlefield portion of the journey is survived) is higher than the expected utility of going west on the mountain road (again given that the battlefield portion is survived), since the battlefield portion is common to both options. However, research suggests that a commander who does survive the battlefield portion will often change plans, decide not to take any more chances, and turn west on the mountain road to the small clinic, violating expected utility.

Why does this occur? Initially, when both probabilities are close to even chance, the difference in gains overwhelms the probability difference, resulting in a preference for the northern route. Later, however, when the probabilities change from certainly to probably safe, the probability difference dominates the difference in gains, resulting in a preference for the western route. Kahneman and Tversky (1979) term this type of result the *certainty effect*. (The actual experiments reported in the literature cited above used simpler and less controversial monetary gambles to establish this type of violation of expected utility theory, rather than a battlefield example.)

Rank-Dependent Utility

To accommodate the above and other empirical facts, new axioms have been proposed to form a revised theory of decision making called *rank-dependent utility* theory (see Tversky and Kahneman, 1992; Luce and Fishburn, 1991; Quiggen, 1982; Yaari, 1987). Rank-dependent utility can be viewed as a generalization of expected utility in the sense that rank-dependent utility theory can explain not only the same preference orders as expected utility, but others as well. The new axioms allow for the introduction of a nonlinear transformation from objective probabilities to subjective decision weights by transforming the cumulative probabilities. The nonlinear transformation of probabilities gives the theory additional flexibility for explaining findings such as the certainty effect described above.

Multiattribute Utility

Military decisions usually involve multiple conflicting objectives (as illustrated in the vignette presented in Chapter 2). For example, to decide between attacking the enemy and withdrawing to safety, the commander would need to anticipate both the losses inflicted on the enemy and the damages incurred by his own forces from retaliation. These two conflicting objectives—maximizing losses to enemy forces versus minimizing losses to friendly forces—usually compete with one another in that increasing the first usually entails increasing the second.

Multiattribute utility models have been developed to represent decisions involving conflicting or competing objectives (see, e.g., Clemen, 1996; Keeney and Raiffa, 1976; von Winterfeldt and Edwards, 1986). There are different forms of these models, depending on the situation. For example, if a commander forces a tradeoff between losses to enemy and friendly forces, a weighted additive value model may be useful for expressing the value of the combination. However, a multiplicative model may be more useful for discounting the values of temporally remote consequences or for evaluating a weapon system on the basis of its performance and reliability.

Game Theory

Game theory is concerned with rational solutions to decisions involving multiple intelligent agents with conflicting objectives. Models employing the principle of subgame perfect equilibrium (see Myerson, 1991, for an introduction) could be useful for simulating command-level decisions because they provide an optimal solution for planning strategies with competing agents. Decision scientists employ decision trees to represent decisions in complex dynamic environments. Although the number of stages and number of branches used to represent future scenarios can be large, these numbers must be limited by the computational capability constraints of the decision maker.

The planning horizon and branching factor are important individual differences among decision makers. For many practical problems, a complete tree representation of the entire problem may not be possible because the tree would become too large and complex, or information about events in the distant future is unavailable or too ill defined. In such cases, it may be possible to use partial trees representing only part of the entire decision. That is, the trees can be cut off at a short horizon, and game theory can then be applied to the truncated tree. Even the use of partial trees may increase the planning capabilities of decision systems for routine tasks.

When decisions, or parts of decisions, cannot be represented easily in tree form, the value of game theory may diminish. Representing highly complex decisions in game theoretic terms may be extremely cumbersome, often requiring the use of a set of games and meta-games.

Concluding Comment

The utility models described in this section are idealized and deterministic with regard to the individual decision maker. He or she is modeled as responding deterministically to inputs from the situation. According to these models, all the branches in a decision tree and all the dimensions of a consequence are evaluated by an individual, and the only source of variability in preference is differences across entities. Individuals may differ in terms of risk attitudes (shapes of utility functions), opinions about the likelihood of outcomes (shapes of decision weight functions), or importance assigned to competing objectives (tradeoff weights). But all of these models assume there is no variability in preferences within a single individual; given the same person and the same decision problem, the same action will always be chosen. As suggested earlier, we believe an appropriate decision model for military situations must introduce variability and adaptability into preference at the level of the individual entity, and means of doing so are discussed in the next section.

INJECTING VARIABILITY AND ADAPTABILITY INTO DECISION MODELS

Three stages in the development of decision models are reviewed in this section. The first-stage models—random-utility—are probabilistic within an entity, but static across time. They provide adequate descriptions of variability across episodes within an entity, but they fail to describe the dynamic characteristics of decision making within a single decision episode. The second-stage models—sequential sampling decision models—describe the dynamic evolution of preference over time within an episode. These models provide mechanisms for explaining the effects of time pressure on decision making, as well as the relations between speed and accuracy of decisions, which are critical factors for military simulations. The third-stage models—adaptive planning models—describe decision making in strategic situations that involve a sequence of interdependent decisions and events. Most military decision strategies entail sequences of decision steps to form a plan of action. These models also provide an interface with learning models (see Chapter 5) and allow for flexible and adaptive planning based on experience (see also Chapter 8).

Random-Utility Models

Probabilistic choice models have developed over the past 40 years and have been used successfully in marketing and prediction of consumer behavior for some time. The earliest models were proposed by Thurstone (1959) and Luce (1959); more recent models incorporate some critical properties missing in these original, oversimplified models.

The most natural way of injecting variability into utility theory is to reformulate the utilities as random variables. For example, a commander is trying to decide between two courses of action (say, attack or withdraw), but he or she does not know the precise utility of an action and may estimate this utility in a way that varies across episodes. However, the simplest versions of these models, called strong random-utility models, predict a property called *strong stochastic transitivity*, a property researchers have shown is often violated (see Restle, 1961; Tversky, 1972; Busemeyer and Townsend, 1993; Mellers and Biagini, 1994). Strong stochastic transitivity states that if (1) action A is chosen more frequently than action B and (2) action B is chosen more frequently than action C, then (3) the frequency of choosing action A over C is greater than the maximum of the previous two frequencies. Something important is still missing.

These simple probabilistic choice models fail to provide a way of capturing the effects of similarity of options on choice, which can be shown in the following example. Imagine that a commander must decide whether to attack or withdraw, and the consequences of this decision depend on an even chance that the enemy is present or absent in a target location. The cells in Tables 6.1a and 6.1b show the percentage of damage to the enemy and friendly forces with each combination of action (attack versus withdraw) and condition (enemy present versus absent) for two different scenarios—one with actions having dissimilar outcomes and the other with actions having similar outcomes.

According to a strong random-utility model, the probability of attacking will be the same for both scenarios because the payoff distributions are identical for each action within each scenario. Attacking provides an equal chance of either (85 percent damage to enemy, 0 percent damage to friendly) or (10 percent damage to enemy, 85 percent damage to friendly) in both scenarios; withdrawing provides an equal chance of either (0 percent damage to enemy, 0 percent damage to friendly) or (0 percent damage to enemy, 85 percent damage to friendly) in both scenarios. If the utility is computed for each action, differences between the utilities for attack and withdraw will be identical across the two tables. However, in the first scenario, the best choice depends on the state of nature (attack if target is present, withdraw of target is absent), but in the second scenario, the best choice is independent of the state of nature (attack if target is present, attack if target is absent). Thus, the first scenario produces negatively correlated or dissimilar outcomes, whereas the second scenario produces positively correlated or similar outcomes. The only difference between the two scenarios is the similarity between actions, but the strong random-utility model is completely insensitive to this difference. Empirical research with human subjects, however, clearly demonstrates a very

TABLE 6.1a Scenario One: Dissimilar Actions

Action	Target Present (.5 probability)	Target Absent (.5 probability)
Attack	(85, 0)	(10, 85)
Withdraw	(0, 85)	(0,0)

NOTE: Cells indicate percent damage to (enemy, friendly) forces.

TABLE 6.1b Scenario Two: Similar Actions

Action	Target Present (.5 probability)	Target Absent (.5 probability)
Attack	(85, 0)	(10, 85)
Withdraw	(0,0)	(0, 85)

large difference in choice probability across these two scenarios (see Busemeyer and Townsend, 1993).

Several different types of random-utility models with correlated random variables have been developed to explain the effects of similarity on probabilistic choice, as well as violations of strong stochastic transitivity. Busemeyer and Townsend (1993) and Mellers and Biagini (1994) have developed models for risky decisions. For multiattribute choice problems, Tversky (1972) and Edgell (1980) have developed one type of model, and De Soete et al. (1989) present an alternative approach.

In summary, more recent random-utility models provide an adequate means for producing variability in decision behavior at the level of an individual entity, provided they incorporate principles for explaining the strong effects of similarity on choice. However, these models are static and fail to describe how preferences evolve over time within an episode. This is an important issue for military decision simulations, in which time pressure is a very common and important consideration. Preferences can change depending on how much time the decision maker takes to make a decision. The next set of models includes examples in which preferences evolve over time within an episode.

Sequential Sampling Decision Models

Sequential sampling models of decision making were originally developed for statistical applications (see DeGroot, 1970). Shortly afterwards, however, they were introduced to psychologists as dynamic models for signal detection that explained both the speed and accuracy of decisions (Stone, 1960). Since that time, this type of model has been used to represent the central decision process for decision making in a wide variety of cognitive tasks, including sensory detection (Smith, 1996), perceptual discrimination (Laming, 1968; Link, 1992), memory retrieval (Ratcliff, 1978), categorization (Nosofsky and Palmeri, 1997), risky decision making (Busemeyer and Townsend, 1993; Wallsten, 1995; Kornbrot, 1988), and multiattribute decision making (Aschenbrenner et al., 1983; Diederich, 1997). Most recently, connectionistic theorists have begun to employ these same ideas in their neural network models (see Chapter 5). There is now a convergence of research that points to the same basic process for making decisions across a wide range of cognitive tasks (see Ratcliff et al., 1998).

The sequential sampling model can be viewed as an extension of the random-utility model. The latter model assumes that the decision maker takes a single random sample and bases his or her decision on this single noisy estimate. The sequential sampling model assumes that the decision maker takes a sequence of random-utility estimates and integrates them over time to obtain a more precise estimate of the unknown expected utility or rank-dependent utility.

Formally, the sequential sampling model is a linear feedback system with stochastic inputs and a threshold output response function. Here we intuitively illustrate how a decision is reached within a single decision episode using this model. Suppose a commander must choose one of three possible actions: attack, communicate, or withdraw. At the beginning of the decision episode, the decision maker retrieves from memory initial preferences (initial biases) for each course of action based on past experience or status quo. The decision maker then begins anticipating and evaluating the possible consequences of each action. At one moment, attention may focus on the losses suffered in an attack, while at another moment, attention may switch to the possibility of enemy detection caused by communication. In this way, the preferences accumulate and integrate the evaluations across time; they evolve continuously, perhaps wavering or vacillating up and down as positive and negative consequences are anticipated, until the preference for one action finally grows strong enough to exceed an inhibitory threshold.

The inhibitory threshold prevents an action from being executed until a preference is strong enough to overcome this inhibition, and this threshold may decay or dissipate over time. In other words, the decision maker may become more impatient as he or she waits to make a decision. The initial magnitude of the inhibitory threshold is determined by two factors: the importance or seriousness of the consequences, and the time pressure or cost of waiting to make the decision.

Important decisions require a relatively high threshold. Thus the decision maker must wait longer and collect more information about the consequences before making a decision, which tends to increase the likelihood of making a good or correct decision. Increasing the inhibitory threshold therefore usually increases the accuracy of the decision. But if the decision maker is under strong time pressure and it is very costly to wait, the inhibitory threshold will decrease, so that a decision will be reached after only a few of the consequences have been reviewed. The decision will be made more quickly, but there will be a greater chance of making a bad decision or an error. Applying these principles, sequential sampling models have provided extremely accurate predictions for both choice probability and choice response time data collected from signal detection experiments designed to test for speed-accuracy tradeoff effects (see Ratcliff et al., 1997)

Not every decision takes a great deal of time or thought according to the sequential sampling model. For example, based on extensive past experience, training, instruction, or a strong status quo, the decision maker may start with a very strong initial bias, so that only a little additional information need be collected before the threshold is reached and a very quick decision is made. This extreme case is close to the recognize-act proposal of Klein (1997).

The sequential sampling model explains how time pressure can drastically change the probabilities of choosing each action (see Figure 6.1). For example, suppose the most important attribute favors action 1, but all of the remaining

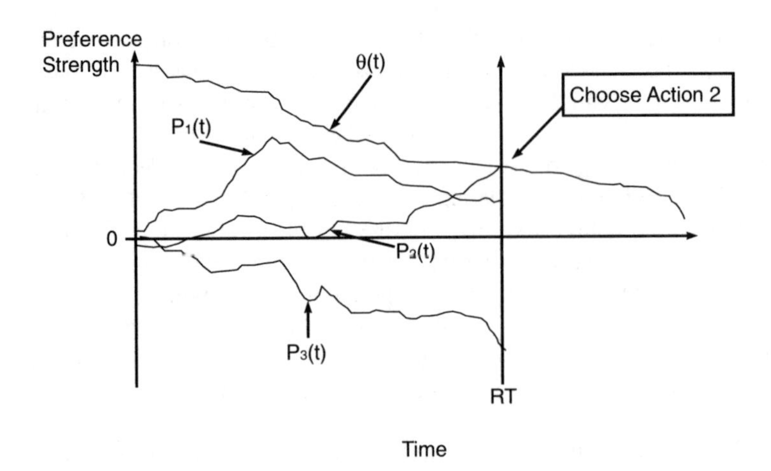

FIGURE 6.1 Illustration of the main ideas of sequential sampling decision models. The vertical axis represents strength of preference, whereas the horizontal axis represents time. The trajectories labeled P1, P2, and P3 represent the evolution of the preference states for actions 1, 2, and 3 over time, and the trajectory labeled theta represents the strength of the inhibitory bound over time. A decision is reached when one of the preference states exceeds the threshold bound, and the first action to exceed this bound determines the choice that is made. In this example, action 2 crosses the threshold first, and thus it is chosen. The vertical line indicated by RT is the time required to make this decision for this example.

attributes favor action 2. Under severe time pressure, the inhibitory threshold is low, and it will be exceeded after a small number of consequences have been sampled. Attention is focused initially on the most important attribute favoring action 1, and the result is a high probability of choosing action 1 under time pressure. Under no time pressure, the inhibitory threshold is higher, and it will be exceeded only after a large number of consequences have been sampled. Attention is focused initially on the most important attribute favoring action 1, but now there is sufficient time to refocus attention on all of the remaining attributes that favor action 2; the result is a high probability of choosing action 2 with no time pressure. In this way, the rank ordering of choice probabilities can change with the manipulation of time pressure.

In summary, sequential sampling models extend random-utility models to provide a dynamic description of the decision process within a single decision episode. This extension is important for explaining the effects of time pressure on decision making, as well as the relation between speed and accuracy in making a decision. However, the example presented above is limited to decisions involving a single stage. Most military decisions require adaptive planning for actions and events across multiple stages (see, for example, the semiautomated force combat instruction set for the close combat tactical trainer). The models discussed next address this need.

Adaptive Planning Models

Before a commander makes a decision, such as attacking versus withdrawing from an engagement, he or she needs to consider various plans and future scenarios that could occur, contingent on an assessment of the current situation (see also Chapters 7 and 8). For example, if the commander attacks, the attack may be completely successful and incapacitate the enemy, making retaliation impossible. Alternatively, the attack may miss completely because of incorrect information about the location of the enemy, and the enemy may learn the attacker's position, in which case the attacker may need to withdraw to cover to avoid retaliation from an unknown position. Or the attack may be only partially successful, in which case the commander will have to wait to see the enemy's reaction and then consider another decision to attack or withdraw.

Adaptive decision making has been a major interest in the field of decision making for some time (see, e.g., the summary report of Payne et al., 1993). Decision makers are quite capable of selecting or constructing strategies on the fly based on a joint consideration of accuracy (probability of choosing the optimal action) and effort or cost (time and attention resources required to execute a strategy). However, most of this work on adaptive decision making is limited to single-stage decisions. There is another line of empirical research using dynamic decision tasks (see Kerstholt and Raaijmakers, forthcoming, for a review) that entail planning across multiple stages decision, but this work is still at an early stage of development and formal models have not yet been fully developed. Therefore, the discussion in the next section is based on a synthesis of learning and decision models that provides a possible direction for building adaptive planning models for decision making. The models reviewed in this subsection are based on a synthesis of the previously described decision models and the learning models described in Chapter 5. More specifically, the discussion here examines how the exemplar- or case-based learning model can be integrated with the sequential sampling decision models (see Gibson et al., 1997).

The first component is an exemplar model for learning to anticipate consequences based on preceding sequences of actions and events. Each decision episode is defined by a choice among actions—a decision—followed by a consequence. Also, preceding each episode is a short sequence of actions and events that lead up to this choice point. Each episode is encoded in a memory trace that contains two parts: a representation of the short sequence of events and actions that preceded a consequence, and the consequence that followed.

The second component is a sequential sampling decision process, with a choice being made on the anticipated consequences of each action. When a decision episode begins, the decision maker is confronted with a choice among several immediately available actions. But this choice depends on plans for future actions and events that will follow the impending decision. The anticipated consequences of the decision are retrieved using a retrieval cue that includes the current action and a short sequence of future actions and events. The consequences associated with the traces activated by this retrieval cue are used to form an evaluation of each imminent action at that moment. These evaluations provide the evaluative input that enters into the updating of the preference state, and this evaluation process continues over time during the decision episode until one of the preference states grows sufficiently strong to exceed the threshold for taking action.

In summary, this integrated exemplar learning and sequential sampling decision model produces immediate actions based on evaluations of future plans and scenarios. Although plans and future scenarios are used to evaluate actions at a choice point, these plans are not rigidly followed and can change at a new choice point depending on recently experienced events. Furthermore, because the evaluations are based on feedback from past outcomes, this evaluation process can include adapting to recent events as well as learning from experience.

The model just described represents only one of many possible ways of integrating decision and learning models. Another possibility, for example, would be to combine rule-based learning models with probabilistic choice models (cf. Anderson, 1993) or to combine neural network learning models with sequential sampling decision models. Clearly more research is needed on this important topic.

INCORPORATING INDIVIDUAL DIFFERENCES AND MODERATING STATES

Decision makers differ in a multitude of ways, such as risk-averse versus risk-seeking attitudes, optimistic versus pessimistic opinions, passive versus aggressive inclinations, rational versus irrational thinking, impulsive versus compulsive tendencies, and expert versus novice abilities. They also differ in terms of physical, mental, and emotional states, such as rested versus fatigued, stressed versus calm, healthy versus wounded, and fearful versus fearless. One way to incorporate these individual difference factors and state factors into the decision process is by relating them to parameters of the decision models. The discussion here is focused on the effects of these factors on the decision-making process; Chapter 9 examines moderator variables in more detail.

Individual Difference Factors

Risk attitude is a classic example of an individual difference factor in expected utility theory. The shape of the utility function is assumed to vary across individuals to represent differences in risk attitudes. Convex-shaped functions result in risk-seeking decisions, concave-shaped functions result in risk aversion, and S-shaped functions result in risk seeking in the domain of losses and risk aversion in the domain of gains (see Tversky and Kahneman, 1992). Individual differences in optimistic versus pessimistic opinions are represented in rank-dependent utility theory by variation in the shape of the weighting function across individuals (see Lopes, 1987). Aggressive versus passive inclinations can be represented in multiattribute expected-utility models as differences in the tradeoff weights decision makers assign to enemy versus friendly losses, with aggressive decision makers placing relatively less weight on the latter. Rational versus irrational thinking can be manipulated by varying the magnitude of the error variance in random-utility models, with smaller error variance resulting in choices that are more consistent with the idealistic utility models. Impulsive versus compulsive (or deliberative) tendencies are represented in sequential sampling models by the magnitude of the inhibitory threshold, with smaller thresholds being used by impulsive decision makers. Expert versus novice differences can be generated from an adaptive planning model by varying the amount of training that is provided with a particular decision task.

State Factors

Decision theorists have not considered state factors to the same extent as individual difference factors. However, several observations can be made. Fatigue and stress tend to limit attentional capacity, and these effects can be represented as placing greater weight on a single most important attribute in a multiattribute decision model, sampling information from the most important dimension in a sequential sampling decision model, or limiting the planning horizon to immediate consequences in an adaptive planning model. Fear is an important factor in approach-avoidance models of decision making, and increasing the level of fear increases the avoidance gradient or attention paid to negative consequences (Lewin, 1935; Miller, 1944; Coombs and Avrunin, 1977; Busemeyer and Townsend, 1993). The state of health of the decision maker will affect the importance weights assigned to the attributes of safety and security in multiattribute decision models.

INCORPORATING JUDGMENTAL ERRORS INTO DECISION MODELS

The models discussed thus far reflect a top-down approach: decision models of great generality are first postulated and then successively modified and extended to accommodate empirical detail. As noted earlier, a second tradition of research on decision making takes a bottom-up approach. Researchers in this tradition have identified a number of phenomena that are related to judgmental errors and offer promise for improving the realism of models of human decision making. Useful recent reviews include Mellers et al. (forthcoming), Dawes (1997), and Ajzen (1996), while Goldstein and Hogarth (1997) and Connolly et al. (forthcoming) provide useful edited collections of papers in the area. Six such phenomena are discussed below. However, it is not clear how these phenomena might best be included in representations of human behavior for military simulations. On the one hand, if the phenomena generalize to military decision-making contexts, they will each have significant implications for both realistic representation and aiding of decision makers. On the other hand, evidence in each case suggests that the phenomena are sensitive to contextual and individual factors, and simple generalization to significant military settings is not to be expected.

Overconfidence

An individual is said to be overconfident in some judgment or estimate if his or her expressed confidence systematically exceeds the accuracy actually achieved. For example, Lichtenstein and Fischhoff (1977) demonstrated overconfidence for students answering general-knowledge questions of moderate or greater difficulty; only 75 percent of the answers for which the subjects expressed complete certainty were correct. Fischhoff (1982) and Yates (1990) provide reviews of such studies. More recent work suggests that such overconfidence may be sensitive to changes in the wording and framing of questions (e.g., Gigerenzer, 1994) and item selection (Juslin, 1994). Others have raised methodological concerns (e.g., Dawes and Mulford, 1996; Griffin and Tversky, 1992). The results are thus interesting, but unresolved ambiguities remain.

It would obviously be useful to know the extent to which assessment of confidence should be relied on in military settings. Should a commander discount a confident report of an enemy sighting or a unit commander's assessment of his probability of reaching a given objective on time? More subtly, would overconfidence be reflected in overoptimism about how long a particular assignment will take or in an overtight range of possibilities (see Connolly and Deane, 1997)? Does the military culture encourage overconfident private assessments, as well as overconfident public reports? And should these discounts, if any, be adjusted in light of contextual factors such as urgency and clarity or individual factors such as personality, training, experience, and fatigue? Existing research is sufficient to raise these as important questions without providing much help with their answers.

Base-Rate Neglect

In many settings, an individual must combine two kinds of information into an overall judgment: a relatively stable long-run average for some class of events (e.g., the frequency of a particular disease in a certain target group) and some specific information about a member of that class (e.g., a diagnostic test on a particular individual). In a classic exploration of this issue by Tversky and Kahneman (1980), subjects were asked to integrate (1) background information on the number of blue and green cabs in town and (2) an eyewitness's report of the color of the cab she saw leaving the scene of an accident. Their primary finding was that the subjects tended to give too much weight to (2) and not enough to (1) in contrast with Bayes' theorem, which the investigators had proposed as the relevant normative rule. As with the overconfidence literature, many subsequent studies have probed variants on the task used by Tversky and Kahneman; see Koehler, 1996, for a review and critique.

Once again, the significance of this phenomenon, if generalizable to military contexts, is obvious. A forward observer may have to judge whether a briefly seen unit is friend or enemy on the basis of visual impression alone. Should he take into account the known preponderance of one or the other force in the area, and if so, to what extent will he do so? Would a decision aid be appropriate? Should training of forward observers include efforts to overcome base-rate neglect, or should such bias be compensated at the command level by training or aiding the commander? Again, it seems unlikely that research will produce simple generalizations of behavioral tendencies that are impervious to the effects of individual differences, context, question wording, training, and so on. Instead, research needs to address to specific settings and factors of military interest, with a view to developing locally stable understandings of phenomena appropriate for representation in models tuned to these specific situations.

Sunk Cost and Escalation Effects

A number of experiments have demonstrated tendencies to persist in failing courses of action to which one has initially become committed and to treat nonrecoverable costs as appropriate considerations in choosing future actions. For example, Arkes and Blumer (1985) showed that theater patrons were more likely to attend performances when they paid full price for their subscriptions than when they received them at an unexpected discount. Staw (1976) used a simulated business case to show that students who made an investment to initiate a course of action that subsequently turned out poorly would invest more in the project at a second opportunity than would those who had not made the initial investment. A considerable amount of laboratory evidence (e.g., Garland, 1990; Heath, 1995) and a modest amount of field evidence (e.g., Staw and Hoang, 1995) are starting to help identify settings in which such entrapments are more and less likely (for reviews see Brockner, 1992, and Staw and Ross, 1989). Despite the obvious relevance to military settings, there is no substantial body of experimental research examining these effects in real or simulated military contexts.

Representation Effects

A central feature of the compendium model known as prospect theory (Kahneman and Tversky, 1979; Tversky and Kahneman, 1992) is that outcomes are evaluated in comparison with some subjectively set reference point and that preferences differ above and below this point. Losses, for example, loom larger than numerically equivalent gains, and value functions in the range of losses may be risk seeking whereas those for gains are risk averse. Since a given outcome can often be framed either as a gain compared with one reference level or as a loss compared with another, outcome preferences are vulnerable to what appear to be purely verbal effects. For example, McNeil et al. (1982) found that both physicians' and patients' preferences between alternative therapies were influenced by whether the therapies were described in terms of their mortality or survival rates. Levin and Gaith (1988) had consumers taste samples of ground beef described as either 90 percent lean or 10 percent fat and found that the descriptions markedly influenced the consumers' evaluations of the product. And Johnson et al. (1993) found that purchase choices for car insurance changed when a price differential between two alternative policies was described as a no-claim rebate on one policy rather than as a deductible on the other. Van Schie and van der Pligt (1995) provide a recent discussion of such framing effects on risk preferences.

The potential relevance to military decision-making contexts is obvious. It is commonplace that a maneuver can be described as a tactical repositioning or as a withdrawal, an outcome as 70 percent success or 30 percent failure, and a given loss of life as a casualty rate or a (complementary) survival rate. What needs to be investigated is whether these descriptive reframings are, in military contexts, anything more than superficial rhetorical tactics to which seasoned decision makers are impervious. Given the transparency of the rhetoric and the standardization of much military language, a naive prediction would be that framing would have no substantial effect. On the other hand, one would have made the identical prediction about McNeil et al.'s (1982) physicians, who also work within a formalized vocabulary and, presumably, are able to infer the 10 percent mortality implied by a 90 percent survival rate, but were nonetheless substantially influenced by the wording used. Framing effects, and problem representation issues in general, thus join our list of candidate phenomena for study in the military context.

Regret, Disappointment, and Multiple Reference Points

A number of theorists (Bell, 1982; Loomes and Sugden, 1982) have proposed decision models in which outcomes are evaluated not only on their own merits, but also in comparison with others. In these models, regret is thought to result from comparison of one's outcome with what would have been received under another choice, and disappointment from comparison with more fortunate outcomes of the same choice. Both regret and disappointment are thus particular forms of the general observation that outcome evaluations are commonly relative rather than absolute judgments and are often made against multiple reference points (e.g., Sullivan and Kida, 1995).

Special interest attaches to regret and disappointment because of their relationship to decision action and inaction and the associated role of decision responsibility. Kahneman and Tversky (1982) found that subjects imputed more regret to an investor who had lost money by a purchasing a losing stock than to another investor who had failed to sell the same stock. Gilovich and Medvec (1995) found this effect only for the short term; in the longer term, inactions are more regretted. Spranca et al. (1991) showed related asymmetries for acts of omission and commission, as in a widespread reluctance to order an inoculation campaign that would directly cause a few deaths even if it would indirectly save many more. The relationships among choice, regret, and responsibility are under active empirical scrutiny (see, for example, Connolly et al., 1997).

Military contexts in which such effects might operate abound. Might, for example, a commander hesitate to take a potentially beneficial but risky action for reasons of anticipated regret? Would such regret be exacerbated by organizational review and evaluation, whereby action might leave one more visibly tied to a failed course of action than would inaction? Studies such as Zellenberg (1996) suggest that social expectations of action are also relevant: soccer coaches who changed winning teams that then lost were blamed; those who changed losing teams were not blamed, even if subsequent results were equally poor. Both internally experienced regret and externally imposed blame would thus have to be considered in the context of a military culture with a strong bias toward action and accomplishment and, presumably, a concomitant understanding that bold action is not always successful.

Impediments to Learning

The general topic of learning is addressed at length in Chapter 5. However two particular topics related to the learning of judgment and decision-making skills have been of interest to decision behavior researchers, and are briefly addressed here: hindsight bias and confirmatory search. The first concerns learning from one to the next in a series of decisions; the second concerns learning and information gathering within a single judgment or decision.

Hindsight bias (Fischhoff, 1975; Fischhoff and Beyth, 1975) is the phenomenon whereby we tend to recall having had greater confidence in an outcome's occurrence or nonoccurrence than we actually had before the fact: in retrospect, then, we feel that "we knew it all along." The effect has been demonstrated in a wide range of contexts, including women undergoing pregnancy tests (Pennington et al., 1980), general public recollections of major news events (Leary, 1982), and business students' analyses of business cases (Bukszar and Connolly, 1987);

Christensen-Szalanski and Fobian (1991) provide an overview and meta-analysis. Supposing the military decision maker to be subject to the same bias, its primary effect would be to impede learning over a series of decisions by making the outcome of each less surprising than it should be. If, in retrospect, the outcome actually realized seems to have been highly predictable from the beginning, one has little to learn. If the outcome was successful, one's skills are once again demonstrated; if it was unsuccessful, all one can do is lament one's stupidity, since any fool could have foreseen the looming disaster. In neither case is one able to recapture the ex ante uncertainty, and one is thus impeded from learning how to cope with the uncertainty of the future.

A second potential obstacle to learning is the tendency, noted by a number of researchers, to shape one's search for information toward sources that can only confirm, but not really test, one's maintained belief. Confirmatory search of this sort might, for example, direct a physician to examine a patient for symptoms commonly found in the presence of the disease the physician suspects. A recruiter might similarly examine the performance of the personnel he or she had actually recruited. Such a search often has a rather modest information yield, though it may increase confidence (see Klayman and Ha, 1987, for a penetrating discussion). More informative is the search for indications that, if discovered, would invalidate one's maintained belief—for example, the physician's search for symptoms inconsistent with the disease initially suspected. Instances of this search pattern might be found in intelligence appraisals of enemy intentions: believing an attack imminent, the intelligence officer might search for, or pay undue attention to, evidence consistent with this belief, rather than seeking out and attending to potentially more valuable evidence that the belief was wrong. Again, there is no specific evidence that such search distortion occurs in military settings, and it is possible that prevailing doctrine, procedures, and training are effective in overcoming the phenomenon if it does exist. As is the case throughout this section, we intend only to suggest that dysfunctions of this sort have been found in other contexts and might well be worth seeking out and documenting in military settings as preparation for building appropriate representations into simulations of military decision makers.

Proposed Research Response

Research on decision behavior embraces a number of phenomena such as those reviewed above, and they presented the panel with special difficulty in forming research recommendations. First, even without a formal task analysis, it seems likely that decision-related phenomena such as these are highly relevant in military contexts and thus to simulations of such contexts. However, research on these topics seems not to fit well with the life-cycle model sketched earlier. That model postulates empirical convergence over time, so that the phenomenon of interest can be modeled with increasing fidelity. Decision behavior phenomena such as those discussed in this section, in contrast, seem often to diverge from this pattern—starting from an initially strong, clear result (e.g., confidence ratings are typically inflated; sunk costs entrap) and moving to increasingly contingent, disputed, variable, moderated, and contextual claims. This divergence is a source of frustration for those who expect research to yield simple, general rules, and the temptation is to ignore the findings altogether as too complex to repay further effort. We think this a poor response and support continued research attention to these issues. The research strategy should, however, be quite focused. It appears unlikely that further research aimed at a general understanding and representation of these phenomena will bear fruit in the near term. Instead, a targeted research program is needed, aimed at matching particular modeling needs to specific empirical research that makes use of military tasks, settings, personnel, and practices. Such a research program should include the following elements:

- **Problem Identification.** It is relatively easy to identify areas of obvious significance to military decision makers for which there is evidence of nonoptimal or nonintuitive behavior. However, an intensive collaboration between experts in military simulations and decision behavior researchers is needed to determine those priority problems/decision topics that the modelers see as important to their simulations and for which the behavioral researchers can identify sufficient evidence from other, nonmilitary contexts to warrant focused research.
- **Focused Empirical Research.** For each of the topics identified as warranting focused research, the extent and significance of the phenomenon in the specific military context identified by the modelers should be addressed. Does the phenomenon appear with these personnel, in this specific task, in this setting, under these contextual parameters? In contrast with most academic research programs, the aim here should not be to build generalizable theory, but to focus at the relatively local and specific level. It is entirely possible, for example, that the military practice of timed reassignment of commanders has developed to obviate sunk cost effects or that specialized military jargon has developed to allow accurate communication of levels of certainty from one command echelon to another.
- **Building of Simulation-Compatible Representations.** Given the above focus on specific estimates of effects, contextual sensitivities, and moderators, there should be no special difficulty in translating the findings of the research into appropriate simulation code. This work is, however, unlikely to yield a standardized behavioral module suitable for all simulations. The aim is to generate high realism in each local context. Cross-context similarities should be sought, but only as a secondary goal when first-level realism has been achieved.

CONCLUSIONS AND GOALS

This chapter has attempted to identify some of the key problems with the currently used decision models for military simulations and to recommend new approaches that would rectify these problems. In the top-down modeling tradition, a key problem with previous decision models is that the decision process is too stereotypical, predictable, rigid, and doctrine limited, so it fails to provide a realistic characterization of the variability, flexibility, and adaptability exhibited by a single entity across many episodes. We have presented a series of models in increasing order of complexity that are designed to overcome these problems. The first stages in this sequence are the easiest and quickest to implement in current simulation models; the later stages will require more effort and are a longer-term objective. Second, the decision process as currently represented is too uniform, homogenous, and invariable, so it fails to display individual differences due to stress, fatigue, experience, aggressiveness, impulsiveness, or risk attitudes that vary widely across entities. We have outlined how various individual difference factors or state factors can be incorporated by being related to parameters of the existing decision models.

We have also identified research opportunities and a research strategy for incorporating work in the bottom-up, phenomenon-focused tradition into models of military decision making. The thrust proposed here is not toward general understanding and representation of the phenomena, but toward specific empirical research in military settings, using military tasks, personnel, and practices. Problem identification studies would identify specific decision topics that modelers see as important to the simulations and behavioral researchers see as potentially problematic, but amenable to fruitful study. Focused empirical research would assess the extent and significance of these phenomena in specific military contexts. And simulation-compatible representations would be developed with full sensitivity to specific effect-size estimates and contextual moderators. Model realism would improve incrementally as each of these phenomenon-focused submodels was implemented and tested. Progress along these lines holds reasonably high promise for improving the realism of the decision models used in military simulations.

Short-Term Goals

- Include judgmental errors such as base-rate neglect and overconfidence as moderators of probability estimates in current simulation models.
- Incorporate individual differences into models by adding variability in model parameters. For example, variations in the tendency to be an impulsive versus a compulsive decision maker could be represented by variation in the threshold bound required to make a decision in sequential sampling decision models.

Intermediate-Term Goals

- Inject choice variability in a principled way by employing random-utility models that properly represent the effect of similarity in the consequences of each action on choice.

- Represent time pressure effects and relations between speed and accuracy of decisions by employing sequential sampling models of decision making.
- Develop models that reflect human decision making biases as suggested by the bottom-up approach.

Long-Term Goals

- Integrate learning with decision models to produce a synthesis that provides adaptive planning based on experience with previous episodes similar to the current situation.

7

Situation Awareness

In everyday parlance, the term *situation awareness*, means the up-to-the-minute cognizance or awareness required to move about, operate equipment, or maintain a system. The term has received considerable attention in the military community for the last decade because of its recognized linkage to effective combat decision making in the tactical environment.

In the applied behavioral science community, the term *situation awareness* has emerged as a psychological concept similar to such terms as *intelligence*, *vigilance*, *attention*, *fatigue*, *stress*, *compatibility*, and *workload*. Each began as a word with a multidimensional but imprecise general meaning in the English language. Each has assumed importance because it captures a characteristic of human performance that is not directly observable, but that psychologists—especially engineering psychologists and human factors specialists—have been asked to assess, or purposefully manipulate because of its importance to everyday living and working. Defining situation awareness in a way that is susceptible to measurement, is different from generalized performance capacity, and is usefully distinguishable from other concepts such as perception, workload, and attention has proved daunting (Fracker, 1988, 1991a, 1991b; Sarter and Woods, 1991, 1995). In fact, there is much overlap among these concepts.

Chapter 4 addresses attention, especially as it relates to time sharing and multitasking. In this chapter we discuss a collection of topics as if they were one. Although each of these topics is a research area in its own right, we treat together the perceptual, diagnostic, and inferential processes that precede decision making and action. The military uses the term *situation awareness* to refer to the static spatial awareness of friendly and enemy troop positions. We broaden this definition substantially, but retain the grouping of perceptual, diagnostic, and inferential processes for modeling purposes. While this approach may be controversial in the research community (see, for example, the discussion by Flach, 1995), we believe it is appropriate for military simulation applications. We review several of the definitions and potential models proposed for situation awareness, discuss potential model implementation approaches, outline connections with other models discussed throughout this report, and present conclusions and goals for future development.

We should also note that considerable research remains to be done on defining, understanding, and quantifying situation awareness as necessary precursors to the eventual development of valid descriptive process models that accurately and reliably model human behavior in this arena. Because of the early stage of the research on situation awareness, much of the discussion that follows focuses on potential prescriptive (as opposed to descriptive) modeling approaches that may, in the longer term, prove to be valid representations of human situation assessment behavior. We believe this is an appropriate research and modeling approach to pursue until a broader behavioral database is developed, and proposed models can be validated (or discarded) on the basis of their descriptive accuracy.

SITUATION AWARENESS AND ITS ROLE IN COMBAT DECISION MAKING

Among the many different definitions for the term *situation awareness* (see Table 7.1), perhaps the most succinct is that of Endsley (1988:97):

> Situation awareness is the perception of the elements in the environment within a volume of time and space, the comprehension of their meaning, and the projection of their status in the near future.

Each of the three hierarchical phases and primary components of this definition can be made more specific (Endsley, 1995):

- **Level 1 Situation Awareness**—perception of the elements in the environment. This is the identification of the key elements or "events" that, in combination, serve to define the situation. This level tags key elements of the situation semantically for higher levels of abstraction in subsequent processing.
- **Level 2 Situation Awareness**—comprehension of the current situation. This is the combination of level 1 events into a comprehensive holistic pattern, or tactical situation. This level serves to define the current status in operationally relevant terms in support of rapid decision making and action.
- **Level 3 Situation Awareness**—projection of future status. This is the projection of the current situation into the future in an attempt to predict the evolution of the tactical situation. This level supports short-term planning and option evaluation when time permits.

TABLE 7.1 Definitions of Situation Awareness

Reference	Definitions
Endsley (1987, 1995)	• Perception of the elements in the environment within a volume of time and space • Comprehension of their meaning • Projection of their status in the near future
Stiffler (1988)	• The ability to envision the current and near-term disposition of both friendly and enemy forces
Harwood et al. (1988)	• Where: knowledge of the spatial relationships among aircraft and other objects • What: knowledge of the presence of threats and their objectives and of ownship system state • Who: knowledge of who is in charge—the operator or an automated system • When: knowledge of the evolution of events over time
Noble (1989)	• Estimate of the purpose of activities in the observed situation • Understanding of the roles of participants in these activities • Inference about completed or ongoing activities that cannot be directly observed • Inference about future activities
Fracker (1988)	• The knowledge that results when attention is allocated to a zone of interest at a level of abstraction
Sarter and Woods (1991, 1995)	• Just a label for a variety of cognitive processing activities that are critical to dyamic, event-driven, and multitask fields of practice
Dominguez (1994)	• Continuous extraction of environmental information, integration of this knowledge to form a coherent mental picture, and the use of that picture in directing further perception and anticipating future events
Pew (1995)	• Spatial awareness • Mission/goal awareness • System awareness • Resource awareness • Crew awareness
Flach (1995)	• Perceive the information • Interpret the meaning with respect to task goals • Anticipate consequences to respond appropriately

Additional definitions are given in Table 7.1; of particular interest is the definition proposed by Dominguez (1994), which was specified after a lengthy review of other situation awareness studies and definitions. It closely matches Endsley's definition, but further emphasizes the impact of awareness on continuous cue extraction and directed perception, i.e., its contribution to attention. Note also, however, the effective nondefinition proposed by Sarter and Woods (1995), which reflects the views of a number of researchers in the field. Further expansion on these notions can be found in Flach (1995).

From the breadth of the various definitions, it should be clear that situation awareness, as viewed by many researchers working in the area, is a considerably broader concept than that conventionally held by the military community. The latter view tends to define situation awareness as merely[1] spatial awareness of the players (self, blue forces, and red forces), and at that, often simply their static positions without regard to their movements. It is also appropriate to point out the distinction between situation *awareness* and situation *assessment*. The former is essentially a *state* of knowledge; the latter is the *process* by which that knowledge is achieved. Unfortunately, the acronyms for both are the same, adding somewhat to the confusion in the literature.

Despite of the numerous definitions for situation awareness, it is appropriate to note that "good situation awareness" in a tactical environment is regarded as critical to successful combat performance. This is the case both for *low-tempo* planning activities, which tend to be dominated by relatively slow-paced and reflective proactive decisions, and *high-tempo* "battle drill" activities, which tend to be dominated by relatively fast-paced reactive decisions.

A number of studies of human behavior in low-tempo tactical planning have demonstrated how decision biases and poor situation awareness contribute to poor planning (Tolcott et al., 1989; Fallesen and Michel, 1991; Serfaty et al., 1991; Lussier et al., 1992; Fallesen, 1993; Deckert et al., 1994). Certain types of failures are common, resulting in inadequate development and selection of courses of action. In these studies, a number of dimensions relating specifically to situation assessment are prominent and distinguish expert from novice decision makers. Examples are awareness of uncertain assumptions, awareness of enemy activities, ability to focus awareness on important factors, and active seeking of confirming/disconfirming evidence.

Maintenance of situation awareness also plays a key role in more high-tempo battlefield activities (e.g., the "battle drills" noted in IBM, 1993). Several studies have focused on scenarios in which the decision maker must make dynamic decisions under " . . . conditions of time pressure, ambiguous infor-

---

[1]Use of the term "merely" is not meant to imply that achieving full military situation awareness is a trivial exercise. Indeed it is not, as much of the military intelligence community is devoted to achieving this state. We do, however, suggest that definitions of situation awareness used frequently by the military be broadened along the lines identified by Endsley (1995).

mation, ill-defined goals, and rapid change" (Klein, 1997:287). These studies span the theoretical-to-applied spectrum and cover many domains. Klein and colleagues (Klein, 1989, 1994, 1997; Klein et al., 1986) have proposed a particular type of decision-making model for these situations, the recognition-primed decision (RPD) model, which incorporates three levels of decisions: (1) "matching," or a simple mapping of a recognized situation to a prescribed decision or action; (2) "diagnosing," or iterative probing of the environment until the situation is recognized, followed by a triggering of the prescribed decision or action; and (3) "evaluating" or mapping of a recognized situation to a set of prescribed decisions, followed by selection of the most appropriate decision through a mental simulation. At all these levels, situation assessment (recognition in the RPD nomenclature) plays a critical role, central to all subsequent decisions or actions.

One aspect of situation awareness, which has been referred to as *crew awareness*, is the extent to which the personnel involved have a common mental image of what is happening and an understanding of how others are perceiving the same situation. The ideas of distributed cognition, shared mental models, and common frame of reference play a role in understanding how groups can be aware of a situation and thus act upon it. Research in distributed cognition (Hutchins, 1990; Sperry, 1995) suggests that as groups solve problems, a group cognition emerges that enables the group to find a solution; however, that group cognition does not reside entirely within the mind of any one individual. Research on shared mental models and common frames of reference suggests that over time, groups come to have a more common image of a problem, and this common image is more or less shared by all participants. What is not known is how much of a common image is needed to enhance performance and what knowledge or processes need to be held in common.

There is currently a great deal of interest in individual and team mental models (Reger and Huff, 1993; Johnson-Laird, 1983; Klimoski and Mohammed, 1994; Eden et al., 1979; Carley, 1986a, 1986b; Roberts, 1989; Weick and Roberts, 1993; Walsh, 1995). Common team or group mental models are arguably critical for team learning and performance (Hutchins, 1990, 1991a, 1991b; Fiol, 1994). However, the relationship between individual and team mental models and the importance of shared cognition to team and organizational performance is a matter requiring extensive research. Although many problems are currently solved by teams, little is known about the conditions for team success.

MODELS OF SITUATION AWARENESS

A variety of situation awareness models have been hypothesized and developed by psychologists and human factors researchers, primarily through empirical studies in the field, but increasingly with computational modeling tools. Because of the critical role of situation awareness in air combat, the U.S. Air Force has taken the lead in studying the modeling, measurement, and trainability of situation awareness (Caretta et al., 1994). Numerous studies have been conducted to develop situation awareness models and metrics for air combat (Stiffler, 1988; Spick, 1988; Harwood et al., 1988; Endsley, 1989, 1990, 1993, 1995; Fracker, 1990; Hartman and Secrist, 1991; Klein, 1994; Zacharias et al., 1996; Mulgund et al., 1997). Situation awareness models can be grouped roughly into two classes: descriptive and prescriptive (or computational).

Descriptive Situation Awareness Models

Most developed situation awareness models are descriptive. Endsley (1995) presents a descriptive model of situation awareness in a generic dynamic decision-making environment, depicting the relevant factors and underlying mechanisms. Figure 7.1 illustrates this model as a component of the overall assessment-decision-action loop and shows how numerous individual and environmental factors interact. Among these factors, attention and working memory are considered the critical factors limiting effective situation awareness. Formulation of mental models and goal-directed behavior are hypothesized as important mechanisms for overcoming these limits.

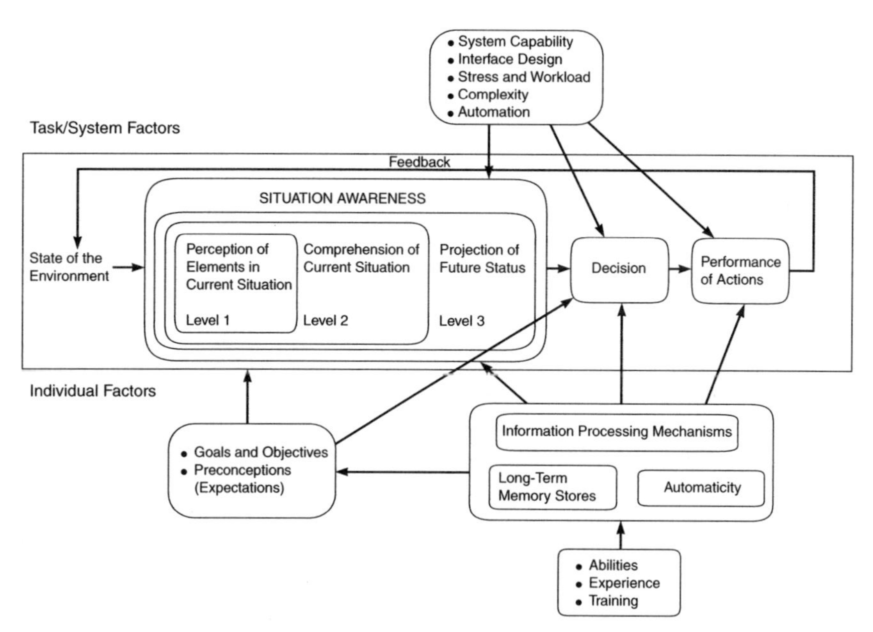

FIGURE 7.1 Generic decision-making model centered on the situation awareness process (Endsley, 1995).

Gaba et al. (1995) describe a qualitative model of situation-based decision making in the domain of anesthesiology. An initial stage of sensory/motor processing is concerned with identifying key physiological "events" as presented on the anesthesiology monitors (Endsley's level 1 situation awareness). A second stage of situation assessment focuses on integrating those events to identify potential problems (Endsley's level 2 situation awareness) and anticipated trends into the future (Endsley's level 3 situation awareness). A third stage of decision making includes a "fast" reactive path of precompiled responses (akin to Klein's RPD strategy [Klein, 1989, 1994]) paralleled by a "slow" contemplative path of model-based reasoning about the situation. Both paths then contribute to the next stage of plan generation and subsequent action implementation.

Although descriptive models are capable of identifying basic issues of decision making in dynamic and uncertain environments, they do not support a quantitative simulation of the process by which cues are processed into perceptions, situations are assessed, and decisions are made. Further, we are unaware of any descriptive model that has been developed into a computational model for actual emulation of human decision-making behavior in embedded simulation studies.

Prescriptive Situation Awareness Models

In contrast to the situation with descriptive models, few prescriptive models of situation awareness have been proposed or developed. Early attempts used production rules (Baron et al., 1980; Milgram et al., 1984). In these efforts, the situation awareness model was developed as a production rule system in which a situation is assessed using the rule "if a set of events E occurs, then the situation is S." Although this pattern-recognition (bottom-up) type of approach is fairly straightforward to implement, it lacks the diagnostic strength of a causal-reasoning (top-down) type of approach. In the latter approach, a hypothesized situation S can be used to generate an *expected* event set E*, which can be compared with the *observed* event set E; close matches between E* and E confirm the hypothesized situation, and poor matches motivate a new hypothesis. Not surprisingly, these early attempts at modeling the situation awareness process using a simple production rule system, going from events to situations, performed poorly because of three factors: (1) they were restricted to simple forward chaining (event to situation); (2) they did not take into consideration uncertainty in events, situations, and the rules linking them; and (3) they did not make use of memory, but simply reflected the current instantaneous event state. More sophisticated use of production rules (e.g., either through combined forward/backward chaining, incorporation of "certainty factors," and/or better internal models) could very well ameliorate or eliminate these problems. Tambe's work on a rule-based model of agent tracking (Taube 1996a, 1996b, 1996c) illustrates just what can be accomplished with this more sophisticated use of production rule systems.

The small unit tactical trainer (SUTT)[2] (see Chapters 2 and 8) includes computer-controlled hostiles that behave as "smart" adversaries to evade and counterattack blue forces. Knowledge of the current situation by the computer-controlled hostiles is specified by a number of fairly low-level state variables defining "self" status: location, speed/direction of movement, posture, weapon state, and the like. No attempt is made to model this type of perception/assessment function; rather, the "true" individual combatant states are used directly. State variables defining the status of others are less extensive and focus primarily on detection/identification status and location relative to self, thus following the conventional military definition of situation awareness—positional knowledge of all players in the scenario at hand. Here, an attempt is made to model the perception of these states in accordance with the level 1 situation awareness process postulated by Endsley (1995). The net result is a list of entities, with attributes specifying the observer's knowledge of each (e.g., undetected, detected but unrecognized) and with parameters specifying the state of each (position, and possibly velocity). In the SUTT human behavior representation, situation awareness is modeled as primarily a low-level collection of "events" (identified and located entities), with no attempt made to assess or infer higher-level situations (e.g., "Is this group of individual entities part of an enemy squad?"). Thus actions or plans for actions are necessarily reflexive at a fairly low level (implemented as either rulebases or decision trees), with little abstraction or generalization involved. While this may be adequate for modeling a wide range of "battle drill" exercises in which highly choreographed offensive and defensive movements are triggered by relatively low-level events (e.g., detection of a threat at a particular location), it is unclear how such battle drills can be selected reliably without the specification of an adequate context afforded by higher-level situation awareness processes (e.g., "We're probably facing an entire platoon hiding behind the building"). Certainly it is unclear how more "inventive" situation-specific tactics can be formulated on the fly without an adequate situation assessment capability.

The man-machine integration design and analysis system (MIDAS) was developed by the National Aeronautics and Space Administration (NASA) and the U.S. Army to model pilot behavior, primarily in support of rotorcraft crew station design and procedural analyses (Banda et al., 1991; Smith et al., 1996; see also Chapters 3 and 8). As described earlier in Chapter 3, an agent-based operator model (comprising three basic modules for representing perceptual, cognitive, and motor processing) interacts with the proximal environment (displays, controls) and, in combination with the distal environment (e.g., own ship, other aircraft), results in observable behavioral activities (such as scanning, deciding, reaching, and communicating). Much effort has gone into developing environmental models, as well as perceptual and motor submodels. Work has also gone

[2]Formerly known as the team target engagement simulatior (TTES).

into developing a number of rule-based cognitive models, each specific to a particular design case and mission specification.

Earlier versions of MIDAS (e.g., as described by Banda et al., 1991) relied primarily on a pattern-matching approach to triggering reactive behavior, based on the internal perceptions of outside world states. Recent work described by Smith et al. (1996) and Shively (1997) is directed at developing a more global assessment of the situation (to be contrasted with a set of perceived states), somewhat along the lines of Reece (1996). Again, the focus is on assessing the situation in terms of the external entities: self, friendly, and threat and their locations. A four-stage assessment process (detection, recognition, identification, and comprehension) yields a list of entities and a numeric value associated with how well each entity assessment matches the actual situation. A weighted calculation of overall situation awareness is made across entities and is used to drive information-seeking behavior: low situation awareness drives the attention allocator to seek more situation-relevant information to improve overall awareness. Currently, the situation awareness model in MIDAS does not drive the decision-making process (except indirectly through its influence on information-seeking behavior), so that MIDAS remains essentially event- rather than situation-driven. The current structure does not, however, appear to preclude development along these lines.

As discussed in Chapter 2, considerable effort has been devoted to applying the Soar cognitive architecture (Laird et al., 1987; Laird and Rosenbloom, 1990) to the modeling of human behavior in the tactical air domain. Initial efforts led to a limited-scope demonstration of feasibility, fixed-wing attack (FWA)-Soar (Tambe et al., 1995);[3] more recent efforts showed how Soar-intelligent forces (Soar-IFOR) could participate in the synthetic theater of war (STOW)-97 large-scale warfighting simulation (Laird, 1996).

As described in Chapter 3, much of the Soar development effort has been focused on implementing a mechanism for goal-driven behavior, in which high-level goals are successively decomposed into low-level actions. The emphasis has been on finding a feasible action sequence taking the Soar entity from the current situation to the desired goal (or end situation); less emphasis has been placed on identifying the current situation (i.e., situation awareness). However, as Tambe et al. (1995) note, identifying the situation is critical in dynamic environments, in which either Soar agents or others continually change the environment, and hence the situation. Thus, more recent emphasis in Soar has been placed on modeling the agent's dynamic perception of the environment and assessment of the situation.

In contrast with conventional "situated action agents" (e.g., Agre and Chapman, 1987), in which external system states are used to drive reflexive agents (rule-based or otherwise) to generate agent actions, Soar attempts to model the

[3]Formerly known as Tactical Air(TacAir)-Soar.

perception and assessment of these states and then use the perceived/assessed states to drive its own goal-oriented behavior. This clear distinction then allows misperceived states or misassessed situations to propagate through the decision/action process, thus supporting a realistic representation of human behavior in uncertain and dynamic environments.

A brief review of the current Soar effort (Tambe et al., 1995; Laird, 1996; Gratch et al., 1996) indicates that this type of "front-end" modeling is still in its early stages. For example, FWA-Soar (Tambe et al., 1995) abstracts the pilot-cockpit interface and treats instrument cues at an informational level. We are unaware of any attempt to model in Soar the detailed visual perceptual processes involved in instrument scanning, cue pickup, and subsequent translation into domain-relevant terms. However, Soar's "perceptual" front end, although primitive, may be adequate for the level of modeling fidelity needed in many scenarios, but also, more important, could support the development of an *explicit* situation assessment module, thus providing *explicit* context for the hundreds or thousands of production rules contained in its implementation.[4] As noted below, this functionality in which low-level perceptual events are transformed into a higher-level decisional context, has been demonstrated by Tambe et al. (1995).

Recognizing that situation assessment is fundamentally a diagnostic reasoning process, Zacharias and colleagues (1992, 1996), Miao et al. (1997), and Mulgund et al. (1996, 1997) have used belief networks to develop prescriptive situation awareness models for two widely different domains: counter-air operations and nuclear power plant diagnostic monitoring. Both efforts model situation awareness as an integrated inferential diagnostic process, in which situations are considered as hypothesized reasons, events as effects, and sensory (and sensor) data as symptoms (detected effects). Situation awareness starts with the detection of event occurrences. After the events are detected, their likelihood (belief) impacts on the situations are evaluated by backward tracing the situation-event relation (diagnostic reasoning) using Bayesian belief networks. The updated situation likelihood assessments then drive the projection of future event occurrences by forward inferencing along the situation-event relation (inferential reasoning) to guide the next step of event detection. This approach of using belief networks to model situation awareness is described at greater length below.

Multiagent Models and Situation Awareness

It is relatively common for multiagent computational models of groups to be designed so that each agent has some internal mental model of what other agents know and are doing; see, for example, the discussion of FWA-Soar in Chapter 10.

[4]Other representation approaches not involving an explicit module for situation assessment are clearly feasible as well, for example, a symbolic mental model maintained jointly by perceptual processes, task-focused knowledge, and so on.

Situation awareness in these models has two parts—the agent's own knowledge of the situation and the agent's knowledge of what others are doing and might do if the situation were to change in certain ways. To date, this approach has been used successfully only for problems in which others can be assumed to act exclusively by following doctrine (preprogrammed rules of behavior), and the agents continually monitor and react to that environment. Whether the approach is extensible to a more mutually reactive situation is not clear.

ENABLING TECHNOLOGIES FOR IMPLEMENTATION OF SITUATION AWARENESS MODELS

A number of approaches can be taken to implement situation awareness models, from generic inferencing approaches (e.g., abductive reasoning, see, for example, Josephson and Josephson, 1994) to specific computational architectures (e.g., blackboard architectures, Hayes-Roth, 1985). We have selected blackboard systems to discuss briefly here, and expert systems, case-based reasoning, and belief networks to discuss in detail below.

Blackboard systems have been used to model all levels of situation awareness as defined by Endsley (1995). Blackboard system models were initially developed to model language processing (Erman et al., 1980), and later used for many situation assessment applications (see Nii, 1986a, 1986b). In the blackboard approach, a situation is decomposed into one or more hierarchical panels of symbolic information, often organized as layers of abstraction. Perceptual knowledge sources encode sensory data and post it on to appropriate locations of the blackboard (level 1 situation awareness), while other knowledge sources reason about the information posted (level 2 situation awareness) and make inferences about future situations or states (level 3 situation awareness), posting all their conclusions back onto the blackboard structure. At any point in time, the collection of information posted on the blackboard constitutes the agent's awareness of the situation. Note that this is a nondiagnostic interpretation of situation awareness. Other knowledge sources can use this situational information to assemble action plans on a goal-driven or reactive basis. These may be posted on other panels. In a tactical application created for the TADMUS program (Zachary et al., forthcoming), a six-panel model in which two panels were used to construct the situational representation and the other four used to build, maintain, and execute the tactical plan as it was developed.

Expert systems or, more generally, production rule systems, are discussed because they have been used consistently since the early 1980s to model situation awareness in computational behavior models. In contrast, case-based reasoning has not been used extensively in modeling situation awareness; it does, however, have considerable potential for this purpose because of both its capabilities for modeling episodic situation awareness memory and the ease with which new situations can be learned within the case-based reasoning paradigm. Finally, belief networks (which implement a specific approach to abductive reasoning, in general) are discussed because of their facility for representing the interrelatedness of situations, their ability to integrate low-level events into high-level situations, and their potential for supporting normative model development.

Expert Systems[5]

An early focus of expert system development was on applications involving inferencing or diagnosis from a set of observed facts to arrive at a more general assessment of the situation that concisely "explains" those observed facts. Consequently, there has been interest in using expert systems to implement situation awareness models.

In typical expert systems, domain knowledge is encoded in the form of production rules (IF-THEN or antecedent-consequent rules). The term *expert system* reflects the fact that the rules are typically derived by interviewing and extracting domain knowledge from human experts. There have been expert systems for legal reasoning, medical diagnosis (e.g., MYCIN, which diagnoses infectious blood diseases [Buchanan and Shortliffe, 1984], and Internist, which performs internal medicine diagnostics [Miller et al., 1982]); troubleshooting of automobiles; and numerous other domains. Expert systems consist of three fundamental components:

- **Rulebase**—a set of rules encoding specific knowledge relevant to the domain.
- **Factbase** (or working memory)—a set assertion of values of properties of objects and events comprising the domain. The factbase encodes the current state of the domain and generally changes as (1) rules are applied to the factbase, resulting in new facts, and (2) new facts (i.e., not derivable from the existing factbase) are added to the system.
- **Inference Engine** (i.e., mechanical theorem prover)—a formal implementation of one or more of the basic rules of logic (e.g., modus ponens, modus tolens, implication) that operate on the factbase to produce new facts (i.e., theorems) and on the rule base to produce new rules. In practice, most expert systems are based on the inference procedure known as *resolution* (Robinson, 1965).

The general operation of an expert system can be characterized as a *match-select-apply* cycle, which can be applied in either a *forward-chaining* fashion, from antecedents to consequents, or a *backward-chaining* fashion, from consequents to antecedents. In forward chaining, a matching algorithm first determines the subset of rules whose antecedent conditions are satisfied by the current set of facts comprising the factbase. A conflict-resolution algorithm then selects one rule in particular and applies it to the factbase, generally resulting in the

[5]This section borrows heavily from Illgen et al. (1997a).

addition of new facts to the factbase. In backward chaining, the matching algorithm determines the rules whose consequent conditions are satisfied by the factbase and then attempts to establish the validity of the antecedent condition. In general, forward chaining is used to derive all the consequences when a new fact is added to the factbase; this is sometimes referred to as *data-driven inference*. In contrast, backward chaining is used when there is a specific problem to solve; this is sometimes referred to as *goal-driven* inference. Many practical expert systems allow for a combination of both strategies (Russell and Norvig, 1995; Ng and Abramson, 1990).

The majority of current military human behavior representations reviewed for this study implicitly incorporate an expert-system-based approach to situation awareness modeling. The implicitness derives from the fact that few attempts have been made to develop explicit production rules defining the situation.

IF [(event 1 is true) AND (event 2 is true) . . .]
THEN [situation 1 is true] (7.1)

Rather, the production rules are more often defined implicitly, in a fairly direct approach to implementing "situated action" agents, by nearly equivalent production rules for action, such as[6]

IF [(event 1 is true) AND (event 2 is true) . . .]
THEN [perform maneuver type 1] (7.2)

An explicit situation awareness approach would include a situational rule of the form given by (7.1) above, followed by a decisional rule of the form

IF [situation 1 is true] THEN [perform maneuver type 1] (7.3)

This approach allows for the generation of higher-level situational assessments, which are of considerable utility, as recognized by the developers of Soar. For example, Tambe et al. (1995) point out that by combining a series of low-level events (e.g., location and speed of a bogey), one can generate a higher-level situational attribute (e.g., the bogey is probably a MiG-29). This higher-level attribute can then be used to infer other likely aspects of the situation that have not been directly observed (e.g., the MiG-29 bogey probably has fire-and-forget missiles), which can have a significant effect on the subsequent decisions made by the human decision maker (e.g., the bogey is likely to engage me immediately after he targets my wingman). Note that nowhere in this situation assessment process is there a production rule of the form given by (7.2); rather, most or all of the process is devoted to detecting the events (Endsley's [1995] level 1 situation awareness), assessing the current situation (level 2), and projecting likely futures (level 3).

[6]Where the events may be "actual" or "perceived," depending on the model.

Case-Based Reasoning[7]

Case-based reasoning is a paradigm in which knowledge is represented as a set of individual cases—or a *case library*—that is used to process (i.e., classify, or more generally, reason about or solve) novel situations/problems. A case is defined as a set of features, and all cases in the library have the same structure. When a new problem, referred to as a *target case*, is presented to the system, it is compared with all the cases in the library, and the case that matches most closely according to a similarity metric defined over a subset of the case features called *index features* is used to solve (or reason further about, respond to) the new problem. There is a wide range of similarity metrics in the literature: Euclidean distance, Hamming distance, the value difference metric of Stanfill and Waltz (1986), and many others.

In general, the closest-matching retrieved case will not match the target case perfectly. Therefore, case-based reasoning systems usually include rules for adapting or tailoring the specifics of the retrieved case to those of the target case. For example, the current battlefield combat scenario may be quite similar to a particular previous case at the tactical level, but involve a different enemy with different operating parameters. Thus, the assessment of the situation made in the prior case may need to be tailored appropriately for the current case. In general, these adaptation (or generalization) rules are highly domain specific; construction of a case-based reasoning system for a new domain generally requires a completely new knowledge engineering effort for the adaptation rules (Leake, 1996). In fact, an important focus of current case-based reasoning research is the development of more general (i.e., domain-neutral) adaptation rules.

A major motivation for using a case-based reasoning approach to the modeling of situation assessment, and human decision making in general, is that explicit reference to individual previously experienced instances or cases is often a central feature of human problem solving. That is, case-specific information is often useful in solving the problem at hand (Ross, 1989; Anderson, 1983; Carbonell, 1986; Gentner and Stevens, 1983). In the domain of human memory theory, such case- or episode-specific knowledge is referred to as *episodic memory* (Tulving, 1972; see also Chapter 5).

The second major motivation for using the case-based approach is that it is preferable when there is no valid domain model (as opposed to a human behavioral model).[8] The state space of most tactical warfare situations is extremely large, thus making construction of a complete and reliable domain model untenable. In the absence of such a domain model, case-based solutions offer an alternative to model-based solutions. Both approaches have their advantages and

---

[7]This section borrows heavily from Gonsalves et al. (1997).

[8]It should be clear, however, that case-based reasoning is not unique in its ability to work in the absence of a valid domain model. In fact, many approaches have this property, including rule-based systems, neural networks, fuzzy logic systems, and genetic algorithms.

disadvantages. Models are condensed summaries of a domain, omitting (i.e., averaging over) details below a certain level. They therefore contain less information (i.e., fewer parameters) and require less storage space and have less computational complexity than case-based approaches. If it is known beforehand that there will never be a need to access details below the model's level of abstraction, models are preferable because of their increased computational efficiency. However, if access to low-level information will be needed, a case-based reasoning approach is preferable.

We are unaware of any military human behavior representations that employ case-based reasoning models of situation awareness. However, we believe this approach offers considerable potential for the development of general situation awareness models and military situation awareness models in particular, given the military's heavy reliance on case-based training through numerous exercises, the well-known development of military expertise with actual combat experience, and the consistent use of "war stories" or cases by subject matter experts to illustrate general concepts. In addition, the ease with which case-based reasoning or case "learning" can be implemented makes the approach particularly attractive for modeling the effects of human learning. Whether effective case-based reasoning models of situation awareness will be developed for future human behavior representations is still an open research question, but clearly one worth pursuing.

Bayesian Belief Networks[9]

Bayesian belief networks (also called belief networks and Bayesian networks) are probabilistic frameworks that provide consistent and coherent evidence reasoning with uncertain information (Pearl, 1986). The Bayesian belief network approach is motivated directly by Bayes' theorem, which allows one to update the likelihood that situation S is true after observing a situation-related event E. If one assumes that the situation S has a *prior* likelihood Pr(S) of being true (that is, prior to the observation of E), and one then observes a related event E, one can compute the *posterior* likelihood of the situation S, Pr(S|E), directly by means of Bayes' theorem:

$$\Pr(S|E) = \frac{\Pr(E|S)\Pr(S)}{\Pr(E)}$$

where Pr(E|S) is the conditional probability of E given S, which is assumed known through an understanding of situation-event causality. The significance of this relation is that it cleanly separates the total information bearing on the

[9]Earlier versions of this section were included in Zacharias et al. (1992) and Gonsalves et al. (1997).

probability of S being true into two sources: (1) the prior likelihood of S, which encodes one's prior knowledge of S; and 2) the likelihood of S being true in light of the observed evidence, E. Pearl (1986) refers to these two sources of knowledge as the predictive (or causal) and diagnostic support for S being true, respectively.

The Bayesian belief network is particularly suitable for environments in which:

- Evidence from various sources may be unreliable, incomplete, and imprecise.
- Each piece of evidence contributes information at its own source-specific level of abstraction. In other words, the evidence may support a set of situations without committing to any single one.
- Uncertainties pervade rules that relate observed evidence and situations.

These conditions clearly hold in the military decision-making environment. However, one can reasonably ask whether human decision makers follow Bayesian rules of inference since, as noted in Chapter 5, many studies show significant deviations from what would be expected of a "rational" Bayesian approach (Kahneman and Tversky, 1982). We believe this question is an open one and, in line with Anderson's (1993) reasoning in developing adaptive control of thought (ACT-R), consider a Bayesian approach to be an appropriate framework for building normative situation awareness models—models that may, in the future, need to be "detuned" in some fashion to better match empirically determined human assessment behavior. For this brief review, the discussion is restricted to simple (unadorned) belief networks and how they might be used to model the military situation assessment process.

A belief network is a graphical representational formalism for reasoning under uncertainty (Pearl, 1988; Lauritzen and Spiegelhalter, 1988). The nodes of the graph represent the domain variables, which may be discrete or continuous, and the links between the nodes represent the probabilistic, and usually causal, relationships between the variables. The overall topology of the network encodes the qualitative knowledge about the domain.

For example, the generic belief network of Figure 7.2 encodes the relationships over a simple domain consisting of five binary variables. Figure 7.2a shows the additional quantitative information needed to fully specify a belief network: prior belief distributions for all root nodes (in this case just A and B) and conditional probability tables for the links between variables. Figure 7.2a shows only one of this belief network's conditional probability tables, which fully specifies the conditional probability distribution for D given X. Similar conditional probability tables would be required for the other links. These initial quantities, in conjunction with the topology itself, constitute the domain model, and in most applications are invariant during usage.

The network is initialized by propagating the initial root prior beliefs downward through the network. The result is initial belief distributions for all vari-

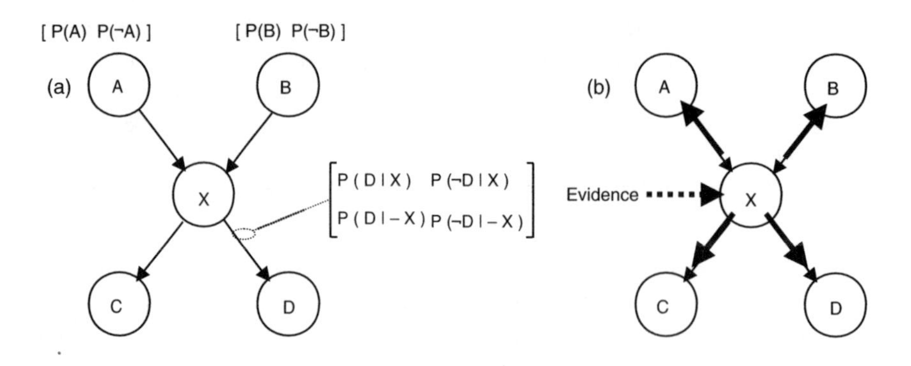

FIGURE 7.2 Belief network components and message passing.

ables and thus for the domain as a whole. When evidence regarding a variable is obtained, it is applied to the corresponding node, which then updates its own total belief distribution, computed as the product of the prior distribution and the new evidence. The node then sends belief revision messages to its parent and children nodes. For example, Figure 7.2b shows the posting of evidence to X and the four resulting belief revision messages. Nodes A, B, C, and D then update their respective belief distributions. If they were connected to further nodes, they would also send out belief revision messages, thus continuing the cycle. The process eventually terminates because no node sends a message back out over the link through which it received a message.

The major feature of the belief network formalism is that it renders probabilistic inference computationally tractable. In addition, it allows causal reasoning from causes to effects and diagnostic reasoning from effects to causes to be freely mixed during inference. Belief network models can be either manually constructed on the basis of knowledge extracted from domain experts or, if sufficient data exist, learned automatically (i.e., built) by sampling the data. Within the artificial intelligence community, belief networks are fast becoming the primary method of reasoning under uncertainty (Binder et al., 1995).

Note that the evidence propagation process involves both top-down (predictive) and bottom-up (diagnostic) inferencing. At each stage, the evidence observed thus far designates a set of likely nodes. Each such node is then used as a source of prediction about additional nodes, directing the information extraction process in a top-down manner. This process makes the underlying inference scheme transparent, as explanations can be traced mechanically along the activated pathways. Pearl's (1986) algorithm is used to compute the belief values given the evidence.

To illustrate how a belief network model could be used to model the situation assessment of an S-2 staff, Figure 7.3 shows a hypothetical battle scenario in which an enemy motorized rifle division will invade its neighbor to the west. The

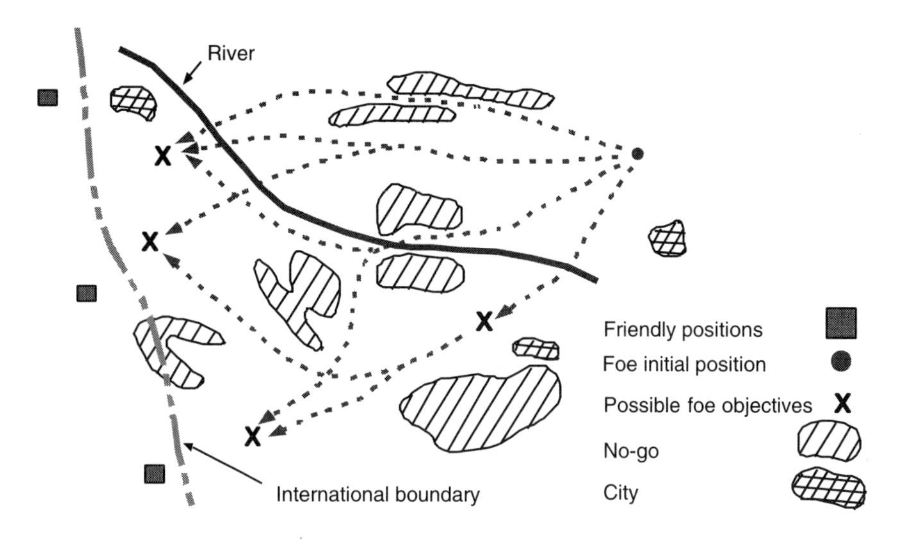

FIGURE 7.3 Hypothetical battle scenario. See text for discussion.

X's and dashed lines represent the friendly S-2's a priori, best hypotheses about possible enemy objectives and paths to those objectives, based on initial intelligence preparation of the battlefield, knowledge of enemy fighting doctrine, and knowledge of more global contexts for this local engagement—for example, that the enemy has already attacked on other fronts, or that high-value strategic assets (e.g., an oil refinery) exist just west of the blue positions. Although not depicted in the figure, the friendly force will, of course, have deployed many and varied sensor assets, such as radio emission monitoring sites, deeply deployed ground-based surveillance units, scout helicopters, and theater-wide surveillance assets (e.g., JSTARS). As the enemy begins to move, these sensors will produce a stream of intelligence reports concerning enemy actions. These reports will be processed by the S-2 staff, who will incrementally adjust degrees of belief in the various hypotheses about enemy positions, dispositions, and so on represented in the system.

Figure 7.4 depicts a simple belief network model supporting blue situation awareness of red activity that might take place under this scenario. It links several lower-level variables that are directly derivable from the sensors themselves to a higher-level (i.e., more abstract) variable, in this case, whether the red force is the main or supporting attack force. The set of values each variable can attain appears next to the node representing that variable.[10] For example, the

[10]These variables are all discrete valued. However, the belief network formalism allows continuous-valued variables as well.

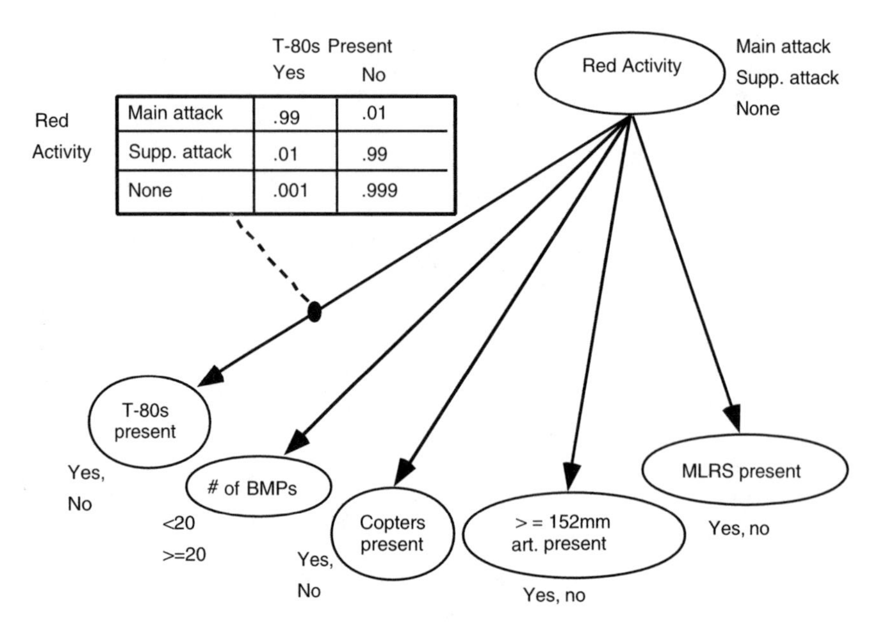

FIGURE 7.4 Belief net linking lower-level and high-level variables. See text for discussion.

binary variable "# of BMPs," representing the number of that particular kind of armored personnel carrier present in the local region, has the two discrete values "≥ 20" and "< 20", since 20 or more BMPs is strongly indicative of the presence of a motorized rifle regiment, which in turn is generally a constituent of the main attack force. The arrows represent direct probabilistic relationships between variables. Thus the arrow from "red activity" to "T-80s present" means that if one knows the value "red activity," one knows, based on doctrine, whether T-80s will be present.

The quantitative aspects of this dependence are encoded in a conditional probability table associated with that link, as shown in the upper left portion of Figure 7.4, based on a priori doctrinal knowledge. In the case of discrete-valued variables, a conditional probability table gives the probability of each value, C_i, of the child variable conditioned upon each value, P_j, of the parent variable; i.e., $P(C_i|P_j)$. The conditional probability table of the figure essentially encodes the doctrinal knowledge that, given that a Russian motorized rifle division (MRD) is attacking, its main attack force will almost certainly (i.e., .99) contain T-80s, any support attacks will almost certainly not have T-80s, and it is very unlikely that a stray T-80 (i.e., one not associated with either a main or support attack) will be encountered. Note that qualitative and quantitative knowledge about the interdependencies among domain variables is represented separately in the belief network formalism; qualitative relationships (i.e., whether one variable directly affects another at all) are reflected in the topology (the links), whereas quantitative details of those dependencies are represented by conditional probability tables.

Gonsalves et al. (1997) describe how this model is used to process low-level data on enemy size, activity, location, and equipment coming in during the course of a red attack and how the situation is continually reassessed during the engagement. A related effort by Illgen et al. (1997b) shows how the approach can be extended to model assessment of enemy courses of action and identification of the most likely course of action selected as the engagement unfolds. Research is now under way, as part of the Federated Laboratory effort supported by the Army Research Laboratory, to develop more extensive belief network models of the situation assessment process for a selected brigade-level engagement. Considerable effort is expected to be devoted to validation of developed models through comparison of model predictions with empirical data.

A number of modeling efforts have also been conducted by Zacharias and colleagues to develop belief network-based situation awareness models of Air Force decision makers. Zacharias et al. (1996) first applied belief networks to model pilot situation awareness in counter-air operations, using them to infer basic mission phases, defensive/offensive strength ratios, and appropriate intercept strategies. Zacharias et al. (1996) built upon this work by including additional factors to incorporate more threat attributes, engagement geometry, and sensor considerations. Work is now under way to refine and validate these situation awareness models and incorporate them into self-contained pilot agents (accounting for other aspects of pilot decision making and flight execution) as part of a larger-scale (M versus N) tactical air combat simulator called man-in-the-loop air to air system performance evaluation model (MIL-AASPEM) (Lawson and Butler, 1995). Parallel efforts are also being conducted to use these situation awareness models for driving tactical displays as a function of the inferred tactical situation (Mulgund et al., 1997) and for providing the battlefield commander with tactical decision aids (Mengshoel and Wilkins, 1997; Schlabach, 1997).

Although there is significant activity regarding the use of belief network-based situation awareness models, it should be emphasized that most of this work has been focused on (1) the knowledge elicitation needed to define the subject matter expert's mental model, (2) the computational model implementation, and (3) the generation of simulation-based traces of the inferencing process and the resulting situation-driven activities of the operator model. Little actual validation has been conducted, and experimental efforts to validate the model output with what is actually observed under well-controlled man-in-the-loop simulator experiments are just now beginning.

We believe that a belief network-based approach to modeling situation assessment has considerable potential, but is not without its drawbacks. First, the graphic representation structure makes explicit, to domain experts and developers alike, the basic interrelationships among key objects and concepts in the domain. However, building these topological structures is not a trivial exercise, nor is the specification of the quantitative probability links between nodes. Second, the event *propagation* algorithm reflects the continuity of situation assessment—an evidence accumulation process that combines new events with prior situation assessments to evaluate events in the context of the assessed situation. Likewise, the event *projection* algorithm reflects the continuity of situation awareness—the projection of future events based on the currently assessed situation. However, it should be clear that both propagation and projection are based on normative Bayesian models, which may, in fact, not accurately reflect actual human behavior in assessment (recall the discussion in Chapter 6 on decision-making models). Finally, it should be pointed out that much of the functionality achievable with belief networks (e.g., continuous updating of the assessed situation, propagation and projection, normative decision-making behavior) may very well be accomplished using other approaches (e.g., forward/backward chaining expert systems incorporating uncertainty factors). Thus it may prove to be more effective to upgrade, for example, legacy production rule systems, rather than to attempt a wholesale replacement using a belief network approach. Obviously the specific details of the model implementation will need to be taken into account before a fair assessment can be made as to the pros and cons of enhancement versus replacement.

RELATIONSHIPS TO OTHER MODELS

This section describes how situation awareness models relate to other models comprising a human behavior representation. In particular, it examines how situation awareness models relate to (1) perceptual models and the way they drive downstream situation awareness, (2) decision-making models and the way situational beliefs can be used to support utility-based decision making, (3) learning models and the way situational learning can be modeled, and (4) planning and replanning models and their critical dependence on accurate situation assessment. The section concludes with a pointer to Chapter 9 for further discussion on how situation awareness models can be adapted to reflect the effects of internal and external behavior moderators.

Perceptual Models and Attention Effects

As noted earlier, the effective "driver" of a realistic situation awareness model should be a perceptual model that transforms sensory cues into perceptual variables. These perceptual variables can be associated with states of the *observer* (e.g., self-location or orientation with respect to the external world), with states of the *environment* (e.g., presence of vegetation in some region), or with states of other *entities* in the environment (e.g., threat presence/absence and location). The perceptual variables may be either discrete (e.g., presence/absence) or continuous (e.g., altitude), and they may be of variable quality, but the key point is to ensure that there is some sort of perceptual model that intervenes between the sensory cues available to the human and the subsequent situation assessment processing that integrates and abstracts those cues. Otherwise, there will be a tendency for the human behavior representation to be omniscient, so that downstream situation assessment processing is likely to be error free and nonrepresentative of actual human assessment behavior. By explicitly modeling perceptual limitations in this fashion, we will be more likely to capture inappropriate assessments of the situation that occur with a "correct" assessment strategy, but an "incorrect" perceptual knowledge base.

A variety of perceptual models could be used to drive a situation awareness model. One that has been used in several modeling efforts conducted by Zacharias and colleagues (1992, 1996) is based on the perceptual estimator submodel of the optimal control model of the human operator developed by Kleinman and colleagues (Kleinman et al., 1971). This is a quantitative perceptual model that transforms continuous sensory cues (e.g., an altimeter needle) into continuous perceived states (e.g., estimated altitude), using a Kalman filter to extract an optimal state estimate from the noisy sensory signal.[11]

Several ongoing modeling efforts could benefit from this estimation theory approach to perceptual modeling. For example, Reece and Wirthlin (1996) introduce the notion of "real" entities, "figments," and "ghosts" to describe concepts that are well understood in the estimation community and would be better handled using conventional estimation theory. Likewise, Tambe et al. (1995) introduce the notion of a "persistent internal state" in Soar; what they mean in estimation theory parlance is simply "state estimate." We suspect that the incorporation of this type of well-founded perceptual model could significantly enhance the realism of Soar's behavior in the face of a limited number of intermittent low-precision cues.

There are, however, at least two problems with this type of perceptual model. The first has to do with its focus on continuous, rather than discrete, states. The continuous-state estimates generated by this type of perceptual model are well suited to modeling human behavior in continuous-control tasks, but not well suited to generating the discrete events typically required by procedure-oriented tasks in which human decision makers engage. Early efforts to solve this problem (e.g., Baron et al., 1980) used simple production rules to effect a continuous-to-discrete transform. More recently, these continuous-to-discrete event transfor-

---

[11]There is considerable treatment of this type of perceptual model in the manual control literature (starting with Kleinman et al., 1971), with extensive validation against empirical data. In addition, several studies and modeling efforts have been directed at accounting for attention sharing across a set of available sensory cues (e.g., separate instruments on a display panel), using the OCM as the basis for postulating an optimal attention-sharing or visual-scanning strategy (Kleinman, 1976).

mation efforts have incorporated fuzzy logic (Zadeh, 1973, 1986, 1997), with its relative declarations of membership and event "trueness," to avoid the inherent brittleness of a production rule definition of events (Zacharias et al., 1996; Mulgund et al., 1997). Once such a transformation into a set of discrete-event declarations has been effected, it becomes straightforward to drive a situation awareness model based on any of the three enabling technologies described in the previous section.

The second problem associated with a perceptual model based on estimation theory is its focus on the informational aspects of the cue environment. Very little emphasis is placed on the underlying visual processes that mediate the transformation of a cue (say, an altimeter pointer) to the associated informational state (the altitude). In fact, these processes are almost trivialized by typically being accounted for with some simple transfer function (e.g., a geometric transform with additive noise).

To overcome this limitation, one must resort to more detailed process-oriented models. Most of the work in this area has been done in the visual modality, and an extensive literature exists. The following subsections briefly describe four visual perception models that offer potential for driving a situation awareness model and accounting for human perceptual capabilities and limitations in a more realistic fashion. All three are focused on visual target detection and recognition, and all have a common functionality: they take two-dimensional images as input and generate probability of detection (P_d) and probability of false alarm (P_{fa}) as output. Another commonality is the concept of separating the probability of fixation (P_{fix}) from the probability of detection (or false alarm) given a fixation ($P_{d|fix}$). P_d and P_{fa} are very relevant to modeling how combatants form a mental picture of what is going on around them because they predict which objects the combatants perceive and how those objects are labeled (target versus clutter).

Night Vision Laboratory Models

The search model developed at the Army Night Vision and Electronic Sensor Directorate (NVESD, formerly known as the Night Vision Laboratory or NVL) is often called the "Classic" model (Ratches, 1976), while the more recent model developed by the Institute for Defense Analysis for NVESD is referred to as the "Neoclassic" model (Nicoll and Hsu, 1995). Further extensions have been developed by others, such as for modeling individual combatants in urban environments (Reece, 1996). The Classic model provides a very simple estimate of $P_{d|fix}$: the longer one spends fixating a target, the more likely one is to detect it. P_{fix} is not modeled (fixation locations are ignored); instead, a single parameter is used to adjust the time-to-fixation manually to fit empirical estimates. The Neoclassic model adds a simple Markov process for modeling eye movements. The eye is assumed to be either in a *wandering state*, characterized by large random saccades, or in a *point-of-interest state*, with fixations clustered around target-like objects. A major drawback of both models is their dependency on empirical estimates, limiting predictive power to the particular situation in which the empirical estimate applies (e.g., a specific type of target, clutter, lighting). Conversely, in simple scenes with no clutter (such as air-to-air or ocean-surface combat), the NVL models can work very well.

Georgia Tech Vision Model

The Georgia Tech Vision model has been in development since the early 1990s at the Georgia Tech Research Institute and has grown to include many specific visual functions relevant to P_{fix} and P_{dlfix}. The most recent version has been integrated into a multispectral signature analysis tool called visual electro-optical (VISEO) for the Army Aviation Applied Technology Directorate (Doll, 1997). One of the early psychophysics experiments under the Georgia Tech Vision research project identified experience as the primary determinant of P_{fix} in visual search tasks for targets in cluttered scenes, even with presentation times as short as 200 milliseconds. While potentially of great importance, this finding needs to be verified independently through more psychophysics experiments, using an eye-tracker and a variety of scenes. The Georgia Tech Vision model also provides a highly detailed representation of the earliest physiological functions of the human visual system, such as dynamics of luminance adaptation due to receptor pigment bleaching, pupil dilation, and receptor thresholds. It is not clear whether this level of detail is appropriate for a model that is ultimately concerned with P_d and P_{fa}, especially given the associated computational complexity: a single input image generates up to 144 intermediate equal-size images. For this reason, it is not clear that the Georgia Tech Vision model is well suited for online simulations unless special image processors are used to host it. Benefits, on the other hand, include the potential to identify P_d and P_{fa} as a function of ambient light level and color and of target attributes such as color, motion, and flicker.

TARDEC Visual Model

The TARDEC Visual Model is a joint effort of the Army Tank-Automotive and Armaments Command Research, Development and Engineering Center and OptiMetrics, Inc. (Gerhart et al., 1995). Its most recent version is known as the National Automotive Center Visual Perception Model (NAC-VPM). The TARDEC Visual Model is a detailed model of P_{dlfix} combined with use of a simple (NVL Neoclassic-style) Markov process approach to set a time-to-fixation parameter. Hence, the model takes as input a subimage containing just the fixated target (manually segmented by the user). Front-end modules simulate temporal filtering and color opponent separation, enabling input with color and motion. Following modules decompose the image into multiple channels through multiresolution spatial filtering and orientation selective filtering. A signal-to-noise ratio is then computed for each channel, and the ratios are combined into a single measure of detectability. A strong point of the TARDEC Visual Model is that it has been calibrated to reproduce typical human observer results under a variety of conditions. Its main shortcoming is the lack of P_{fix} prediction, preventing the use of whole-scene images as input.

Decision-Making Models

Chapter 6 describes how the foundations of utility theory can be used to develop a number of deterministic and stochastic models of human decision making. At its simplest level, utility theory predicates that the decision maker has a model of the world whereby if one takes action A_i, under situation S_j, one can expect an outcome O_{ij} with an associated utility[12] of U_{ij}, or:

$$\text{action } A_i \text{ under situation } S_j \text{ yields outcome } O_{ij}, \text{ with utility } U_{ij}. \quad (7.4)$$

If the decision maker has a belief B_j that situation S_j is true, then the expected utility of action A_i across the ensemble of possible situations (S_j) is

$$EU_i = \frac{\Sigma B_j U_{ij}}{j} \quad (7.5)$$

Simple utility theory models the rational decision maker as choosing the "best" action A_i* to yield the maximal EU_i. As described in Chapter 6, a number of variants of this utility-based model have been developed to account for observed human decision making.

It should be recognized that there is a close connection between this model of decision making and the situation assessment process, since it is this process that provides the belief values (B_j) for use in the above expected utility calculation, which in turn is the basis for the selection of a best-choice decision option. Clearly, a misassessment of the situation in terms of either the set of possible situations (S_j) or their corresponding likelihoods (B_j) can easily bias the computation of the expected utilities, leading to the inappropriate selection of a non-optimal action A_i.

As discussed in Chapter 6, the human decision maker also often makes decisions consistent with Klein's (1994) RPD, in which action selection immediately follows situation recognition. The majority of human behavior representations reviewed for this study model this process using a simple production rule of the form:

$$\text{IF } (S_j \text{ is true}) \text{ THEN } (A_i \text{ is true}) \quad (7.6)$$

---

[12]The net benefit of an action, less the cost of effecting that action.

or simply "Select action A_i in situation S_j." As pointed out in Chapter 6, this is merely a special case of the more general utility-based decision model.

Whatever approach is taken to modeling the actual decision-making process of action selection, it seems clear that accurate modeling of the situation awareness process deserves close attention if human decisions in the kind of complex military environments under consideration are to be reliably accounted for.

Learning Models

With one exception, the models reviewed here do not incorporate learning. The one exception is Soar, with its chunking mechanism (see Chapter 5). However, as noted in Chapter 5, this capability is not enabled in any of the tactical applications of Soar reviewed for this study. Research in simpler domains has produced Soar models that learn situational information through experience. For example, one system learns new concepts in a laboratory concept-acquisition task (Miller and Laird, 1996), one learns the environmental situations under which actions should be taken in a videogame (Bauer and John, 1995), one deduces hidden situational information from behavioral evidence in a black-box problem-solving task (Johnson et al., 1994), and one learns episodic information in a programming task (Altmann and John, forthcoming). Thus, in principle, chunking can be used for situation learning and assessment, but it remains to be proven that this technique will scale up to a complex tactical environment.

As discussed in Chapter 5, exemplar-based or case-based reasoning models are particularly well suited to the modeling of learning because of their simplicity of implementation: new examples or cases are merely added to the case memory. Thus with increasing experience (and presumed learning), case-based reasoning models are more likely to have a case in memory that closely matches the index case, and recognition is more likely to occur in any given situation as experience is gained and more situational cases populate the case library.[13] (See the discussion of exemplar-based learning in Chapter 5 for more detail on learning within the context of a case-based reasoning model.)

Finally, recall that the earlier discussion of belief network models assumed a fixed belief network "mental model"—a fixed-node topology and fixed conditional probability tables characterizing the node links. Both of these can be learned if sufficient training cases are presented to the belief network. To learn the conditional probability tables for a fixed structure, one keeps track of the frequencies with which the values of a given variable and those of its parents occur across all samples observed up to the current time (Koller and Breese,

[13]However, it should be recognized that this ease of learning comes at the cost of increased retrieval time as experience grows (to be contrasted with observed human behavior), and a lack of abstraction, rule generation, or "chunking," depending on the flavor of alternative learning protocols one might consider.

1997). One can then compute sample conditional probabilities using standard maximum likelihood estimates and the available sample data. Infinite memory or finite sliding-window techniques can be used to emphasize long-term statistics or short-term trends, but the basic approach allows for dynamic conditional probability table tracking of changing node relations, as well as long-term learning with experience.

Planning Models

Planning is described more fully in Chapter 8. The discussion here focuses on how a situation awareness model would support a planning model during both initial planning and plan monitoring.

During the *initial planning stage* there is a need to specify three key categories of information:

- **Desired Goals**—In standard planning jargon, this is the desired goal state. In the U.S. Army, it is doctrinally specified as part of the METT-T (mission, enemy, troops, terrain, and time) planning effort (recall the vignette of Chapter 2). In this case, the desired goal state is specified by the mission and the time constraints for achieving the mission.
- **Current Situation**—In standard planning jargon, the current situation is the initial state. In the U.S. Army, it is doctrinally specified by an assessment of the enemy forces (the E of METT-T, e.g., strength, location, and likely course of action), an assessment of friendly troop resources (the first T of METT-T, e.g., their location and strength), and an assessment of the terrain (the second T of METT-T, e.g., trafficability and dependability). Clearly, all these elements are entirely dependent on a proper assessment of the current situation. A failure to model this assessment function adequately will thus result not only in failure to model adequately the initial planning efforts conducted by commander and staff, but also failure to set the proper initial state for the planning process proper.

Once the above categories of information have been specified, the actual planning process can proceed in a number of ways, as discussed in Chapter 8. Whichever approach is taken, however, additional situation awareness needs may be generated (e.g., "Are these resources that I hadn't considered before now available for this new plan?" or "For this plan to work I need more detailed information on the enemy's deployment, specifically, ..."). These new needs would in turn trigger additional situation assessment as a component activity of the initial planning process.[14]

During the *plan monitoring* stage, the objective is to evaluate how well the plan is being followed. To do this requires distinguishing among three situations:

---

[14]In practice, the U.S. Army formalizes this process through the generation of Priority Information Requirements (PIRs).

(1) the actual situation that reflects the physical reality of the battlefield; (2) the assessed situation that is maintained by the commander and his/her staff (and, to a lesser extent, by each player in the engagement); and (3) the planned situation, which is the situation expected at this particular point in the plan if the plan is being successfully followed.

Stressors/Moderators

The panel's review of existing perceptual and situation awareness models failed to uncover any that specifically attempt to provide a computational representation linking stressors/moderator variables to specific situation awareness model structures or parameters. We have also failed to identify specific techniques for modeling cognitive limitations through computational constructs in any given situation awareness model (e.g., limiting the number of rules that can be processed per unit time in a production rule model; or limiting the number of belief nodes that can reliably be "remembered" in a belief network model, etc.). Clearly, further work needs to be done in this direction if we are to realistically model human situation assessment under limited cognitive processing abilities. This lack, however, does not preclude the future development of models that incorporate means of representing cognitive processing limitations and/or individual differences, and indeed Chapter 9 proposes some approaches that might be taken to this end.

CONCLUSIONS AND GOALS

This chapter has reviewed a number of descriptive and prescriptive situation awareness models. Although few prescriptive models of situation awareness have been proposed or developed, the panel believes such models should be the focus of efforts to develop computational situation awareness submodels of larger-scope human behavior representations. As part of this effort, we reviewed three specific technologies that appear to have significant potential in situation awareness model development: expert systems, case-based reasoning, and belief networks. We noted in this review that early situation awareness models, and a surprising number of current DoD human behavior representations, rely on an expert systems approach (or, closely related, decision tables). In contrast, case-based reasoning has not been used in DoD human behavior representations; however, we believe there is a good match between case-based reasoning capabilities and situation awareness behavior, and this approach therefore deserves a closer look. Finally, recent situation awareness modeling work has begun to incorporate belief networks, and we believe that there is considerable modeling potential here for reasoning in dynamic and uncertain situations.

We concluded this review of situation awareness models by identifying the intimate connections between a situation awareness model and other models that are part of a human behavior representation. In particular, situation awareness models are closely related to: (1) perceptual models, since situation awareness is critically dependent on perceptions; (2) decision-making models, since awareness drives decisions; (3) learning models, since situation awareness models can incorporate learning; and (4) planning models, since awareness drives planning.

Short-Term Goals

- Include explicitly in human behavior representations a perceptual "front end" serving as the interface between the outside world and the internal processing of the human behavior representation. However trivial or complex the representation, the purpose is to make explicit (1) the information provided to the human behavior representation from the external world model, (2) the processing (if any) that goes on inside the perceptual model, and (3) the outputs generated by the perceptual model and provided to other portions of the human behavior representation. Incorporating this element will help make explicit many of the assumptions used in developing the model and will also support "plug-and-play" modularity throughout the simulation environment populated by the synthetic human behavior representations.
- In future perceptual modeling efforts undertaken for the development of military human behavior representations, include a review of the following:

—Informational modeling approaches in the existing human operator literature.

—Visual modeling approaches to object detection/recognition supported extensively by a number of DoD research laboratories.

—Auditory modeling approaches to object detection/recognition. Developers of perceptual models should also limit the modeling of low-level sensory processes, since their inclusion will clearly hamper any attempts at implementation of real-time human behavior representation.

- In the development of human behavior representations, make explicit the internal knowledge base subserving internal model decisions and external model behaviors. A straightforward way of accomplishing this is to incorporate an explicit representation of a situation assessment function, serving as the interface between the perceptual model and subsequent stages of processing that are represented. Again, the representation may be trivial or complex, but the purpose is to make explicit (1) the information base upon which situation assessments are to be made, (2) the processing (if any) that goes on inside the situation awareness model, and (3) the outputs generated by the situation awareness model and provided to other components of the human behavior representation. At a minimum, incorporating this element will specify the information base available to the decision-making and planning functions of the human behavior representation to support model development and empirical validation. This element will also serve as a placeholder to support the eventual development of limited-scope situation awareness models focused primarily on location awareness, as well as the development of more sophisticated models supporting awareness of higher-level tactical situations, anticipated evolution of the situation, and the like. These limited-scope models, in turn, will support key monitoring, decision-making, and planning functions by providing situational context. Finally, this explicit partitioning of the situation awareness function will support its modular inclusion in existing integrative models, as described in Chapter 3.

• Consider taking several approaches to developing situation awareness models, but be aware of the limitations of each. Consider an initial focus on approaches based on expert systems, case-based reasoning, and Bayesian belief networks, but also assess the potential of more generic approaches, such as abductive reasoning, and enabling architectures, such as blackboard systems.

• For situation awareness models based on expert systems, review techniques for knowledge elicitation for the development of rule-based systems. When developing the knowledge base, provide an audit trail showing how formal sources (e.g., doctrine expressed in field manuals) and informal sources (e.g., subject matter experts) contributed to the final rulebase. In addition, consider using an embedded expert system engine, rather than one developed from scratch. Finally, pay particular attention to how the expert system-based model will accommodate soft or uncertain facts, under the assumption that a realistic sensory/perceptual model will generate these types of data.

• For situation awareness models based on case-based reasoning, consider using an embeddable case-based reasoning engine. As with an approach based on expert systems, identify how uncertainty in the feature set will be addressed. Finally, consider a hybrid approach incorporating more abstract relations to represent abstract non-case-specific military doctrine, which must somehow be represented if realistic situation assessment behavior is to be achieved in human behavior representations.

• For situation awareness models based on belief networks, consider using an embeddable belief network engine. As with model development based on expert systems, maintain an audit trail of the way knowledge information sources contributed to the specification of belief network topology and parameter values. A potentially fruitful approach to belief network development at higher echelons may be the formal intelligence analysis process specified by Army doctrine. The top-down specification of intelligence collection planning and the bottom-up integration of intelligence reports suggest that belief networks could be put to good use in modeling the overall process, and perhaps in modeling the internal processing conducted by the human decision makers.

• Validate all developed situation awareness models against empirical data. Although not reviewed here, there are a number of empirical techniques for evaluating the extent and precision of a decision maker's situation awareness, and this literature should be reviewed before a model validation effort is planned.

Model validation efforts have only begun, and considerable work is still needed in this area.

Intermediate-Term Goals

- Explore means of incorporating learning into situation awareness models as a means of both assessing the impact of expertise on situation awareness performance and "teaching" the model through repeated simulation exposures. For models based on expert systems, explore chunking mechanisms such as that used in Soar. For models based on case-based reasoning, learning comes as a direct result of recording past cases, so no technology development is required. For models based on belief networks, explore new learning techniques now being proposed that are based on sampled statistics derived from observed situations.
- Explore means of incorporating the effects of stressors/moderators in situation awareness models. Conduct the behavioral research needed to identify across-model "mode switching" as a function of stressors, as well as within-model structural and/or parametric changes induced by stressors. Conduct concurrent model development and validation to implement stressor-induced changes in situation awareness behavior.
- Expand model validation efforts across larger-scale simulations.

Long-Term Goals

- Conduct the basic research needed to support the development of team-centered situation awareness models, looking particularly at distributed cognition, shared mental models, and common frames of reference. Conduct concurrent model development for representing team situation awareness processes.

8

Planning

Effective planning is often the key determinant of mission success in the military, and conversely, mission failure can often be traced back to poor planning. Planning's proper representation (both good and bad) within a military human behavior representation is therefore essential for the generation of realistic behaviors in the battlespace. This chapter reviews military planning in some detail, describing both doctrinal and nondoctrinal behaviors and their implications for model development. It then examines approaches taken to modeling planning—first in military planning models, and then in the work of the artificial intelligence and behavioral science communities. The chapter concludes with conclusions and goals in the area of planning.

PLANNING AND ITS ROLE IN TACTICAL DECISION MAKING

In an attempt to identify key implications for modeling the tactical planning process, this section describes the planning process itself in the context of the vignette presented in Chapter 2. Doing so allows us to identify essential attributes of this type of *doctrinally correct* planning behavior. This discussion is then balanced by an examination of *observed* planning behavior, which differs significantly from that prescribed by doctrine. The section concludes with a summary of the modeling implications of both doctrinally correct and observed planning behavior as context for the ensuing discussion of existing models and potential development approaches.

The Tactical Planning Process

A plan can be defined as "any detailed method, formulated beforehand, for doing or making something" (Guralnik, 1986). More specifically, an *activity* plan (for doing something) usually begins with a system at some initial state, specifies some desired final or goal state, and identifies constraints on the allowable sequence of actions that will take the system from the initial to final state. A valid plan is one that specifies an allowable action sequence (formulated beforehand). If it does no more than this, it is sufficing; if it optimizes some utility function of the actions and the states, it is optimizing.

Planning, or the generation of a plan, is critical to successful operations—it plays a key role in the tactical decision-making process across all services and throughout all echelons. Capturing the substance of the planning process in a realistic human behavior representation is therefore essential to developing behavioral models that realistically reflect actual tactical decision-making behavior.

Because of the limited scope of the present study, the focus of this chapter is limited to the U.S. Army planning process. Based on brief reviews presented to the panel by the other services, we believe the basic activities comprising this process are similar across the services, although they have different names, so the findings presented here should be generalizable. We believe similar generalization holds across echelons, although we will have more to say about this below.

The vignette in Chapter 2 describing the planning and execution of a hasty defense operation by a tank platoon incorporates a planning process that closely follows the doctrinally specified process detailed in Army publication FM 101-5, *Staff Organization and Operation*, and described more tutorially in Command and General Staff College publication ST 100-9, *The Tactical Decisionmaking Process*. As described in the latter publication, planning is part of a five-stage process:

1. Mission analysis
2. Intelligence preparation of the battlefield
3. Development of courses of action
4. Analysis of courses of action
5. Decision and execution

The paragraphs below provide a brief description of each of these stages. More complete descriptions can be found in FM 101-5 and ST 100-9.

The *mission analysis* stage begins with receipt of an order from the unit's command and proceeds to more complete definition of the initial state (or, equivalently, current situation), as well as a definition of the final goal state (or, equivalently, the mission objectives). Consideration is also given to operational constraints that will apply during the course of the operation. In the platoon-level vignette of Chapter 2, this process is formalized by expanding the mission, enemy, terrain, troops, time available (METT-T) process, which makes explicit the

TABLE 8.1 Relating METT-T Components to Initial and Goal States of a Plan

METT-T	Initial State	Goal State
Mission		X
Enemy	X	
Terrain	X	
Troops	X	
Time		X

consideration of initial and final state planning components (see Table 8.1). This stage thus includes situation assessment activities (initial state specification) and planning activities (goal state specification).

The *intelligence preparation of the battlefield* stage then focuses on a detailed assessment of the situation, covering three key components (FM 101-5):

- Environment in the area of operations (e.g., terrain, weather)
- Enemy situation (e.g., disposition, composition, strength)
- Friendly situation

In our platoon-level vignette, the environmental assessment is formalized by expanding the observation, cover and concealment, obstacles, key terrain, and avenues of approach (OCOKA) process, with consideration also given to the weather. Assessment of the enemy and friendly situations attempts to go beyond simple disposition and strength estimates (recall Endsley's [1995] level 2 situation awareness as described in Chapter 7) and to focus on expectations of operational behavior (level 3 situation awareness). As noted in ST 100-9, the OCOKA process may be adequate for the brigade level and below, but a more sophisticated preparation-of-the-battlefield process is typically undertaken at higher echelons based on current tactical intelligence.

The *course-of-action development* stage is the generative stage of the planning process,[1] in which alternative sufficing plans are generated to accomplish the mission and take the unit from its current to its goal state. Each course of action is a candidate plan developed at a relatively high level, addressing such things as the type of operation and when it will happen. In our vignette, the results of this stage, which are typical, are three candidate courses of action; Army doctrine calls for the generation of "several suitable" courses of action for every enemy course of action under consideration.

The *course-of-action analysis* stage is the evaluative stage of the planning process,[2] in which candidate courses of action are elaborated, wargamed against likely enemy courses of action (through mental simulations), and

[1]Designated the "art" of command by ST 100-9.

[2]Designated the "science" of control by ST 100-9.

evaluated across multiple dimensions. Courses of action are then scripted out in more detail, showing (through a synchronization matrix) the anticipated temporal sequence of subordinate, support, and enemy unit activities. The wargaming itself is fairly shallow in terms of blue moves and red countermoves, covering a three-step sequence of action, reaction, and counteraction. Evaluation criteria for each course of action are specified by each commander, and each course of action is scored on the criteria according to the outcome of the wargame.

The *course-of-action selection* stage is the decision and execution stage, in which the commander selects the highest-rated course of action, refines it to ensure clarity (commander's intent), and generates the plans and orders for unit execution.

Key Attributes of Doctrinally Correct Tactical Planning

Even from the above brief overview of tactical planning, it is possible to identify several key attributes of the overall doctrinally prescribed process:

- Planning is doctrinalized and knowledge intensive.
- Echelon affects the planning focus and process.
- Planning strongly reflects the resource context.
- Planning strongly reflects the task context.

Planning is Doctrinalized and Knowledge Intensive

Because of the complexity of the domain, the risk/reward ratios of the outcome, the limited resources available for finite-time plan generation, and, perhaps most important, individual differences in planning and decision-making abilities, the planning process has been highly doctrinalized.[3] As noted, step-by-step procedures are designed to take the planner from the current situation to the mission end state. This is true for all echelons, with the complexity of the planning process increasing with echelon (and time available). The planning process also relies on a detailed and extensive knowledge base of the domain,[4] covering the environment, enemy capabilities, and friendly capabilities. At mid-level and higher echelons, the planning process is so knowledge intensive that its accomplishment demands a staff of specialists. It is perhaps as far from reasoning by first principles as one can imagine.

---

[3]This last point is brought out by Falleson (1993:9), who notes that the Prussian Army formalized the tactical decision-making process in the early 1800s so as "not to leave success to the rare chance of tactical genius."

[4]Provided in the U.S. Army by an extensive collection of field manuals, augmented by a broad array of training material and formal courses.

Echelon Affects the Planning Focus and Process

Although this review does not include instances of planning at different echelons, it is clear that the focus of the planning differs with echelon. At the lower echelons, much of the emphasis is on path planning and movement to an objective. For example, an individual may spend time planning a route that maximizes cover under military operations in urban terrain (MOUT), or, as in our vignette, a platoon commander will spend time planning a best route to an objective that his unit is to defend. At the middle echelons, say the company to brigade levels, route planning is clearly important, but greater emphasis is placed on coordinating the subordinate and collateral units in the operation, so that issues of synchronization, timing, and communication become more important. At still higher echelons, the division level and above, the same also holds, but now logistics planning takes on greater importance because of the longer time spans involved. It should be emphasized that we are not suggesting that lower-echelon planning excludes logistics concerns (since clearly the individual combatant is concerned with ammunition reserves) or that higher echelons are unconcerned with movement along possible avenues of approach. Rather, we are saying that the *focus* changes with echelon, longer-term coordination and logistics issues taking on greater weight for the higher echelons.

Planning at different echelons is also characterized by different time windows afforded the operation and a corresponding change in the planning process itself. At lower levels, the operation and the planning for it may take from minutes to hours, as illustrated by our vignette. At higher levels, the operation may take days, and planning for it may take weeks to months. Because fewer resources are available at the lower echelons, this time compression serves to change not only the focus but also the nature of the planning process. Planning at the lower echelons becomes more reflexive and more driven by past experience in similar situations. Planning at the higher echelons is more contemplative and is supported by extensive rational analyses of the options.[5] This suggests that different planning techniques and processes may be employed by the tactical decision maker at each echelon.

Planning at different echelons also affects how current situations (initial state), missions (final goal state), and actions (plan activities) are viewed by the planner. Planning and execution occur within a heavily hierarchical organization, so that one unit's planned tasks or activities become the plan goals or missions of that unit's subordinates. This hierarchical goal/task decomposition starts at the highest operational level, say the corps level, and proceeds down the organization to the lowest individual combatant. The result, in theory, is a com-

[5]Although the next section describes research that clearly disputes this viewpoint, the focus here is on "doctrinally correct" planning.

mon planning process at each echelon, but one whose objects (unit situation, unit mission, and subordinate tasks/missions) change to reflect a unit's location in the hierarchy. In the artificial intelligence planning literature, this process is viewed as abstraction planning (Sacerdoti, 1974).

Planning Strongly Reflects the Resource Context

Planning is not conducted in an abstract world of perfect and complete information (as in, for example, a chess game), nor is it conducted as an open-ended exercise, free of time constraints. Effective planning relies on a proper interpretation of the mission plan goal specified by headquarters. As discussed at a recent Workshop on Commander's Intent (Anonymous, 1997), the goal may not be clear because of a failure to properly specify the intent, a failure in the communication channels, or a failure to properly interpret the intent. Effective planning also relies on an accurate assessment of the current situation. As discussed in Chapter 7, this accurate assessment depends, in turn, on an accurate perception of key tactical events or cues, which may be circumscribed and noisy and thus lead to erroneous planning decisions. Finally, the planning process itself is time constrained, so that the planning effort must be scaled down to the time available. This meta-planning skill of time management is illustrated in our vignette: one of the first of the platoon leader's activities after receiving the mission is to define a timeline for all activities leading up to occupation of the battle position, including planning, logistics preparation, and rehearsal. In this case, plan preparation takes somewhat less than 2 hours. In more dynamic situations, there may be considerably less time available, so that an abbreviated planning process occurs, with less extensive consideration of alternative courses of action, less wargaming, and so forth. ST100-9 provides guidelines on how to speed up the decision-making process under time stress.

Planning Strongly Reflects the Task Context

Planning is only one activity that occurs in the overall military operational context as part of the following sequence of activities:

1. Reception of orders from the command unit, defining the plan goal and plan constraints (ST 100-9: mission analysis).
2. Specification of the environment, the enemy situation, and the friendly situation (ST 100-9: intelligence preparation of the battlefield generation) and selection of a high-level plan for the unit and subordinate units (ST 100-9: course-of-action development, analysis, and decision).
3. Communication of operational plans and operational orders to subordinate units for recursive plan elaboration (ST 100-9: decision and execution).
4. Execution of plan (operations).

5. Continued situation assessment to support monitoring of plan compliance through comparison of the assessed and the planned situation.

6. Replanning whenever noncompliance exceeds a decision threshold or when an opportunity exists for improving on the planned outcome (opportunistic planning).

Note that only the first four steps are covered in detail in FM 101-5 (U.S. Army, 1997). The latter three steps, which actually occur in parallel and in the heat of battle, are clearly critical to the overall success of the operation. Being able to determine when a plan is failing and replan accordingly under adverse conditions and time stress would appear to be key to successful tactical decision making. From our review, however, it is not clear that this class of plan monitoring/modification activities is as well covered by doctrine as are the preoperations planning activities.

Key Attributes of Tactical Planning in Practice

To this point, the discussion has focused on tactical planning as prescribed by U.S. Army doctrine. If the ultimate objective of future efforts in human behavior representation is to model doctrinally correct planning behavior, this should be an adequate starting point for a requirements definition effort. If, however, future efforts in human behavior representation are to be aimed at modeling *actual* tactical planning, a serious knowledge engineering effort is needed to build a behavioral database describing planning during combat decision making. This conclusion follows from the observations of many researchers that actual planning behavior observed and reported differs markedly from that prescribed by doctrine.

Fallesen (1993) conducted an extensive review of the relevant literature in battlefield assessment and combat decision making over a 5-year span, covering hundreds of studies, reports, and research summaries. The interested reader is referred to this review for an in-depth discussion of the disconnect between doctrine and practice. The discussion here highlights some of the key points relevant to the planning process.

Management of the Planning Process

Overall management of the planning process and staff is often poor. There is inadequate involvement by the commander; Fallesen (1993:15) notes that this may be because "current doctrine establishes an unclear role for the commander." There is inadequate coordination across the staff specialties, with engineers and fire support officers (FSOs) often being left out. Finally, there is inadequate meta-planning over the scheduling of the planning activities and the allocation of staff resources. The net result is that the process defined by doctrine "*typically* is not followed closely" [emphasis added] (Fallesen, 1993:17).

Information Exchange for Planning

The underlying information needed for good decisions is often not sought or not specifically related to the decision process. There is inadequate exchange of critical information across staff members, planners fail to seek actively the information they need (Thordsen et al., 1989), information generators fail to present interpretations or implications of the briefed information, and finalized plans are often not presented to the commander for final review (Metlay et al., 1985).

Situation Assessment to Support Planning Assumptions

Situation assessment for planning suffers from a number of shortcomings, including (Fallesen, 1993) failure to consider an adequate number of facts for accurate situation awareness, failure to verify assumptions made for planning purposes, failure to weight information according to its quality, failure to interpret information (Endsley's [1995] Level 2 situation awareness), and failure to make predictions (level 3 situation awareness). Although these shortcomings are discussed in detail in Chapter 7, they are noted here because of the critical importance of correct situation awareness for planning.

Course-of-Action Development

The creative heart of the planning process, course-of-action development, deviates along three key dimensions from what is doctrinally specified (Fallesen, 1993). First, the management and tracking of course-of-action alternatives, another meta-planning task, is poorly done, with no audit trails and inadequate documentation of assumptions and concepts. Second, course-of-action generation does not appear to follow the doctrinally prescribed model:

> The tactical decision-making model of the [doctrinally prescribed] process indicates that multiple options should be generated, and that options should be distinct from one another. Findings have shown that multiple options are often not generated, and that the options are not always unique. When three courses of action are produced, they are sometimes called the 'best', the 'look-alike', and the 'throw-away.' (Fallesen, 1993:24).

Fallesen goes on to say that this deviation from doctrine is not necessarily bad. He points out that Klein (1989, 1994) and Thordsen et al. (1989) term this behavior recognition-primed decision-making (see Chapter 7) and believe it to be the normal operating mode of experienced decision makers:

> [Thordsen et al. (1989)] concluded that multiple options were not considered as a matter of course nor were [the planners] compelled to conform to the traditional decision analytic model. Planners considered alternatives out of necessity if the first alternative proved infeasible (Fallesen, 1993:24).

This view clearly suggests a sufficing rather than an optimizing mode of plan generation. Finally, a third deviation from doctrine concerns the failure of planners to generate sufficiently detailed courses of action, including contingency plans for dealing with deviations from the plan.

Course-of-Action Analysis and Selection

The actual analysis and selection of courses of action also deviates significantly from doctrine. Doctrine prescribes extensive use of wargaming, concurrent evaluation of the different courses of action, and avoidance of premature decisions. What is observed, however, is a different matter (Fallesen, 1993). First, wargaming is often simply not conducted, or is done superficially or using noncomparable wargaming techniques (Fallesen and Michel, 1991). Second, course-of-action evaluation tends to be serial; that is, it is conducted depth first, rather than the doctrinally prescribed breadth first:

> We found planners tended to employ a process where they would evaluate an option or idea by gradually examining deeper and deeper branches of the idea for workability. . . . If [an idea] is rejected the decision maker either moves on to a totally different option or idea or goes back up the deepening chain to a point (theoretically) above the source of the flaw and then follows another branch. (Thordsen et al., 1991:2)

In addition, the method for evaluating courses of action does not follow the recommended multiattribute decision matrix approach (Thordsen, 1989). Fallesen et al. (1992) explain:

> Using a decision analytic approach, as complicated as a weighted, multiattribute utility matrix or as simple as a summary column of pluses and minuses, can be misleading for complex, dynamic tactical problems (Fallesen et al., 1992:87).

Finally, doctrine cautions against reaching early decisions about the desirability of one course of action over another. In practice, however, the course of action first generated is often selected (Geva, 1988), and it has been found that early decisions do not degrade performance in selection of the proper course of action (Fallesen et al., 1992):

> Making early decisions about courses of action are contrary to a formal, analytical process, but making decisions early has been shown not to impact solutions adversely in one study and to promote better solutions in another (Fallesen, 1993:30).

As Lussier and Litavec (1992:17) point out, the time saved by making an early decision can be put to good use in detailed plan refinement, contingency planning, and rehearsal:

> Other commanders said they must just bite the bullet and decide quickly. They emphasized that the important thing is how well planned and executed the mission is, not which course of action is chosen.

Plan Monitoring and Replanning

Once the operation is under way, ongoing battlefield monitoring is key to determining how closely the actual situation is following the plan. This situation awareness function, unfortunately, is not well carried out:

> The 1992 analysis of NTC [National Training Center], JRTC [Joint Readiness Training Center], and CMTC [Combat Mamzuren Training Center] trends reported that 59 percent of brigade, battalion task forces, and company/teams did not track the battle quickly and accurately. The failure to do so created conditions for fratricide and being unaware of available combat power. (Fallesen, 1993:35)

This failure to track the battle also clearly created conditions for an inability to see when a plan was failing or, for that matter, succeeding beyond the staff's expectations. If adequate plan monitoring can be achieved, then plan failures can be dealt with quickly, provided that contingency plans have anticipated the failure involved. Otherwise, a replanning exercise must take place concurrently with operations. However, as Lussier and Litavec (1992:16) note:

> Commanders distinguish two situations: limited time situations, with only a few hours of planning time available, and execution situations, where mission planning is occurring at the same time as execution. In the latter case, the changing tactical environment makes the doctrinal decision making process even less applicable. Most commanders believe that they are not given much doctrinal help in doing that truncation; each must develop his own techniques and planning processes.

Clearly, one can expect the real-time replanning effort to deviate significantly from the doctrinally specified preoperations planning process.

Implications for Modeling the Tactical Planning Process

The above discussion has a number of key implications for modeling the tactical planning process, related to the difference between theory and practice; the need for domain knowledge in model development; the need to model the various stages of planning; and the effects of echelon, resources, and context.

Theory vs. Practice

The preceding discussion makes a clear distinction between planning activities that are prescribed by doctrine and those observed in practice. Table 8.2 summarizes the implications in a matrix of procedural options (doctrinal, nondoctrinal) and execution[6] options (perfect, imperfect) for the modeling of planning. The option selection by the modeler will clearly depend on the purpose for which the human behavior representation is to be applied. For example, in the

[6]By execution, we mean execution of the planning activity, not execution of the plan.

TABLE 8.2 Options for Modeling Planning

	Execution Options	
Procedural Options	Perfect	Imperfect
Doctrinal	Doctrinally correct with error-free execution (blue subordinate HBR and/or red adversary HBR used for training novice commanders)	Doctrinally correct with error-prone execution (novice blue subordinate HBR used to train expert blue commander)
Nondoctrinal	Nondoctrinal with error-free execution (innovative red adversary HBR used to train expert blue commanders)	Nondoctrinal with error-prone execution (realistic blue subordinate HBR and/or red adversary HBR used for mission rehearsal)

NOTE: HBR, human behavior representations.

training area, one might take the upper left option (doctrinal, perfect) to represent blue subordinates during initial training of blue commanders; a transition to the upper right option (doctrinal, imperfect) might be made for advanced training in dealing with less competent subordinates, and a transition to the lower right option (nondoctrinal, imperfect) for commander evaluation under realistic conditions. Alternatively, the upper left option might be used to represent red adversaries during early blue force training, with a transition to the lower left option (nondoctrinal, perfect) for advanced training in dealing with innovative adversaries. Clearly, other options abound, especially if the scope is broadened beyond training to include systems evaluation, development of tactics, and the like.

Need for Domain Knowledge

Whether a doctrinal or nondoctrinal modeling approach to planning is taken, it is clear that any model will rely on a detailed and extensive knowledge base of the service-specific domain, covering the environment, friendly and enemy unit capabilities, operational constraints, general doctrine, and the like. It is evident that any computational planner must be capable of dealing with a highly complex domain, far removed from first-principle planners typifying early artificial intelligence approaches to planning.

Planning Stages

If a planning model is to generate planning behaviors that somehow mimic those of a human planner, the modeler must attempt to replicate the various stages of planning, especially if the goal is to achieve some measure of doctrinal fidelity. Specifically, a planning model[7] should attempt to represent the planning stages discussed in the previous section:

---

[7]That is, a model suited for the Army.

1. *Information exchange/filtering/retrieval*, a front-end active perception process, responsible for generating the scenario-specific knowledge base for assessment and decision making.

2. *Situation assessment*, another front-end process responsible for generating informational abstractions and situational context to support the planning process.

3. *Course-of-action development*, the process responsible for generating candidate plans. A decision-theoretic approach might be taken for doctrinally correct planning, whereas a recognition-primed decision making approach, relying on episodic memory, might be better suited to the modeling of actual planning behaviors. Whichever approach is taken, attention needs to be paid to whether an optimizing or a satisficing approach is used.

4. *Course-of-action analysis and selection*, the process responsible for selecting the best among candidate plans. A decision-theoretic approach would provide a natural environment for incorporating multistage wargaming and evaluation activities. A recognition-primed decision making approach would bypass this stage.

5. *Monitoring and replanning*, the processes responsible for assessing the situation and any deviations from the plan, and then developing or calling up new plans to compensate for those deviations.

Echelon Effects

The strong dependence of plan focus and process on echelon was noted earlier. Table 8.3 provides a qualitative overview of this echelon effect. It seems clear that any realistic planning model must account for this echelon dependence through the development of either separate echelon-specific models (as is the current practice) or a more generic planner that reflects echelon-dependent changes in goals, scope, staff resources, time planning windows, and the like. The latter approach may be feasible because of the heavily hierarchical command and control structures employed by the services, which lead to a natural hierarchical goal decomposition by echelon.

Resource Effects

Planning's strong dependence on the resources available, including information, time, and staff, was also noted earlier. Plans are based on incomplete, uncertain, and outdated information, and planning models need to reflect this fact. The best way of ensuring that they do so is to create a planning model that has no access to "outside" information, but instead relies solely on perceived states and assessed situations, which, if modeled correctly, will realistically intro-

TABLE 8.3 Qualitative Trends in Planning Due to Echelon

Echelon	Focus	Process
High	Logistics, overall operation	Contemplative
Mid	Coordination, communication	(Not available)
Low	Route planning, cover	Reflexive

duce uncertainty and untimeliness.[8] Plan generation is also time and resource (staff) constrained, and any realistic planning model must reflect this fact as well. One approach is to incorporate meta-planning capabilities in the model to budget time and allocate staff (subplanners) for the planning process, change planning strategies as a function of time/staff available,[9] monitor plan generation progress, and the like. An added benefit of this approach is that the resulting planning model can begin to reflect the actual meta-planning activities of the human tactical decision maker, clearly adding to the face validity of the model.

Contextual Effects

Finally, it has also been noted that planning is only one activity that occurs in the overall operational context. It is not an isolated process, but depends highly on—and often competes for resources with—many other activities (e.g., information gathering, situation assessment). In modeling planning, it is therefore necessary to represent in-context planning, interruptions caused by higher-priority activities, attention allocation across concurrent tasks, and a variety of other issues that arise in multitask environments.

MODELS FOR PLANNING IN MILITARY HUMAN BEHAVIOR REPRESENTATIONS

The following three subsections (1) review planning models in existing military human behavior representations; (2) examine military decision-aiding systems that contain a planning component, since they may have potential use in the development of computational planning modules for future military human behavior representations; and (3) summarize the key findings in both areas. Table

[8]If some meta-knowledge of this uncertainty/untimeliness is held by the planner, then it seems clear that a similar meta-knowledge base should be maintained by a model. This meta-knowledge base would support modifications in plan goals, planning strategies, and the like as a function of the uncertainty of the planning knowledge base (i.e., high uncertainty should lead the planner to choose from robust, but possibly nonoptimal plans; low uncertainty should lead to the choice of more optimal plans, which may, however, be more likely to fail when the planning assumptions are violated).

[9]For example, using shortcuts under time stress.

8.4 provides an overview of the human behavior representations, decision aids, and simulations reviewed in this section.

Planning Models in Military Human Behavior Representations

Adaptive Combat Model

The Marine Corps Modeling and Simulation Management Office has a program in Adaptive Combat Modeling, aimed at modeling individual and unit-level behaviors in three separate phases (Fisk, 1997):

- Phase 1: Route Planning
- Phase 2: Control of Movement
- Phase 3: Terrain Exploitation

The objective is to develop adaptive autonomous synthetic agents for use in tutoring systems and command trainers, and possibly as a knowledge-base source in advanced decision aids for fire support. For Phase 1, route planning in the face of multiple constraints (e.g., terrain obstacles, time, fuel) is being modeled through the use of genetic algorithms that serve to generate, evaluate, and evolve quasi-optimal route solutions satisfying the overall route constraints (risk, personal communication). To date, no effort has been made to match the agent's performance with that of a human route planner, although the possibility exists—if there is more of a focus on human behavior model development, as opposed to tactical decision-aid development.

Commander's Visual Reasoning Tool

A high-level descriptive model of the staff course-of-action planning process is presented by Barnes and Knapp (1997). Although not properly a model, it provides an overview (through a command/staff dependency matrix) of the course-of-action planning activities undertaken by the brigade commander and his intelligence officer (S2), operations officer (S3), and fire support officer (FSO). It thereby provides a framework for ensuring that all individual commander/staff functions are represented in any subsequent command and control (C^2) human behavior representation development effort. It also identifies the key points of interaction among team members (e.g., "provide commander's guidance to staff," "coordinate with division G2"), thus identifying communication channels and information flow within and outside of the commander/staff team.

Barnes and Knapp (1997) also describe an experimental tool—the commander's visual reasoning tool (CoVRT)—for prototyping future brigade C^2 environments by means of a three-workstation network supporting the battle management functions of the commander, S2, and S3. Such a research environment could be used for future model validation efforts in developing scenarios and recording the detailed course-of-action planning process conducted by individual staff members. These records could then serve as a real-world database for evaluating the realism of comparable model-generated behaviors.

Dismounted Infantry Computer-Generated Force

The Hughes Research Laboratory has developed a computer-generated force (CGF) representation for the Marine Corps (see also Chapter 2), intended to model individual fire-team members, fire-team leaders, and squad leader to support the training of their respective superiors (Hoff, 1996). Individual behavior is simulated by a production rule system that generates a plan comprising a number of "task frames" that are eventually implemented by the modular semiautomated forces (ModSAF) dismounted infantry "action" module (see Chapter 3) operating within the distributed interactive simulation (DIS) environment.

The CGF plan generation module consists of 22 separate modules or rulesets, each containing 10 to 100 production rules, which encode a subject matter expert's ranking of decision alternatives for a number of key decisions, such as attack or abort, avoid contact, select assault point, and select route. For example, a select route decision rule might be:

IF (route A distance < 50 m)
AND (route A terrain is slow-go)
AND (route B distance is 100 m)
AND (route B terrain is go)
THEN (select route B)

The distance measures are left "fuzzy" to allow for variations.

Ruleset antecedents are essentially features of the tactical situation (e.g., "enemy posture is dug-in"), which are obtained directly by querying the ModSAF module(s) responsible for maintaining the state of the simulation. Thus, no attempt is made to model the individual's cognitive functions of information gathering, perception, correlation, or situation assessment. Once the necessary state information has been gathered from the ModSAF modules, one frame of a plan is generated and sent to ModSAF for execution (e.g., "attack the objective in a wedge formation"). It is unclear how often the plan can be updated to account for unplanned circumstances, but the state-driven production rule approach appears to allow for almost continuous updating (e.g., "situated cognition").

Although components of the tactical planning function are segregated by ruleset, it is not clear whether there is any attempt at hierarchical planning (e.g., decide to attack first and *then* decide the route) and/or modeling of the interaction of interdependent or conflicting goals (e.g., attack quickly, but minimize casualties). Finally, it is unclear whether this is much more than a one-step-ahead planner, and thus effectively a purely reflexive agent, or if it can generate a phased sequence of frames defining the overall evolution of the plan from start to finish.

TABLE 8.4 Models for Planning in Military Human Behavior Representations (HBR)

Name	Sponsor	Service (Domain)	Level
Adaptive Combat Model	Marine Corps Modeling and Simulation Office	Infantry/Marines (HBR models)	Individual and unit
Commander's Visual Reasoning Tool (CoVRT)	Army Research Laboratory (ARL)	Army (decision support)	Brigade
Marine Computer-Generated Force (CGF)	Defense Advanced Research Projects Agency (DARPA), Hughes Research	Marine (HBR models)	Individual and unit
Computer-Controlled Hostiles for Team Target Engagement Simulator (TTES)	Institute for Simulation and Training/ University of Central Florida (IST/UCE)	Marine (HBR models)	Individual
Fixed-wing Attack (FWA)-Soar and Soar-Intelligent Force Agents (IFOR)	DARPA, Air Force, University of Michigan (UMich), University of California/Los Angeles (UCLA), Carnegie Mellon University	Air Force (HBR models)	Individual and unit
Rotary-Wing Attack (RWA)-Soar	DARPA, Army, UMich, UCLA	Army Aviation (HBR models)	Individual and unit (company)
Man-Machine Integration Design and Analysis System (MIDAS)	National Aeronautics and Space Administration, Army	Army Aviation (HBR models)	Individual and unit

Functions/ Activities Modeled	Implementation/ Architecture	Comments
Route planning	• Genetic algorithm optimization	• Currently an exploratory study
Course-of-action generation	• Decision-aiding workstation with graphic depiction of military entities	• Underlying command/staff dependency matrix provides overall framework for information flow among HBR agents
Route planning	• Production rule system driven by external state variables	• Multiple rulesets for different decision alternatives • Generation of single plan frame for ModSAF execution
Short-term planning for individual military operations in urban terrain activities	• Hierarchical goal decomposition • Decision-theoretic goal selection • Situation-driven rules	• Appears to be single-step planner, but could be expanded to multiple steps for individual course-of-action generation
Full activities of tactical pilots across a wide range of aircraft and missions	• Soar architecture with hierarchical goal decomposition • Efficient production rule system to deal with large rulebase • Situation-driven rules	• Planning not explicitly represented, as Soar supports only single-step planning • Could be expanded to support an explicit planning module
Full activities of rotorcraft pilots and company commander for RWA mission	• Live battalion commander • Soar-CFOR company commander that: —Generates mission plan —Monitors progress —Replans as necessary • Soar-IFOR RWA pilots • ModSAF vehicle entities	• Plan generation and elaboration done through tactical templates and standard operating procedures • Plan refinement done through checks on task interdependencies, timing
Full activities of tactical rotorcraft pilots	• Symbolic operator model architecture with hierarchical mission activity decomposition • Production rule system to implement procedures • Situation-driven productions	• Similar to Soar in its top-down decomposition from mission phases to low-level activities • Single-step planning, but could be expanded to support an explicit planning module

continued

TABLE 8.4 Models for Planning in Military Human Behavior Representations (HBR) (continued)

Name	Sponsor	Service (Domain)	Level
Naval Simulation System (NSS)	Navy	Navy (HBR models)	Command and control (C^2) and unit
Automated Mission Planner (AMP)	IST/UCF, University of Florida	Army (HBR models)	Unit, company
	International Advisory Group-EUropean Cooperation for the Long Term In Defence Consortium (ISAG EUCLID)	Army and Navy (decision support)	Unit
Battlefield Reasoning System (BRS)	ARL Federated Laboratory, University of Illinois at Urbana-Champaign	Army (decision support)	C^2
Decision Support Display (DSD)	ARL Federated Laboratory, North Carolina Agricultural and Technical State University	Army (decision support)	C^2 Unit

Computer-Controlled Hostiles for SUTT

The small unit tactical trainer (SUTT)[10] is used to train Marine rifle squads in MOUT clearing (Reece, 1996; see also Chapter 2). The virtual world is populated by computer-controlled hostiles, which have been developed to behave as "smart" adversaries that evade and counterattack blue forces. The key state variables are fairly low level (e.g., soldier position, heading, posture), and perception and situation awareness are based on simple detection and identification of enemy forces.

Most of the activity appears to be fairly reflexive, but there is an attempt at situation-driven planning using hierarchical goal decomposition. High-level goals

[10] Formerly known as the team target engagement simulator (TTES).

Functions/ Activities Modeled	Implementation/ Architecture	Comments
Full activities of a Navy task force at entity level	• Decision tables or prioritized production rules • Future state predictor that supports some planning	• Future state predictor could be used to support reactive
Course-of-action generation	Four stage process: 1. Terrain analysis for route planning 2. Course-of-action generation 3. Course-of-action simulation 4. Course-of-action selection	• For use in ModSAF company commander entities • Development status unknown • Key use of simulation-based plan evaluation
Course-of-action generation, maneuver and FS planning. AGW planning	• Multiagent architecture	• Broad Europe-wide effort in decision-support tools • Planning modules may be useful in HBR modeling
Course-of-action generation	• Blackboard architecture with multiple specialist agents • Current focus on course-of-action generation	• ARL-sponsored effort in decision-support tools • Course-of-action planning modules may be useful in HBR models
Course-of-action generation and logistics planning	• Multiple distributed rulesets	• Currently an exploratory study • Rulesets may be useful for future model development

are decomposed into subgoals, which can be either concurrent or alternative. For example, a concurrent goal set for attack and move is to attack *and* move simultaneously; an alternative goal set, say, for the attack subgoal, is to perform a quick fire *or* aimed fire. For each high-level goal, a full tree of candidate goals and subgoals is constructed, down to the leaf action nodes, prior to any action generation. Deconflicting of conflicting concurrent activities and selection of nonconflicting alternative activities are performed by calculating the effective utility of each path in the goal tree in a fashion similar to the usual decision-theoretic approach based on priority values assigned to each goal/action node (see Chapter 6). The action associated with the maximum utility goal path is then selected, and this action specifies the entity's behavior in the next time frame.

This type of hierarchical goal decomposition approach to action selection could subserve a tactical planning function, but in the current implementation it appears that only a single action is selected at any one time frame and not a *sequence* of actions, as one would expect in any sort of plan that spanned a finite time window (from current situation to future goal). What has been implemented is, in effect, a situation-driven decision-theoretic reflexive action generator, operating over a planning horizon of a single time frame. It may be possible to convert the current planner into a full-fledged multistep plan generator, although it is unclear how this would be achieved in the current implementation.

Fixed-Wing Attack (FWA)-Soar and Soar-Intelligent Forces (IFOR)

Considerable effort has been devoted to applying the Soar cognitive architecture (Laird et al., 1987; Laird and Rosenbloom, 1990) to the modeling of human behavior in fixed-wing tactical missions across three services: the Air Force, the Navy, and the Marines (see Chapter 2). Initial efforts have led to a limited-scope demonstration of feasibility for FWA-Soar (Tambe et al., 1995); more recently, FWA-Soar was used in the synthetic theater of war (STOW)-97 large-scale warfighting simulation (Laird, 1996).

The standard Soar approach to hierarchical goal decomposition is taken, so that a high-level goal (e.g., "conduct intercept") is broken down into its component subgoals (e.g., "adjust radar," "achieve proximity"). Each of these subgoals is then successively broken down until some implicit "atomic-level" goal is reached (e.g., "press trigger"), beyond which no further subgoals or activities are generated. The successful achievement of any given goal in the hierarchy typically requires the satisfaction of one or more associated subgoals, so that a tree structure naturally arises. Thus in the tactical air domain, one might see the following tree structure (modified from Laird, 1996):

Intercept
Search
Select threat
Achieve proximity
  Employ weapons
  Select missile
  Get threat in weapons employment zone (WEZ)
    Launch missile
    Get steering circle
    Get lock-on
    Push launch button
    Do FPOLE maneuver

In the above, indenting is used to indicate the level in the tree, and at each level, only one node (shown in bold) is expanded. Satisfaction of any given goal may require the satisfaction of all its subgoals (for example, the launch missile goal shown above) or of only some subset of its subgoals (for example, perform evasive turn could require a left *or* a right turn). In addition, subgoal sequencing may be specified by a need to satisfy antecedent goal conditions (e.g., in the launch missile sequence, push launch button must be preceded by achieving lock-on). Finally, arbitration or deconflicting of concurrent conflicting goals (e.g., evade threat 1 *and* attack threat 2) is accomplished by assigning fuzzy preferences (e.g., worst, poor, neutral, good, best) and, on the basis of these symbolic preferences, selecting the most preferable.

Implementation is accomplished by means of a highly efficient production rule system, with all triggered rules being fired until quiescence, followed by a deconflicting/prioritization stage for selecting the behavior(s) in which to engage for that particular time frame (nominally every 50 milliseconds, but frame times can be less frequent with more complex scenarios). In FWA-Soar, approximately 400 goals are represented, with about 4,000 rules specified on how to achieve them (Laird, 1996).

In generating a path through the goal tree and subsequently pruning off lower-priority paths, Soar has the capability of generating a linear sequence of goals and subgoals to be followed over some finite time horizon, starting with the current time frame. This linear sequence of goals and their associated behaviors is effectively a plan: it is a linear sequence of behaviors that will transform the current state of the system into some desired goal state through a path of sequential system states defined by the subgoals along the way. It is an open-loop plan in that it has been generated for the current state of the system and the current goal; any uncontemplated change in either, or any uncontemplated failure to reach a particular subgoal in the path, may cause the plan to fail. However, this is not a problem in the Soar implementation since replanning occurs every time frame, so that plan deviations can be detected and accounted for via a (possibly) different path through the goal space. By replanning every time frame, Soar effectively converts an open-loop plan into a closed-loop plan.

Essentially, then, Soar acts as a purely reflexive agent[11] in the cited Air Force applications, with no clearly specified planning function (although the latter could be implemented through the subgoal path training just noted) (see also Tambe et al., 1995). Clearly, additional development is needed, and a likely candidate for that development would appear to be the subgoal path-tracing approach outlined here. Implementation of this approach would require: (1) the specification of an explicit planning task, which would generate the open-loop subgoal path for some finite time window into the future; (2) the storage of that plan in memory;

[11]Tambe et al. (1995) distinguish Soar from pure "situated-action" systems (Agre and Chapman, 1987) in that the latter reflexive systems are typically driven by external world states (e.g., actual threat position), whereas Soar is driven by internal *estimates* of those states (e.g., perceived threat position). Although the distinction can lead to significantly different behavioral outcomes, it is irrelevant with regard to planning functionality: neither incorporates planners, and both are thus purely reflexive.

(3) the generation of appropriately time- or event-driven behaviors selected to achieve each subgoal in sequence; and (4) the monitoring of conformance with the plan in memory. Nonconformance would trigger a replanning function, starting the process over again as a reflexive function driven, say, by inadequate planning, unanticipated changes in the world state, and the like. In this fashion, Soar could balance a purely closed-loop reflexive production rule approach to behavior generation with a purely open-loop plan-based and scripted approach to behavior generation. The former approach clearly imposes greater workload per computation cycle (with full replanning every frame), but is much more responsive to unanticipated changes in world state; the latter is a lower-workload solution with replanning occurring only as required, but is less responsive to plan failures or unanticipated world state changes.

Rotary Wing Attack (RWA)-Soar

Soar has also been used to model an Army rotary-wing attack (RWA) mission, specifically an attack aviation company commander and subordinate pilots flying the attack helicopters (Gratch et al., 1996; Gratch, 1996; see also Chapter 2). A layered architecture is used, consisting from the top down of the following:

1. A "live" battalion commander, specifying the overall mission by means of the structured command language CCSIL, communicating with:

2. a simulated company commander modeled as a Soar-command forces (CFOR)[12] entity, which receives the battalion order and generates a detailed mission plan, and then communicates the detailed operations order, again in CCSIL, to each:

3. RWA pilot, modeled as a Soar-IFOR[13] entity that receives individual task orders and generates the appropriate entity behavioral goals (e.g., vehicle flight path parameters, weapon employment activities) to command:

4. the corresponding ModSAF entity (the pilot's rotorcraft) to pursue the mission plan within the DIS operating environment.

Naturally, there is also an upward information flow of hierarchically aggregated status information from the DIS environment, to the ModSAF entity, to its Soar-IFOR controller (the pilot), to the Soar-CFOR commander (the company commander), and finally to the live battalion commander.

---

[12]A further description of the CFOR program is provided by Hartzog and Salisbury (1996) and Calder et al (1996); see also Chapter 2. The planning component uses an approach based on constraint satisfaction to link together component activities, which are doctrinally specified, and which, when fully linked, form an activity plan, taking the command entity from the current situation to the desired end state. Specific details of how activities are linked efficiently to avoid a combinatorial explosion of options through the planning space are described in Calder et al. (1966).

[13]See the earlier description of the Soar-IFOR program for fixed wing pilot modeling.

From the planning point of view, the key component of the RWA-Soar architecture is the mission planning function executed by the Soar-CFOR representation of the company commander. The basic process involves three stages:

1. Obtain the CCSIL battalion order, which specifies (a) the tactical situation, enemy and friendly; (b) the mission and the sequence of tasks to be executed; (c) the general means of execution; and (d) any coordination requirements.
2. Generate the detailed company order for the individual RWA entities, using (a) prestored templates (actually rules) for elaborating the battalion orders, and (b) standard operating procedures (SOPs) or rules for filling in unspecified components of the battalion orders.
3. Refine the company order by looking for inconsistencies and ensuring that intertask dependencies and timing constraints are met.

Note that the second stage of this planning process—essentially open-loop plan generation/elaboration—incorporates some general planning capabilities along the lines originally proposed by the general problem solver (GPS) of Newell and Simon (1963), in particular, step addition when necessary to accomplish the preconditions for an individual in an overall plan. As noted by Gratch (1996), Soar-CFOR adopted a plan representation approach based on hierarchical task networks (Sacerdoti, 1974; Tate, 1977; Wilkins, 1988)—a graphical means of hierarchical task decomposition with a specification of preconditions required for individual task execution. Plan modification occurs through refinement search (Kambhampati, 1997), which adapts plan segments to yield a complete sequence of steps, taking the planner from the initial to the goal state. Replanning occurs in the same fashion as does initial plan generation, so that dynamic unanticipated changes in the battlespace can be accounted for.

Finally, although the above discussion focuses on plan generation/elaboration, it should be noted that RWA-Soar also performs the key functions of plan monitoring and replanning. Plan monitoring occurs during execution of the plan and requires a comparison of the (perceived) tactical situation with the tactical situation anticipated by the plan. If the two do not match sufficiently closely, replanning is triggered. A production rule framework is used to check for antecedent satisfaction for plan monitoring and to maintain a memory of the plan goal for subsequent backward chaining and replanning.

Man-Machine Integration Design and Analysis System (MIDAS)

MIDAS is an agent-based operator model used in helicopter crew station design and procedure analysis (Banda et al., 1991; Smith et al., 1996; see also Chapters 3 and 7). A considerable amount of work has gone into developing a number of rule-based cognitive models, each specific to a particular design case and mission specification.

Development of these cognitive models begins with the specification of a nominal scenario and then proceeds through a process of mission decomposition akin to the hierarchical goal decomposition performed in applying Soar to any realistic modeling task. With MIDAS, one would typically decompose a flight mission into mission phases, say, pretaxi, taxi, takeoff, ... The pretaxi phase would be broken down further into Soar-like procedures or goals, as follows:

Pretaxi
Preflight check
Takeoff clearance
  Engine start check
  Exhaust gas temperature check
  Oil pressure check
  Oil temperature check
  Engine start light check
Subsystems check
Release brakes

As with Soar, the above shows the hierarchical nature of the procedure structure and how some procedures (in bold) can be broken down into subprocedures. Also as with Soar, a production rule system is used to represent these procedures, with the antecedents chosen to ensure proper sequencing relations and priorities. There are other similarities between MIDAS and Soar (specifically FWA-Soar and Soar-IFOR), such as relying on internal perceptions of external states, but it is this common approach to goal/procedure decomposition that suggests, at least in the planning realm, that MIDAS and Soar are effectively equivalent representational approaches: Both are reflexive production rule systems,[14] in which the production rules representing behaviors at various levels of abstraction are carefully crafted to generate the appropriate behavioral time/event sequences for the scenarios and environment for which they were designed. Both are one-step or single-frame planners, but, as discussed earlier for Soar, both could be transitioned to multiframe planners, apparently without great difficulty.

Naval Simulation System

The naval simulation system (NSS) was developed to support the simulation of Navy operations in support of tactical analyses, decision-support applications, and training (Stevens and Parish, 1996; see also Chapter 2). The NSS represents tactics through what is termed a decision table, although it might more properly be termed a prioritized production rule system. A typical rule looks like the following:

[14]Again, reflexive with respect to internal or perceived world states.

```
IF (situation x is true)
        AND (event y  is true within criteria z)
        THEN (do action a with priority p)
```

The situational antecedent shown in the first line above ensures that only situationally relevant (e.g., for the right mission, in the right phase of the mission) rulesets are considered. The event trigger shown in the second line provides a "fuzzy" Boolean check by allowing for variable closeness in matching, through the parameter Z. Finally, the consequent in the third line specifies the action and also assigns it a priority, so that deconflicting or scheduling can occur when multiple actions are triggered in a given frame.

Operational plans in the NSS are represented either as open-loop plans, scripted by the user, or as reflexive, generated in the fashion just described by applying the appropriate tactical ruleset to the current tactical situation. Thus instead of "do action a with priority p," we might see "generate plan a with priority p." However, it is unclear whether actual plan generation is triggered by this production rule, or prestored contingency plans are retrieved from memory and issued by the command entity. An intermediate possibility, however, is suggested by the existence of another NSS module, the future enemy state predictor. Its function is to project the future battle status of the enemy some finite time into the future. Given this predicted future tactical situation, one could then apply the appropriate future tactical response. Since this situation would be in the future, it would become a response goal. Repetitive applications at multiple future time frames could then be used to generate multiple response goals, in effect producing a sequential plan for future actions and goals. Thus there would appear to be a mechanism (implemented or not) within the NSS for generating a tactical plan based on the anticipated evolution of the current tactical situation.

Automated Mission Planner

The University of Florida, in cooperation with the Institute for Simulation and Training at the University of Central Florida, is developing an automated mission planning (AMP) module for eventual incorporation into a command entity representing a company commander, as part of a ModSAF development effort within the DIS community (Lee and Fishwick, 1994; Karr, 1996). A four-stage planning process is outlined:

1. A terrain analyzer generates possible routes based on the current situation, the mission orders, and the terrain database.
2. A course-of-action generator uses "skeletal plans" (i.e., plan templates) to generate candidate courses of action identifying specific unit roles, routes, and tactical positions.
3. A course-of-action simulation runs through each candidate course of

action, evaluating it in terms of critical factors (e.g., how long a unit is exposed to enemy fire).

4. A course-of-action selector selects the best course of action based on the scores from the evaluation.

AMP would appear to have many functions in common with the battlefield reasoning system (BRS) discussed below and currently being developed as a battlefield decision aid.

A key feature of the AMP approach is the introduction of simulation-based planning, wherein generated plans are evaluated through fast-time simulation in order to predict expected outcomes and provide a basis for ranking candidate courses of action according to their anticipated outcomes. Note the similarity to the doctrinally specified wargaming stage of course-of-action analysis, in which commanders conduct mental simulations or verbal walkthroughs of candidate courses of action to anticipate problems and evaluate outcomes. It would appear that this type of simulation-based evaluation should be a key component of any planning model considered for human behavior representation.

A concurrent effort at the Institute for Simulation and Training is the development of a unit route planning module for modeling route planning at multiple echelons (battalion, company, platoon) (Karr, 1996). The effort is currently focused on route finding using an Astar-search algorithm designed to find the optimal path based on a cost function defined over path length, trafficability, and covertness.

Planning Components of Military Decision Aids

International Simulation Advisory Group (ISAG) Effort in Planning Decision Aids

ISAG, a European consortium aimed at demonstrating how artificial intelligence and human-computer interaction tools can be applied to command, control, communications, and intelligence (C^3I) systems development, has efforts under way in three areas (Ryan, 1997):

- Automated report analysis
- Army decision support, planning, and tasking
  - —Terrain analysis
  - —Course-of-action development
  - —Maneuver and fire support planning
- Navy decision support, planning, and tasking
  - —Tactical threat evaluation
  - —Antisubmarine warfare planning
  - —Engagement coordination

The objective is to develop decision aids, not human behavior representations per se, but the overall approach is based on a multiagent architecture in which "intelligent" autonomous agents specialize in particular subtasks and coordinate their efforts through interagent communication protocols. Development appears to be in the early stages, but several of the focus areas may hold promise for the development of agent-based planning models for human behavior representation, particularly in Army course-of-action development, Army maneuver and fire support planning, and Navy antisubmarine warfare planning.

Battlefield Reasoning System

The battlefield reasoning system (BRS) is currently under development by the Beckman Institute of the University of Illinois, Urbana-Champaign, as part of the Army Research Laboratory's Federated Laboratory in Interactive Displays. The BRS is an overall architecture designed as a decision aid to support a number of reasoning tasks occurring in battlefield management, including (Fiebig et al., 1997):

- Terrain analysis
- Course-of-action generation
- Wargaming
- Collection planning
- Intelligence analysis

The architecture is built around five agents, each specializing in one of the above tasks, a common data bus for interagent communication, a number of shared blackboards (e.g., offensive and defensive courses of action), and a number of shared databases (e.g., map data). The BRS is intended to serve as a prototype decision aid at the battalion through corps echelons, but two aspects of the system suggest that it may have potential for the development of future models for human behavior representation: (1) the development of each task agent is founded on the protocol analysis of a subject matter expert solving the same task, so that, at least on the surface, the agent should follow a reasoning process that somewhat mirrors that of the human decision maker; and (2) the agent-oriented architecture itself may serve as a good (normative) model of how to compartmentalize the multiple concurrent tasks facing the battlefield decision maker, providing a basis for subsequent matching of observed human behavior in comparable multitasking situations.

The current focus of BRS development is on the course-of-action generation agent, which reduced to its essentials, operates in accordance with the following process:

1. Obtain the output of the terrain analysis agent, showing the go/no-go regions and other salient terrain attributes within the area of interest.

2. Generate a modified combined obstacles overlay that populates the ter-

rain map with the red and blue forces, the attack objectives, and the potential avenues of approach.

3. Specify the commander's initial guidance, which essentially identifies the mission, defines key subobjectives, and identifies key constraints to be met while accomplishing the mission (e.g., "use armored units only in reserve").

4. Generate a candidate course of action by matching some combination of subordinate units to some combination of avenues of approach. Filter out any combination that fails to meet the commander's initial guidance specified in the previous step. Methodically cycle through all possible combinations (i.e., generate all possible courses of action) for the given set of units and avenues of approach.

5. Present the filtered (i.e., commander-acceptable) courses of action to the G2 and G3 planners for further evaluation and narrowing down using the war-gaming and collection agents.

As currently implemented, the operational planning component of the BRS focuses on course-of-action generation through a simple combinatorial generation of potential unit/avenue-of-approach matches, followed by subsequent filtering out of courses of action that do not comply with the commander's initial guidance. Clearly other planning factors need to be taken into account (e.g., fire support, logistical support), as well as other attributes of plan "goodness" beyond compliance with the commander's initial guidance (e.g., robustness to uncertainty). In addition, realistic courses of action need to account for time-phased activities of both blue and red units (e.g., through a synchronization matrix), so that considerable branching can occur even in one "baseline" course of action. Whether the exhaustive combinatorial approach now implemented in the BRS can handle the true complexity involved in the course-of-action planning task is not clear at this point. If so, it still remains to be seen whether this approach maintains some degree of psychological validity with respect to what is seen with the operations staff.

Decision Support Display

The effort to develop the decision support display (DSD) is described by Ntuen et al. (1997), who are members of the Federated Laboratory in Interactive Displays. The DSD is designed to support the commander and his/her staff as they deal with information at different steps during the following four-stage process:

- Perception
- Situation awareness and recognition
- Situation analysis
- Decision making

No explicit attempt has been made to model the commander's planning function, although the system codifies a number of rules covering components of course-of-action development, such as rules for resource use, rules for combat readiness, rules for battle engagement, and so on. It is unclear how the DSD will use these component rules to aid the commander, and in particular how they can be used for development of a cohesive battle plan over space and time. However, the DSD knowledge base may be useful in other applications, particularly planning models for human behavior representation, if the rulebase effectively codifies current tactical guidelines, rules, and decision heuristics. As this is an ongoing project under the auspices of the Federated Laboratory program, it should be rereviewed again at a later date.

Summary

For the set of models reviewed, the following four general approaches to model development have been taken.

Production Rule or Decision Table. The most common approach relies on production rules or decision tables. The rules/tables often follow doctrine closely, thus yielding greater face validity, at least for doctrinally correct behavior. Rules can be generated hierarchically to support a goal decomposition (Soar) or mission decomposition (MIDAS) hierarchy. The approach suffers the usual problem of brittleness, although fuzzy logic may offer a way around this problem. Also, the approach is unlikely to generate novel or emergent behavior, especially in situations not well covered by the rulebase.

Combinational Search or Genetic Algorithm. Used in some planning decision aids, this approach can be used for exhaustive generation of option sequences (plans) when the option spaces are well defined. Rapid course-of-action generation followed by constraint-based filtering of undesirable courses of action is one technique for developing sufficing plans. A related approach is to use genetic algorithms to generate plans through replication, mutation, and selection based on a plan "fitness function." Experience with other genetic algorithm applications suggests that this approach may prove quite useful in developing novel plans for novel situations, but it is unclear how difficult it will be to enforce doctrinally correct behavior when called for. In addition, a genetic algorithm approach may be quite computationally expensive in complex planning domains, precluding its use in real-time human behavior representations.

Planning Templates or Case-Based Reasoning. Planning templates have been used to expand mission orders into more detailed plans in accordance with well-specified doctrine. SOPs can fill in mission plan details, again in accordance with doctrine. These details can then become mission orders for subordinate units, thus supporting a hierarchical application of the approach at each echelon of human behavior representation. Although no specific examples of a comparable approach based on case-based reasoning were reviewed here, it would appear that a case-based reasoning planning model could also have considerable potential because of both its capabilities for encoding extensive domain knowledge and its ability to model the experienced commander's apparent reliance on episodic memory for rapid plan development.

Simulation-Based Planning. Although not a purely generative technique, simulation-based planning relies on fast-time simulation of candidate plans to support rapid evaluation, modification, elaboration, and optimization of plans. Our review indicates that this is not a popular approach, either as a decision aid or as a planner for human behavior representation, but it may show more promise as better simulation models are developed and computational costs decrease. Simulation-based planning could also be used more in a final evaluative mode to model doctrinally specified wargaming of candidate courses of action.

The review of military planning models in this section reinforces a key finding of the previous section on tactical decision making: military planning is extremely knowledge intensive. Although early artificial intelligence planners emphasized general problem-solving capabilities in an effort to develop domain-independent "planning engines," it seems clear that the military planning domain, especially Army planning, requires that planning be conducted in accordance with an extensive doctrine covering operations and intelligence at all echelons in the military hierarchy. Thus, the method by which planning is to be conducted is specified, and it seems highly unlikely that any approach based on generic algorithms or artificial intelligence would converge on the existing set of planning techniques embraced by the military.

The result is a need for a highly knowledge-based approach to the modeling of existing military planning functions, particularly the process of transforming incoming operations orders into detailed plans for subordinate units. Thus, it would appear that wherever planning functions are specified explicitly by doctrine, a production rule or decision table approach could be used effectively to represent the process, especially at the lower unit levels, where course-of-action activities appear to be much more explicitly characterized. The same may also be true at higher unit levels, but here we suspect that a template or case-based approach may be more effective because of the additional complexity involved at these levels and the limited potential for solving complex problems with a simple rule-based system. In any case, it seems clear that to model military planning behavior, much of which is doctrinally specified (by either rules or cases), it is necessary to model the process with methods that explicitly embed that doctrine. A general-purpose planner would seem doomed to failure in this effort.

The military planning models reviewed here also conform to our earlier findings regarding echelon effects. At lower levels (individual to company), planning focuses on path planning and movement. It may be that more algorithmic approaches are appropriate here (e.g., genetic algorithm optimization, dynamic programming) to solve essentially skill-level problems. At middle levels (company to brigade), path planning and unit movement are still key, but so are communication and coordination. Perhaps approaches that better represent explicit formalized doctrine are better here (e.g., production rule or case-based approaches) to solve what appear to be more rule-level problems. Finally, at higher levels (division and above), more emphasis is needed on conceptualizing the overall operation and specifying the mission. Whether the latter are best modeled by a higher-level decision-theoretic approach, say, or a case-based approach drawing on thousands of years of military history, is unclear at this point. It should also be recognized that the military hierarchy and the command and control information flow within that hierarchy (i.e., operational orders *to* subordinates and situational status *from* subordinates) are an ideal match for the development of hierarchical human behavior representation planners distributed across echelons. Thus at high echelons, only highly abstracted plans are formulated. As these plans are distributed downward, each subordinate unit elaborates on them (possibly through templates or production rules) and redistributes them downward as a new set of orders. Thus, plan granularity grows increasingly finer as one progresses downward in echelon, much as it does in a number of abstraction planners in the artificial intelligence community (e.g., Abstraction-Stanford Research Institute Planning System [ABSTRIPS]) or goal/task decomposers in the human behavior representation community (e.g., Soar or MIDAS).

The review of models in this section also reinforces our earlier findings about how planning relies critically on perception and situation assessment. The planning function must be given not only a goal, but also a specification of the starting point. That is, it must be given a description of the current situation and a specification of any relevant problem variables that might affect the planning solution. This is the function of "upstream" perceptual and situation assessment submodels, which are represented at varying levels of fidelity in the models reviewed here:

- In some cases, neither the perceptual nor situation assessment functions are explicitly represented (e.g., FWA-Soar), so that environmental state variables (e.g., the identity and location of a bogey) are directly available in an error-free fashion to support either proactive planning or reactive behavior. This is clearly a low-fidelity representation of the sensors, the communications links, and the human perceptual apparatus that all serve to process and distort the information upon which plans are based.
- In other cases, the perceptual functions are represented at some level, but no attempt is made to model higher-level situation assessment activities (e.g., computer-controlled hostiles for TTES). Thus one might model line-of sight

limitations in "seeing" a potential threat, but not model the abstraction of "seen" information (such as recognizing an attack formation that needs to be defended against). Unfortunately, by modeling only low-level perceived events (e.g., threat present at location A), one can generate only a low-level description of the current situation (e.g., three threats at locations A, B, C). Thus any subsequent planning can take place only at low levels, since the specification of the current situation is not sufficiently abstracted.

- In yet other cases, both the perceptual and situation assessment functions are represented, so that low-level events are incorporated (in an appropriately errorful manner depending on sensor, communications, and perceptual limitations), as well as higher-level situations. This situation assessment function (e.g., recognizing that the three threats at locations A, B, C are in a defensive posture and unlikely to attack in the near future) then allows for situationally relevant rather than event-driven planning (e.g., select a course of action that is best for a dug-in threat group). This capability supports top-down hierarchical planning, or planning in context, clearly something in which most expert military planners engage.

Finally, the review in this section reinforces the requirement to regard the planning process as only one activity that occurs in the overall military operational context. The models reviewed here address, to one degree or another, the need to represent several stages of planning, starting with receipt of orders and proceeding to situation assessment of enemy and friendly dispositions, generation/evaluation/selection of a plan, communication of the plan to subordinates, plan execution, plan monitoring, and finally replanning. Modeling approaches for dealing with multitasking of this sort are discussed further in Chapter 4, and the findings presented there are clearly appropriate to the modeling of in-context military planning.

PLANNING MODELS IN THE ARTIFICIAL INTELLIGENCE AND BEHAVIORAL SCIENCE COMMUNITIES

An extensive literature on planning models exists in the artificial and behavioral science communities, and it is beyond the scope of this study to do much more than provide the reader with pointers to the relevant literature.

Excellent although slightly out-of-date reviews of artificial planners are provided by Georgeff (1987) and by Tate et al. (1990). They describe the basic chronology, starting with general problem solver (GPS) means-end analysis (Newell and Simon, 1963) and proceeding through a series of ever more sophisticated planners operating in increasingly complex domains. Starting with the essential definition of a plan as a sequence of operators transforming a problem's initial state into a goal state, Tate et al. (1990) outline the essential characteristics of artificial intelligence planners as follows.

Planning is a search problem. Generating an operator sequence is a search in operator/state space, so search algorithms/heuristics are paramount. Tate et al. (1990) review and critique a dozen techniques, including Newell and Simon's (1963) early work with GPS as a formalization of means-end analysis, Sacerdoti's (1975) work with procedural nets in nets of action hierarchies (NOAH), and Tate's (1977) work with partial plans in the nonlinear planner (NONLIN).

Plan goals/tasks can be hierarchized. Goals have subgoals, and some goals can be independent of others. The way goals are organized is also key to planning. Tate et al. (1990) review a number of techniques, including the hierarchical planner ABSTRIPS by Sacerdoti (1974); work by Waldinger (1977) that deals with multiple simultaneous goal task satisfaction, and the system for interactive planning and execution monitoring system (SIPE) by Wilkins (1984); which supports the generation of plans that are both hierarchical and partially ordered.

Plans can be partial and nonlinear. Some planners generate a linear sequence of operators; others generate "partial-order" sequences, with segments to be joined later for a complete plan. There are almost a dozen ways to do this, starting with the goal interference checks conducted by Stanford Research Institute problem solver (STRIPS) (Fikes and Nilsson, 1971), and proceeding to more sophisticated interaction corrections accomplished by NOAH (Sacerdoti, 1975) and NONLIN (Tate, 1977).

Plan goals/tasks can have conditionals. Plans can have conditions for task initiation, termination, or branching, and there are several ways of dealing with this issue.

Effective planners depend on a proper domain representation. Different planners represent their domains differently, with different formalizations for capturing the information about the applications domain. Effective planners have built on the success of earlier planners. It is fair to say, however, that most of the planners reviewed by Tate et al. (1990) deal with deliberately abstracted and simplified domains in order to focus on the key planning issues and mechanisms. (See Wilensky [1981] for a general overview of how more realistic and complex domains might be dealt with through meta-planning.)

Planning is time constrained and resource dependent. Several planners have begun to attack the problems imposed by time and resource constraints in recognition of the need for "practical" solutions to real-world planning problems. Much of the work in efficient search algorithms has supported the implementation of "practical" planners. Although motivated by memory and learning considerations, case-based planners such as Chinese meal planning (CHEF)

(Hammond, 1986) may have significant potential for the eventual development of practical planners in knowledge-intensive domains.

Planning is situation dependent. Early planners were "open-loop" in that the world remained static once the plan was initiated. More recent planners deal with change during plan execution, as well as uncertainty in specifying the state of the environment. Interleaving of planning and execution is described by McDermott (1978) and replanning by Hayes (1975); general issues concerning situation-dependent planning are covered by Georgeoff and Lansky (1990) in their description of the procedural reasoning system (PRS).

Planners can learn. Most planners generate plans from scratch, but the better ones learn from experience. Clearly these are the planners most appropriate for a military human behavior representation. As noted above, case-based planners such as CHEF (Hammond, 1986) show potential for performance improvement with experience. The interested reader is directed to Tate et al. (1990) for a more in-depth discussion of these essential characteristics of artificial planners.

A slightly more recent review of artificial intelligence planners, again covering approximately 150 citations, is provided by Akyurek (1992), who proposes the basic planner taxonomy shown in Table 8.5. The key point here is the distinction between search-based and case-based planners: the former treat planning as a search in an appropriately structured state space characterizing the domain, while the latter treat planning as a retrieval and adaptation of prior plans and plan segments, using an appropriate feature space for case retrieval and an appropriate "adaptation" scheme for plan adjustment. Akyurek (1992) argues that the case-based approach appears to be better supported by the psychological literature and proposes the development and incorporation of a case-based planner within the Soar architecture (Laird et al., 1987). Fasciano (1996) builds upon

TABLE 8.5 Taxonomy of Planners (Akyurek, 1992)

Family	Type	Example	Reference
Search-based	Linear	STRIPS	Fikes and Nilsson (1971)
		HACKER	Sussman (1975)
	Abstraction	ABSTRIPS	Sacerdoti (1974)
		NOAH	Sacerdoti (1975)
		MOLGEN	Stefik (1981a, 1981b)
	Opportunistic	OPM	Hayes-Roth and Hayes-Roth (1979)
Case-based		CHEF	Hammond (1986)
		JULIA	Hinrichs (1988); Kolodner (1987)
		PLEXUS	Alterman (1988)

SOURCE: Akyurek (1992).

the concept of case-based planning with his "Sim City" planner; here, the planner effects case adaptation by means of a qualitative causal model of the external world relations, which in turn is learned online through observation of past cause/effect correlations.

One other review of artificial intelligence planners should be noted, primarily because of its recency, if not coverage. A review by Kambhampati (1995:334) notes that "most implemented planners settle on some variant of the STRIPS action model," where the world is represented in a state space, actions transform the world from one state to another, and a plan is a series of such actions (Fikes and Nilsson, 1971). This approach runs into problems when the world is only partially observable, dynamic, and stochastic. Table 8.6 summarizes Kambhampati's (1995) observations on how current planners are attempting to deal with these problems.

TABLE 8.6 Advanced Artificial Intelligence Planner Techniques

World Characteristics	Planner Strategy
Fully observable, static, deterministic	Use classical planners
Partially observable	Specify information-gathering strategy
Dynamic	Interleave planning and acting
Stochastic	Use Markov models

A more recent although not as extensive review of cognitive architectures that includes planning functions is to be found at the following University of Michigan website: *CapabilLists/Plan.html.* Table 8.7 summarizes the architectures reviewed at this site, and the interested reader is referred there for further information.

TABLE 8.7 Example Planners as Part of Existing Cognitive Architectures

Planner	Reference
Gat's ATLANTIS	Gat (1991)
Theo	Mitchell et al. (1991)
Icarus	Langley et al. (1991)
Prodigy	Carbonell et al. (1991)
Meta reasoning architecture fax	Kuokka (1991)
Architecture for intelligent systems	Hayes-Roth (1991)
Homer	Vere and Bickmore (1990)
Soar	Newell (1990)
RALPH	Ogasawara and Russell (1994)
Entropy reduction engine	Drummond et al. (1991)

SOURCE: University of Michigan (1994).

From our review of these and other artificial intelligence planners, it would seem that many of the artificial intelligence planners are limited by a number of basic problems, including the following:

- The environment with which the planner deals is often very simple (e.g., a blocks world), and the plan objectives and constraints are fairly trivial (e.g., stack the blocks in some order).
- The environment is static, not dynamic, in the sense that it changes only in response to the planner's actions (e.g., route planning). Often, the planner is the only agent inhabiting the environment.
- The environment is deterministic, not stochastic, so that the effects of the planner's actions are the same every time.
- The planner has perfect information about the environment, with complete omniscience, error-free sensors, and perfectly timely feedback.
- The planner works from first principles (e.g., means-end analysis), has little explicit domain knowledge, avoids useful domain-specific heuristics, and fails to learn from past successes and failures.

Clearly, the above limitations are most applicable to early planner designs; as noted earlier by Kambhampati (1995), more recent work has focused on ameliorating one or more of these problems. One last key shortcoming of artificial intelligence planners in general is due to the artificial intelligence planning community's interest in developing computationally effective planners, and not necessarily in developing *models* of human planners. Thus, most artificial intelligence specialists working on planner development efforts pay scant attention to how actual human planners solve problems. Modeling of human planning behavior per se is left to those in the behavioral science community who are interested in planning and problem solving. Some computational modeling efforts outside the military domain have been conducted, however, and the interested reader is referred to the behavioral science planning literature.

Our review of the literature has, unfortunately, uncovered remarkably few attempts at developing computational models of human planning behavior. Perhaps the earliest and best-known effort is the analysis and modeling of human errand-planning behavior performed by Hayes-Roth and Hayes-Roth (1979). They observed a mix of strategies used by humans, involving different levels of abstraction; partial planning; and, most clearly, *opportunistic planning*, a behavior that involves situation-based triggering of new goals, subgoals, and/or partial plans (e.g., "Well, as long as I'm in the neighborhood, I might as well go to the pet store nearby, even though it wasn't a high-priority item on my errand list."). The resulting opportunistic planning model (OPM) developed to reflect the observed planning behavior in the errand-planning domain makes use of a blackboard architecture (Hayes-Roth, 1985), which is reviewed by specialized (rule-based) agents to see if they can contribute to the plan, and if so, to which they post their plan components. The key architectural contribution of the OPM is the five conceptual levels of the blackboard (Hayes-Roth and Hayes-Roth, 1979):

• Executive—determines goal priorities, attentional focus, and resolution across competing agents.

• Meta-plan—determines the planning objective, problem representation, policies for constraint satisfaction, and plan evaluation criteria.

• Plan abstraction—determines more specific plan intentions, schemes for meeting these intentions, overall strategies, and specific tactics to be used.

• Plan—determines specific plan goals, a plan design to meet these goals, general procedures for action, and detailed operations for finer action.

• Knowledge base—maintains key problem-specific planning information, such as subgoal priorities and route geometry.

An implementation of this architecture, populated with about 40 rule-based agents, allowed Hayes-Roth and Hayes-Roth (1979) to compare OPM-generated protocol and plans with those generated by human subjects. As noted by the authors, some sections of the protocol (i.e., the behavior engaged in while generating the plan) showed remarkable similarities, while others diverged; the resulting plans, however, were not too far apart. Overall, the OPM seems to be a reasonable candidate for modeling human behavior, at least in this domain. Further validation work would appear necessary, however.

A partial validation of this approach is provided by Kuipers et al. (1988) in their study of how pulmonary physicians develop diagnosis and treatment plans (i.e., information-gathering and action plans) for treating a patient in a high-risk, uncertain medical scenario. Kuipers et al. conducted a protocol analysis during treatment plan generation ("thinking aloud") and after specification of a plan ("cross-examination") with a number of medical specialists to develop a behavioral database. They then attempted to match the protocol and treatment plan using a classical decision-theoretic model approach, which provides a rational means for combining diagnostic uncertainty with treatment utility (see Chapter 6 for further discussion). They found, however, that this model fails to capture the planning behaviors they observed. Instead, Kuipers et al. (1988:193) found:

> The decision process we observed in the protocols is well described as one of planning by successive refinement of an abstract plan (Sacerdoti, 1977), combined with opportunistic insertion of plan steps (Hayes-Roth and Hayes-Roth, 1979).

Kuipers et al. go on to suggest (but not implement) a simple rule-based model to explain the protocols they observed and to argue, more generically, that such a planning model:

• Is simpler to effect by a human than one requiring an estimate of probabilities, a specification of utilities, and a computation of expected value.

• Is simpler to remember, at least at high abstraction levels (e.g., the Hippocratic dictum "First, do no harm.").

• Can be explained and defended (to patients and colleagues) most easily through a step-by-step syllogistic explanation.

- Can be built up from experience through rule abstraction from past cases.

We are unaware of related follow-up work focusing on the development of a computational planning model. It would appear, however, that the behavioral data and initial modeling efforts of Kuipers et al. (1988) would support such work.

CONCLUSIONS AND GOALS

Short-Term Goals

- Make explicit the difference between doctrinal and actual planning behaviors. Modeling the former behaviors may be easier since they are well documented, but the results may not be representative of actual behavior. Modeling the latter may yield a more accurate representation of what occurs, but several years of research will be required to measure, describe, formalize, and formulate such models.
- Begin the above research process to build the behavioral database needed to support follow-on development of nondoctrinal planning models.
- In parallel, for the short term, develop planning models based on doctrine. Some of the observed (nondoctrinal) behaviors may be reproduced by these models because of limitations in "upstream" process models (e.g., perception, situation assessment); other behaviors might be modeled by injecting deliberate deviations from doctrine through, say, a fuzzy logic implementation of a "crisp" doctrinal rule.
- Consider expanding the scope of most military planners to encompass the entire planning process, from receipt of command orders, to issuance of subordinate orders, to plan monitoring and replanning.
- Give greater consideration to the next generation of planning decision aids that will be deployed on the battlefield. Some of the lower-level planning functions (e.g., route optimization) may not need to be modeled within the human behavior representation because of their automation (analogous to the drop-off in pilot modeling research as autopilots were introduced).
- Consider expending effort on developing models in the tactical domain from the ground up, rather than searching in other domains for candidate models to adapt to the military case. It is not clear that any particular community has the answer to the development of sophisticated planning models, algorithms, or decision aids, especially at the higher conceptual levels of plan generation (although there are many domain-specific tools to support the more tedious plan elaboration tasks).
- Consider expending development effort on codifying the large numbers of doctrinal (or nondoctrinal) planning behaviors once those behaviors have been agreed upon. The conceptual model of the mission space (CMMS) development effort may be an appropriate place to start. It does not appear likely that a "first-

principles" planning architecture emphasizing process sophistication over knowledge content will succeed in the highly knowledge-intensive and options-constrained military environment.

Intermediate- and Long-Term Goals

- Devote effort to developing planners that incorporate, at the least, an extensive domain-specific knowledge base and, more important, a learning capability to build upon past planning and operations experience.
- Develop planning models that are more reactive to account for plan failures, dynamically changing environments, and changes in plan goals. The focus should be on developing planners that are more robust to failures, develop contingency plans, and can rapidly replan on the fly.
- Develop planning models that account for a range of individual differences, skill levels, and stressor effects. Since the military plan is such a key determinant of overall mission success, it is critical that this behavior be represented accurately across the population of military planners.

9

Behavior Moderators

INTRODUCTION

Previous chapters have reviewed the current state of the science and state of computational modeling regarding various human characteristics with a view to identifying ways in which models of human behavior in military simulations could be made more realistic and therefore more useful. The analysis has covered basic human mechanisms associated with attention and multitasking, memory and learning, decision making, situation awareness, and planning. This chapter examines the role of a class of variables we refer to as behavior moderators and how such variables might be used in engagement simulations. These are the individual difference variables (e.g., task stress, emotion) about which the least is known, so that they cannot at this point be encoded directly into a model. Until sufficient information about the structure and function of these variables is obtained, one approach is to introduce them by assigning values to parameters of various behavior models on the basis of informed judgment. Tentatively, such values could, for example, shift the probabilities associated with a range of responses generated by the triggering of a production rule.

There are a number of important questions to be addressed about variables that moderate behavior. For example, what is the desirability and feasibility of developing models of humans that vary with respect to personalities, emotions, cultural values, skills, and intellectual capabilities? How is behavior influenced by external stressors such as extremes in temperature, long hours of continuous work, and exposure to toxins? Can the effects of fear and fatigue on soldier performance reasonably be represented? At the unit level, what are the effects on performance of average level of training, the extent to which standard operating procedures are followed, and the degree of coupling between procedures? Although there is some discussion of unit-level behavior moderators in this chapter, the focus is primarily at the individual level.

Interest in incorporating moderator variables such as task stress into military engagement simulations emerged in the late 1960s (Kern, 1966; Siegel and Wolf, 1969). In retrospect, these efforts were probably ahead of their time. For example, Siegel and Wolf's single-operator model incorporated a complex amplifying interaction between the level of operator capability and performance degradation due to time stress. Likewise, their multiperson model included such features as the costs of coordination as a function of the relative appropriateness of task allocation to team members. One of the factors that probably inhibited rapid elaboration of their models was the limits of computer capacity at that time.

There was renewed interest in moderator variables in the late 1980s. All branches of the military services supported associated research and system development efforts (for Army-supported efforts, see Van Nostrand, 1986; for Air Force programs, see Davis, 1989). Van Nostrand in particular opened several useful lines of inquiry. For example, she undertook to assemble and interpret data from Army field exercises that characterized the decline in soldier performance as a consequence of continuous operations, fatigue, and fear. In this regard, she was among the first to see the potential contribution of empirical data on soldier behavior under stress conditions to the development of a computational model of human behavior under battlefield conditions.

Van Nostrand also gave some quantitative specificity to the idea that morale can influence performance. She noted from various sources of military data that specific conditions, such as extremes in temperature, noise, the rate of friendly casualties observable by individual soldiers, and the turnover rate among unit members, had specific adverse effects on both individual and unit performance. Her work represents an attempt to examine and assess the state of the data from military studies on the relationship between external stressors and human behavior on the battlefield.

Davis' work at the RAND Strategic Assessment Center (1989) had a broader scope—incorporating theater-wide, strategic operations. This work was conducted in response to the hypothesis of military historians (see, e.g., Woldron, 1997) and users of war games that observable variations in large-unit performance can be linked to the character of commanders or the philosophical directives of particular cultures (see, e.g., Bennett et al., 1994).[1] Davis' work addressed three broad factors: (1) fighting quality (different forces with equivalent equipment), (2) the friction of actual combat, and (3) the processes of command decision making in a particular political climate.

---

[1]The successes of the Israeli Defense Forces against larger and similarly equipped opponents are often cited as cases in point (see, e.g., Bond, 1996).

Davis introduced these factors into strategic simulations by applying weights to standard quantitative indices of performance. The weights were derived from estimates of characteristics such as level of unit training, level of logistic support, and whether a combat unit was defending its home territory or fighting on foreign ground. These early models were simple and based mainly on common sense; however, as results from psychological research studies and feedback from simulation became available, these primitive models were elaborated and refined.

More recently, several investigators (Lockett and Archer, 1997; Allender et al., 1997a; Hudlicka, 1997; Fineberg et al., 1996) have worked toward developing computational models that incorporate some of these moderator variables. In her work on MicroSaint and IMPRINT (improved performance integration tool), Archer (1997) has built on Van Nostrand's efforts by introducing the effects of various levels of environmental stressors (such as heat, cold, and the presence of toxins [Anno et al., 1996]) into computational models representing human behavior (see also Conroy and Masterson, 1991). Fineberg and colleagues (1996) have demonstrated, in the context of a military exercise, a model that incorporates the level of anxiety for dismounted infantry confronted with indirect fire. In this model, behaviors such as taking a prone position, moving while prone, and returning fire are affected by the frequency of incoming rounds, the pattern of intermittency, the apparent caliber of the rounds, and their proximity. These latter factors are linked to various anxiety levels. Hudlicka (1997) has proposed a general approach to modeling the effects of a wide range of individual differences, including cognitive styles, individual preferences, training, personality, and emotion; the approach moderates specific components of the overall cognitive architecture so as to indirectly affect explicit behavioral outcomes.

There are many behavior moderators and many human behaviors that might be influenced by these moderators. Examining them all is beyond the scope of this chapter. Our intent here is to select a subset to illustrate the current state of knowledge in this area. Based on our review, we believe these variables have the potential to impact the performance of soldiers and commanders in the field and therefore should be considered for further development. We address two broad categories of moderators: those that are imposed on the human externally and those that are part of the human's internal makeup. External moderators include physiological stressors (e.g., heat, toxins, noise, vibration), physical workload and fatigue, and cognitive/task workload. Internal moderators include such factors as intelligence, level and type of expertise, personality traits, emotions/affective factors, attitudes/expectancies, and cultural values. Each of these moderators may have both positive and negative effects, depending on the value of the variable. For example, soldiers who received high scores on the military entrance examination and have the required expertise will probably perform better than soldiers who received lower scores on the entrance examination or have had less training.

The following sections provide a review of the major moderator variables—

external and internal—in terms of what is known about their effects on human behavior. This review is followed by a discussion of efforts to build computational models incorporating these variables. The final section presents conclusions and goals in the area of behavior moderators.

EXTERNAL MODERATORS OF HUMAN BEHAVIOR

Most of the behavior moderators that originate outside the human are stressors and as a result have a negative influence on performance. For example, if a unit is filled with rested soldiers who are perfectly trained and in ideal physical condition, external factors such as extreme heat, long hours of continuous work, and exposure to toxins or the requirement to operate in protective gear to guard against toxins can only serve to degrade performance.

Physiological Stressors

Environment

Several environmental factors, including temperature, toxic substances, noise, and vibration, can degrade both the mental and physical performance of soldiers. Although some data suggest tolerable limits for these factors beyond which work is not safe or cannot be performed, there is unfortunately little research on performance decrements expected during exposure. As Van Nostrand (1986) suggests, it is unlikely that all performance is 100 percent effective until workload reaches maximum and performance drops to zero. However, data on intermediate stages are sparse (Hart, 1986).

Heat in relation to heavy work (Parsons, 1996) has been studied extensively by the Army over the years. The results of these studies have been translated into guidelines giving the maximum safe time exposures for various levels of heat (Technical Bulletin, Medical 507, 1980). As shown in Figure 9.1, heavy work can extend somewhat beyond 3 hours at a temperature of 82 degrees before heat exhaustion is expected; at 94 degrees, the work time is reduced to 90 minutes before exhaustion is expected. The figure also shows, for 2 hours of work, the relationship between type of work and recommended safe temperature.

Military Standard 1472D (1981) provides tolerance levels for heat and cold in terms of the number of hours of exposure for both physical and cognitive tasks. According to this standard, at 102 degrees the effective work time is limited to 30 minutes, while at 85 degrees work can be performed effectively for up to 12 hours. For cold, the limits are 30 minutes at 32 degrees and 4 hours at 59 degrees. Hancock and Vercruyssen (1988) have proposed recommended exposure time limits at various temperatures for cognitive performance; these limits are similar to those suggested in the military standard.

A second source of stress from the environment is toxic substances. One

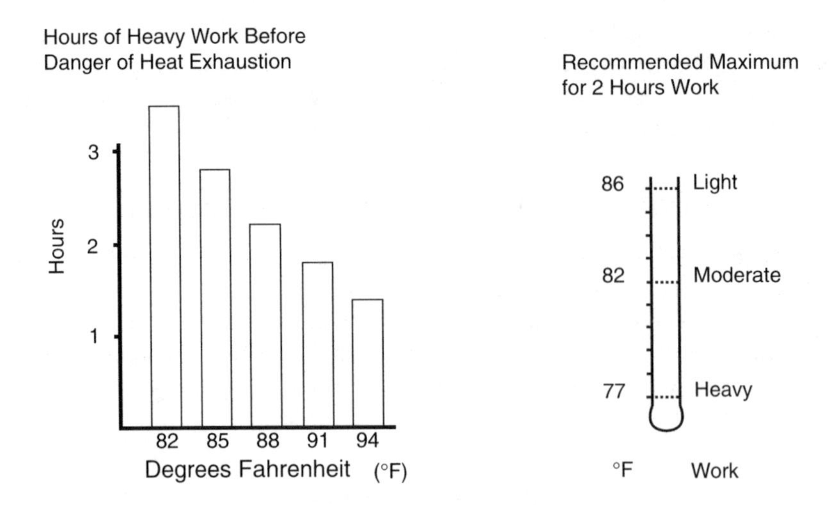

FIGURE 9.1 Maximum exposures for heat, measured with "wet bulb temperature."

effect of toxic substances is to reduce the oxygen supply to the blood. This condition, referred to as hypoxia, has a straightforward physiological effect on the human body and mental processes. There is a near-linear correlation between oxygen partial pressure and rate of breathing, cardiac output, and degradation in some sensory functions. Specifically, by the time arterial oxygen saturation has dropped only 15 percent, visual acuity has already dropped 30 percent (Fulco and Cymerman, 1988). Physical power output and endurance are also promptly depressed, proportionally to the drop in arterial saturation. In contrast with these direct physiological effects, the performance of higher cognitive functions is only moderately depressed, particularly during the early stages of oxygen deprivation. Although compensatory adjustments operate in this domain as in other areas of physical stress, there appears to be a threshold that, once passed, signals a precipitous drop in performance. Also, changes in manifest mood can appear early.[2]

Another important consideration from the standpoint of modeling variations in performance concerns soldiers' ability to perform their tasks in the protective gear they must wear in the presence of a toxic threat. Similar issues arise for soldiers wearing heavy clothing to protect them from extreme cold. In her review of the literature on toxic substances, Van Nostrand (1986) notes one project

[2]The whole process of change in physical, cognitive, and affective behavior appears to parallel the effects of alcohol intoxication. Since the relationship between blood alcohol level and performance has been studied intensively and good quantification has been achieved, it might be useful to employ the performance models from this source for a generic representation of stress effects until focused research on stress and performance has been completed.

(Hamilton et al., 1982) that showed no performance differences due to protective gear for aircrew chemical defense ensembles and another (Banks, 1985), on combat tasks for the Army, in which the performance of soldiers in protective gear was completely ineffective. Based on her review, Van Nostrand (1986:2-10) states:

> Apparently the major problems are heat buildup inside the clothing, the inability to perform work requiring manual dexterity with heavy gloves on, and to see well with the face mask in position. The end result of prolonged exposure to heat is extreme fatigue, and the result of fatigue is degradation of thinking and decision making skills. Therefore the tests that require soldiers to perform tasks that are so well practiced that they can do them without thinking about them will show less decrement than tasks which require the soldier to decide what to do next. Any task which requires vigilance will probably show large decrements in performance.

A third source of stress from the environment is ambient noise. According to Haslegrave (1996), people vary enormously in their response to noise and can adapt well to a continuous background of sound as long as it is below the intensity level that results in physiological damage and actual hearing impairment. The core response to a noisy environment is annoyance. Noise in combat environments, however, is likely to be of a class that is maximally disruptive—sudden, unexpected, and intense. Yet there do not seem to be quantitative (i.e., programmable) models for the performance effects of even extraordinary noise conditions.

A fourth source of environmental stress is vibration (Bonney, 1996). Studies of the relationship between vibration and human performance generally use three parameters to characterize vibration: acceleration, frequency, and duration. There are several modes by which vibration may be experienced, but the most commonly studied is whole-body vibration, as a soldier would experience in a moving tank, for example. Unfortunately, studies of whole-body vibration do not link the vibration parameters directly to measures of performance. Outcomes are typically expressed as tolerance limits. For example, the tolerance limit for accelerations of 3 meters per second per second at 5 cycles per minute is 1 minute. That is, the average soldier can tolerate this acceleration and frequency for 1 minute before performance begins to decline. In contrast, an acceleration of 0.3 meters per second per second at the same frequency can be tolerated for 8 hours. The standards information does not include a function showing the rate of decline once the threshold has been exceeded (Air Force Systems Command, 1980).

In the past, studies of environmental stressors were designed with the goal of specifying limits for effective performance, rather than demonstrating performance decrements over time as a function of various levels of the stressor. The results of these studies were used in making decisions about equipment design, task design, and mission length; they were not intended for use in developing models of human behavior. With this new focus in mind it is important to recognize that collecting accurate data on performance decrements is both difficult and challenging given the range of environmental conditions and variety of task performance requirements.

Physical Work and Fatigue

The U.S. military has conducted several studies of the effects of continuous operations (and the resulting fatigue) on performance. These studies have demonstrated some quantitative changes in performance that might be used in building computational models of troops operating over extended periods of time. For example, Pfeiffer et al. (1979) and Siegal et al. (1980, 1981) developed a model for the Army Research Institute that calculated the performance over time of a soldier in a small unit. This model—performance effectiveness for combat troops (PERFECT)—combines task analysis of Army jobs with results from laboratory research on sleep loss, noise, visual acuity, and reasoning abilities.

In another study, Dupuy (1979) analyzed real battles in World War II and the Arab and Israeli wars from the standpoint of continuous operations involving enemy contact and recovery periods following combat. Fatigue was measured in terms of degradation in the ability to produce casualties. Three battle intensity levels were specified, based on the percentage of days during the operation that the unit was in contact with the enemy. If the unit was in combat 80 percent of the time, the degradation in performance per day was determined to be 7 percent; if the unit was in combat between 50 and 80 percent of the time, performance was determined to drop off at a rate of 2 percent per day. It was estimated that during recovery periods, performance capability was regenerated at a rate of 6 percent per day.

Additional research on continuous work was conducted at Walter Reed Army Institute (Belenky, 1986). The results of this effort suggest that a 7 percent degradation in performance per day for all tasks is a reasonable estimate for troops with 5 hours of sleep per night.

Figures 9.2 and 9.3 show the expected performance degradation over 5 days of continuous work under two conditions of sleep deprivation (U.S. Department of the Army, 1983). The soldiers in Figure 9.2 were deprived of sleep for 24 hours before they began continuous operations; the soldiers in Figure 9.3 were not deprived of sleep before continuous operations began. In both cases, the soldiers received approximately 4 hours of sleep each night during operations.

Cognitive Workload Stressors

When an individual is subjected to variations in information input rate and composition, performance changes are not monotonic, but follow an inverted U-shaped function (Hancock and Warm, 1989).[3] A consistent human adaptation to

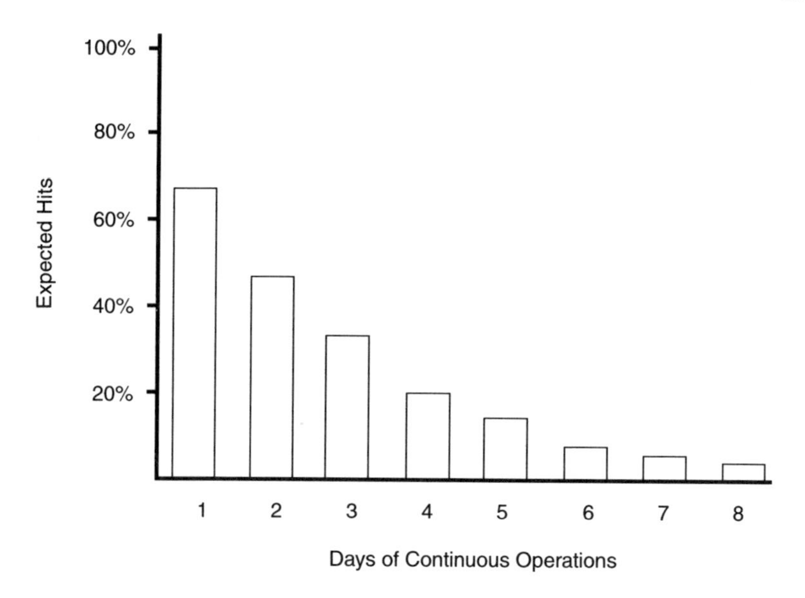

FIGURE 9.2 Hits during first model cycle after 24 hours of sustained operations followed by continuous operations.

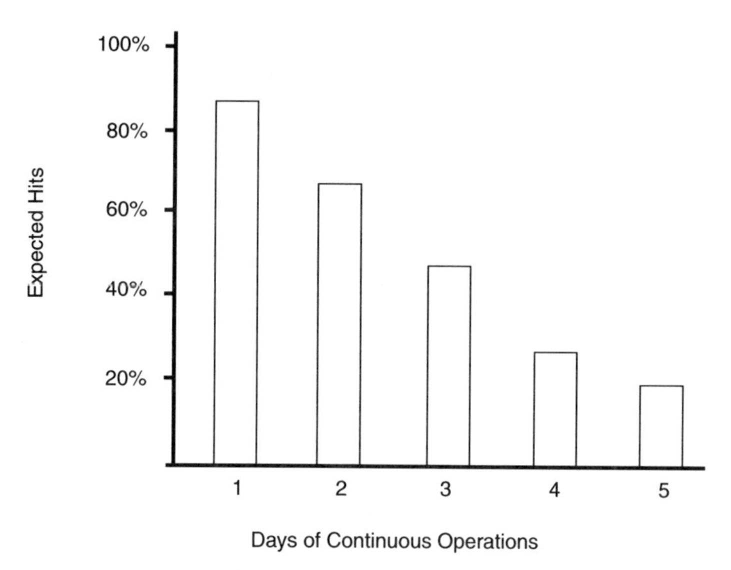

FIGURE 9.3 Hits during first model cycle after 1-5 days of continuous operations.

overwhelming task loads in information processing tasks (such as command and control) is to exclude some portion of the signal flow or postpone the processing response until the peak period has passed (Hancock and Chignell, 1987, 1988). The programmer intent on representing such behavior in an engagement simulation must address the question of the information flow rate threshold for such adaptive responses. It should be noted that this question in turn raises complex measurement issues concerning the level at which task elements are defined and their relationships specified.

Some responses to cognitive overload are dysfunctional. For example, sequence sense is often lost (Hancock et al., 1994). That is, some events are sequentially related, such as the cues a tank battalion commander will receive regarding unit losses. When confronted with a high signal flow load, the commander may forget the initial signals and not keep an accurate tally of losses, thus being unaware when a threshold for withdrawal has been reached.

The relating of actions to consequences can also be faulty, particularly if there is a time delay between action initiation and outcome or between outcome and feedback (Hancock et al., 1994). Such decoupling may seem intentional when a decision maker does not perceive negative outcomes; that is, he or she adopts a form of denial.

INTERNAL MODERATORS OF HUMAN BEHAVIOR

Internal moderators of human behavior include variables such as intelligence, level and type of expertise, personality traits, emotions/affective factors, attitudes/expectations, and cultural values. These moderators are complex, they interact, and they can influence performance in multiple directions. For example, individuals with a low level of expertise and aggressive personality characteristics will probably select different strategies than individuals with a high level of expertise and nonaggressive personalities. However, the form of this relationship cannot currently be predicted. With regard to emotions, fear may result in a tendency to interpret a situation as dangerous, while anger may result in an interpretation that others are hostile. Furthermore, an individual with an aggressive personality who is fearful may behave differently than an aggressive individual who is self-confident.

Intelligence

The Department of Defense conducted an extensive research project on linking selection criteria to job performance (Wigdor and Green, 1991). This project, known as the Job Performance Measurement Project, involved researchers from

[3] There is an extensive literature on cognitive workload, including major works by Wickens (see, for example, Hart and Wickens, 1990; Huey and Wickens, 1993; Wickens et al., 1997).

all branches of the services. Specifically, data were collected on the relationship between performance on selected military tasks and the mental category classification of enlistees as measured by the Armed Forces Qualification Test, which places recruits into mental categories ranging from Category I (high) to Category V (low). Project A, conducted by the Army, also developed a database linking mental category to job performance (Campbell and Zook, 1991).

Two other studies, one conducted on M-60 tankers and the other on M-1 tankers (Scribner et al., 1986; Wallace, 1982), demonstrated a reduction in the number of target hits in a gunnery exercise as mental category declined. For the M-60 the reduction was from 108 hits for Category I to 61 for Category IV; for the M-1 it was from 102 hits for Category I to 86 for Category IV. The smaller decrement shown for the M-I tankers was attributed to the increased automation in this tank. Data such as these could be used to introduce variability into models of human performance.

Level and Type of Expertise and Cognitive Abilities

Rather than discussing education, training, and level and type of experience as internal behavior moderators, we focus here on level and type of expertise (which should be a function of those factors). According to Van Nostrand (1986), it appears that if a task is within the physical capability of the soldier, the cognitive aspects of the task will display greater variation in performance than will physical aspects that are frequently practiced. This conclusion was supported by testing the use of human data in the combined arms task force effectiveness model (CASTFOREM) (Miller, 1985).

As noted by Hudlicka (1997), recent research has identified a number of differences among individuals in terms of level of expertise, cognitive abilities (specifically the ability to perform mental "what-if" simulations), the ability to visualize situations from multiple perspectives, the ability to abstract relevant information, and preferred style of information presentation—textual or visual, and abstract or concrete (Hammond et al., 1987; Deckert et al., 1994; Badre, 1978). The Meyers-Briggs Inventory (MBI) has been used to measure individual differences on the cognitive style scale. The validity of such scores for characterizing students has been problematic (Bjork and Druckman, 1991). However, better success has been achieved when adult decision makers have been tested (Richards and Giminez, 1994) and their cognitive style scores have been correlated with performance.

Studies of tactical planning and command decision making identify features that characterize expert performance as distinguished from the performance of novice tactical planners (Fallesen and Michel, 1991; Lussier et al., 1992; Tolcott et al., 1989). According to these studies, experts have better situation awareness than novices in that they are aware of uncertain assumptions, make better use of available information, actively seek confirming and disconfirming evidence, and have greater awareness of enemy activities. Experts also make better decisions than novices because they can develop flexible plans, conduct more elaborate wargaming, and make better predictions of battle events and outcomes.

Personality, Emotions, Attitudes, and Cultural Values

Although this section briefly addresses personality, emotions, attitudes and expectancies, and cultural values separately, it is clear that there is great overlap among these variables. For example, the personality trait of neuroticism has many of the same defining characteristics as depression and anxiety, the latter being treated by some authors as emotions. As another example, the personality trait of agreeableness results in a predisposition toward a positive attitude. Since distinctions among these classes of variables are difficult to make, and the nature of the data describing their impact on behavior is less than rigorous, the most useful approach to modeling them may be to treat them as a single class. Work on incorporating these variables into models of human behavior is discussed in a later section of this chapter. Here we review the some of the relevant research literature.

Personality

During the first part of the twentieth century, scientific interest in personality variables as descriptors and predictors of behavior was focused on the relationship between these variables and the quality of managerial decision-making performance. After World War II, however, simulation was introduced as a means of studying organizational processes (Bass, 1964), and interest shifted from the study of personality variables to the study of the influence of situational variables on managerial performance (Mischel, 1968). After some years it was recognized (see, e.g., Janis and Mann, 1977) that both situational and the personality variables may have important influences on decision-making behavior, and thus it would be useful to examine their interactions (Byrn and Kelley, 1981).

In recent years, the measurement of personality has matured to the point where there is broad abstract agreement among researchers that the nature of personality can be captured by relatively simple self-descriptive questionnaires (Campbell, 1996). Such questionnaires are usually constructed "rationally" to reflect a particular theory or to support personality profiling (i.e., diagnostics) for certain groups, such as alcoholics or students. The result has been the specification of a finite set of traits that are extracted from the test data by means of factor analysis. Profiles based on the pattern of test-taker traits answer specific questions, such as whether this individual's personality is congruent with the requirements of this job (Hayes, 1996) or whether this individual will benefit from this particular therapeutic regimen (Parker et al., 1996).

There are quantitative scales for various traits, but most interpretations of these traits in the context of military engagement situations are anecdotal (Dixon, 1976). Furthermore, there are as yet no up-to-date instruments for assessing the traits specific to soldiers or their commanders. However, certain traits are frequently judged by observers of military affairs to have a high likelihood of salience for the military character. At the top of any such list is "need achievement" or "n-Ach" (Bond, 1996). Other potentially relevant traits include risk taking and innovativeness. While position in communication networks and other situational factors influence awareness and acceptance of innovations (Valente, 1995), individual differences are also significant (Rogers, 1983). Some individuals are clearly more innovative and more accepting of innovations than others. Innovation can be a powerful factor in a battle command situation. Another factor for consideration is general alertness or vigilance.

One approach to characterizing personality traits is known as the five factor theory. This approach has been used by several investigators over the years, including Hogan and Hogan (1992). Although different investigators use different names for the factors, they can be generally classified as follows (Barrick and Mount, 1991):

- *Openness*: curious, broad interests, creative, original, imaginative, untraditional
- *Conscientiousness*: organized, reliable, hard-working, self-disciplined, honest, clean
- *Extroversion*: sociable, active, talkative, optimistic, fun-loving, affectionate
- *Agreeableness*: good-natured, trusting, helpful, forgiving, gullible, straightforward
- *Neuroticism*: worrying, nervous, emotional, insecure, inadequate, hypochondriac

One possibility is to create a typology based on these factors that could be used to assign numerical values to specific agents in an engagement simulation.

Janis (1989) suggests that for the specific purpose of predicting and understanding command behavior, only three of the five factors are essential: openness, conscientiousness, and neuroticism. He asserts that these three traits can explain executive decision making in that the scores on these traits for individual managers will predict the manner in which each will fail.

At the organizational level, it is clear that personality plays a role in shaping behavior (Argyris, 1957). It appears likely, for example, that individuals in an organizational context tend to behave in ways that are congruent with their public personas. In particular, people attempt to present a consistent pattern of behavior that represents their mode of fulfilling an organizational role (Heise, 1992).

Role assignments and personality do not fully determine behavior in organizations. One reason for this is that the command, control, and communications (C^3) structure and the task can constrain individual decision making to the point where individual differences become secondary. A second reason is that in large numbers of units or very large units, the effects of cognition and individual differences often counteract each other to the point that the only main effect is that caused by the task or the C^3 architecture.

In recent years there has been a strong push to merge concepts of personality with concepts of emotion (Plutchik and Conte, 1997). This merger is seen as a means of reducing the complexity of concepts and enhancing the power to predict behavior.

Emotions[4]

Hudlicka (1997) has developed a cogent approach to the problem of fitting concepts from research on human emotions into a framework of military operations—and the simulation thereof. She notes, for example, that emotions have until recently had a somewhat marginal status in both cognitive science and neuroscience. Over the past 10 years, however, important discoveries in neuroscience and experimental psychology have contributed to an interest in the scientific study of emotion. A growing body of evidence from neuroscience research points to the existence of neural circuitry processing emotionally relevant stimuli (i.e., stimuli that threaten or enhance the survival of the organism or its species) (LeDoux, 1992, 1989, 1987). Studies in experimental psychology have identified memory systems with distinct processing characteristics—explicit and implicit memory (Schachter, 1987; see also Chapter 5), analogous to the characteristics of the neural circuitry identified by LeDoux of fast, less differentiated processing and slower, more refined processing. Cognitive psychologists have described a variety of appraisal processes involved in inducing a particular emotional state in response to a situation (Frijda and Swagerman, 1987; Lazarus, 1991), and several models of these processes have been proposed (Ortony et al., 1988). The link between cognition and emotion has been strengthened by the findings of Cacioppo and Tassinary (1990), who show that some individuals have an emotional need to engage in cognitive activities.

A number of studies in cognitive and organizational psychology have documented the differential impact of various emotional states on cognition. While there are still disagreements among researchers regarding the number of basic emotions (Ortony et al., 1992b), three affective states and traits—anxiety, obsessiveness, and depression (Oatley and Johnson-Laird, 1987; Ortony et al., 1988)—have been studied extensively; findings of these studies are briefly summarized below. These findings illustrate the impact of emotion on cognitive processing and the central role of emotion in the control of behavior. They also begin to blur the distinction between what have traditionally been thought of as the separate realms of cognition and emotion. Modern examples of conformance to group norms in highly charged situations, such as protest marches, exemplify the point.

[4]This section borrows heavily from Hudlicka (1997).

Participation in such an event is a deliberated decision; behavior during the march is determined partly by emotive situations and the behavior of others.

Anxiety The primary impact of anxiety is on attention. Specifically, anxiety narrows the focus of attention and predisposes toward the detection of threatening stimuli and the interpretation of ambiguous stimuli as dangerous (Williams et al., 1997; Williams et al., 1988; Mineka and Sutton, 1992).

Obsessiveness A number of studies have documented the impact of obsessiveness, characterized by "checking" behavior, on cognitive processing. Among the primary effects identified are lack of confidence in one's own attention apparatus to capture salient features in the environment (Broadbent et al., 1986); narrow conceptual categories (Reed, 1969; Persons and Foa, 1984); poor memory for previous actions and a general lack of certainty about one's own ability to distinguish between events that occurred and those that were planned or imagined, and slower decision-making speed related to obsessive gathering of confirming information (Sher et al., 1989).

Depression The primary impact of depression is on memory. Perhaps the best-documented phenomenon is mood-congruent recall in memory (Bower, 1981; Blaney, 1986), whereby a particular affective mood induces recall of similarly valenced memories (e.g., a depressed mood enhances recall of past negative experiences, including negative self-appraisals). Depression has also been studied in the context of particular inferencing tasks, such as judgment and decision making, in which depression appears to lower estimates of the degree of control (Isen, 1993). Even more to the point, a positive disposition appears to be correlated with high-quality performance among managerial-level employees (Straw and Barsade, 1993).

Attitudes and Expectancies

Attitudes were once defined as predispositions to respond in certain rigid ways. That is, an individual with a negative attitude toward a particular situation would always respond negatively or with avoidance. Current conceptualizations, however, distinguish between fundamental and transient attitudes (Heise, 1979) and allow for a change in position. This approach includes feedback controls that continually rebalance the relationship between the two types of attitude. Thus behavior congruent with fundamental attitudes can result in a change in the transient state, and if a difference in tone between fundamental and transient attitudes is detected, behavior that will minimize the discrepancy is initiated. These are core propositions of affect control theory (Heise, 1992)

In prior, stage-setting work, Osgood (1967) used the techniques of semantic differential in a multinational survey approach to test the idea that attitude response sets are aligned along three affective dimensions: evaluation (good or bad), potency (strong or weak) and activity (dynamic or static). Under this classification, a commander's attitude toward an enemy force might be bad, strong, and dynamic, and these tendencies would guide his/her actions in confronting the enemy force.

Expectancies are a product of the interaction between beliefs and attitudes. Hypothetically, beliefs are purely cognitive—based on past objective experience. When a belief and an attitude come together, an expectancy is established. Such a state is not only a subjective estimate of the probability of a given occurrence, but also a perception moderator such that low subjective probability of occurrence is associated with a form of perceptual denial. If an event falls outside the beliefs and attitudes of the perceiver as manifested by his/her expectations, the event is susceptible to misperception and misidentification. The linkage of attitudes to the quality of military performance probably should be through the concept of expectancies. That is, there appear to be no a priori grounds for suggesting that any specific pattern of attitudes will generate any particular impact on military performances (Cesta et al., 1996). However, if an attitude influences expectancies—particularly if it is not supported by objective facts—it can influence decisions made about the object.

Cultural Values

Cultural values and their impact on behavior are of critical importance to the military. Although the study of these values and the influence of nationality on behavior is an active area of behavioral research (Berry et al., 1992), the efforts to date have been limited, and most of the research is not centrally relevant to military operations. While anthropological studies reveal that cultural values have a profound effect on the behavior of the individual warrior (Thomas, 1994) and the planning of war strategy (Pospisil, 1994) most such research has been carried out in the setting of preindustrial societies (Carneiro, 1994).

Cross-cultural studies that compare behaviors in technologically advanced societies tend to focus on either industrial/commercial operations (Carey, 1994) or educational settings (Rosen and Weil, 1995). The Rosen and Weil report is an example of a study of nationality and performance in which the researchers looked at relative "computer anxiety" among college students in 10 countries. The result was a pattern of three clusters: high anxiety on the part of the Japanese and Czechoslovakians and low anxiety on the part of the Israelis, with the remaining (including U.S.) students falling in a large middle cluster. In support of their work, Rosen and Weil suggest that despite its industrial strength and technological expertise, Japan shows a low level of utilization of information processing systems, and the Japanese demonstrate a general lack of individual exposure to information technology that could be the cause of their anxiety. Although the Rosen and Weil study provides no observations on the relative competence in the use of computer technology that might be shown by representatives of different nationalities, it does suggest the possibility of some variation among nationalities, particularly in operators and beneficiaries of computer-based command and control systems.

In contrast to the limited data on nationality/culture and the performance of militarily relevant tasks, conceptual modeling of cultural values has an elaborate history. The key empirical contribution to the formulation of cross-cultural psychology was made by Hofstede (1980), who conducted an extensive study, using questionnaires, of the employees in a multinational firm. Citizens of 67 nation-states were participants. Factor analytic techniques revealed a set of four major dimensions: power distance, uncertainty avoidance, individualism, and masculinity/femininity. Power distance is interpreted as a characteristic of the culture, rather than an individual differences dimension. Basically, it corresponds to the degree to which societal organizations stress status hierarchies. The higher the score, the more transactions between individuals in different strata are restricted. A possible implication is that procedures such as adherence to the formal chain of command will be enforced more rigorously in a culture with high power distance.

Hofstede does not link power distance scores (or other factor scores) to variations in performance. Possible performance outcomes can only be inferred on the basis of other behavioral ramifications. For example, it is relevant in a military engagement context to know the extent to which subordinates can discuss the desirability of a given tactic with the command authority who issues a tactical order. One might infer that for nationalities having high power distance, the amount of discussion of tactical orders would be minimal. However, building a performance model on the basis of such inferences is difficult.

Helmreich and colleagues (Helmreich et al., 1996; Sherman and Helmreich, 1996; Merritt and Helmreich, 1996) took Hofstede's approach further toward explicit treatment of performance in the setting of advanced systems operations. Specifically, they looked at a large set of pilots and copilots of commercial aircraft from many countries. Their main focus was on pilot attitudes toward automation, in particular their affective responses to an advanced training program in cockpit resource management. In light of Hofstede's findings, they also considered behavior such as the acceptance or rejection of messages from subordinates about system status. Such behavior could be relevant in a military context, but direct linkages between scores on the cultural value dimensions and performance were not identified.

Even taking a very broad view of culture and nationality does not produce many findings of specific relevance (see, e.g., Berry et al., 1992). Matters such as the powerful value commitments of soldiers in the subcultures of elite units are rarely studied. Present sociological and anthropological studies focus on issues such as the effects of the representation of events of the Gulf War to civilian populations by the mass media (Mann, 1988; also see Berry et al., 1997). Some connections appear in the area of leadership styles of various nationalities and the contrast between vertical and horizontal organizational structures (see Chapter 10). Whether there are important variations in the organization of military formations as a direct consequence of cultural values is an open question. One factor that holds promise for varying in a systematic way across nationalities is risk taking. Brown (1989) suggests, for example, that risk taking is valued much more in western European and North American societies than in other cultures. If this is true for military commanders, it could be a crucial variable for some forms of simulation.

Possibly the most cogent basis for research directed at the influence of cultural values on the performance of military combatants comes from work on culture and cognition. Much of the momentum in this research area is centered at the University of Chicago (see, e.g., Stigler et al., 1990). The movement has satellite centers at the University of Pennsylvania (Heine and Lehman, 1997) and Kyoto University in Japan (Kitayama and Markus, 1994) that are producing research findings at a basic level. However, the work that is beginning to take on the shape of a computational model is that of Forgas (1985, 1995), which links the affective state (being culturally defined) to strategic decision making and high-level problem solving. While it is still not possible to assert that a soldier from a particular culture will be culture bound to respond in a particular way to a given battlefield event, the impact of culture is being investigated in a progressively more rigorous manner.

Other Intrinsic Variables

In addition to personality, emotion, and cultural values, a number of other intrinsic behavior moderators may impact performance and can be parameterized (Hudlicka, 1997). These include, at the individual level, the following:

- Cognitive styles (attentional factors, metacognitive skills, susceptibility to cognitive bias, assimilating versus accommodating, analytic versus initiative, goal-directed versus data-directed)
- Individual preference for or avoidance of situations, activities, types of units, types of targets, and previous operations
- Training/education, including doctrinal emphasis on units and types of maneuvers, doctrinal preferences for timing of operations, and specific areas of competency

Additional moderators include social factors such as demographic homogeneity, educational homogeneity, cultural diversity, and cultural stability. We do not discuss these variables here, but merely mention them to show the range. As noted earlier in this chapter, our purpose is to select a few variables for discussion to illustrate the state of knowledge about individual differences and how this knowledge might be applied in representing human behavior in computational models.

MODELING BEHAVIOR MODERATORS

This section reviews various attempts to incorporate behavior moderators into computational models of human behavior. The first is the Army's work on the improved performance research integration tool (IMPRINT) (Archer and Lockett, 1997; Allender et al., 1997a), to which physical environmental stressors, mental workload, personnel characteristics, and training are added to make the human component more realistic. A primary use of this model is to provide human factors input for models designed to test and compare the performance capabilities of various systems considered for acquisition. We then review efforts to develop models that incorporate affective variables into synthetic agents (Moffat, 1997; Elliot, 1994; Cesta et al., 1996). The focus of this work has been twofold: to understand human behavior and to build realistic synthetic actors for entertainment. We then examine an approach to the specific problem of incorporating emotion into the simulation of command decision making proposed by Hudlicka (1997). Finally, we describe some alternative approaches to modeling the effects of intrinsic moderator variables on perception and situation awareness. We include these examples to show that even with imprecise definitions and weak quantification, some progress has been made toward incorporating such variables in models of human behavior.

IMPRINT

As described by Allender et al. (1995, 1997a, 1997b) and Lockett and Archer (1997), IMPRINT operates as an event-based task network in which a mission is decomposed into functions that are further decomposed into tasks. The tasks are linked together in a network that represents the flow of events. Task performance time and accuracy, along with expected error rates and failure to perform, are entered for each task. These data, obtained from military research studies (field tests, laboratory tests, and subject matter experts), are assumed to be representative of average performance under typical conditions.

In addition to basic task network simulation, IMPRINT also provides the capability to incorporate the effects of training, personnel characteristics, cognitive workload, and various environmental stressors (see, e.g., Kuhn, 1989). Stressors degrade performance, whereas training can both increase and degrade performance.

The stressors include protective clothing (mission-oriented protective posture, or MOPP), heat, cold, noise, and hours since sleep. Each of these stressors can have an impact on one or more of several classes of human behavior, such as visual, numerical, cognitive, fine motor, and gross motor (referred to as taxons). Taxons are basic

TABLE 9.1 Environmental Stressors and Taxon Effects in IMPRINT

Taxon	Military Operations Protective Posture	Heat	Cold	Noise	Sleepless Hours
Visual	T	A	T		
Numerical		A			TA
Cognitive		A			TA
Fine Motor Discrete	T	A	T		
Fine Motor Continuous					
Gross Motor Light	T		T		
Gross Motor Heavy					
Communications (Read and Write)		A			
Communications (Oral)	T	A		A	

NOTE: T, affects task time; A, affects tack accuracy; TA, affects both.

task categories. The nine taxons used in IMPRINT were derived from the task taxonomies described by Fleishman and Quaintance (1984). Table 9.1 provides an overview of the proposed effects of the environmental stressors on the various taxons. Taxons are used to characterize tasks by means of a weighing system, and since stressors are linked to taxons, they are similarly weighted. Stressors can have two types of effects—they can increase performance time or reduce performance accuracy. The data used to determine these values for each taxon are drawn from several military publications, including Mil STD 1472D (1981), Mil-HDBK-759-A (1981), Belenky et al. (1987), Ramsey and Morrissey (1978), and Harris (1985). A user of the IMPRINT tool can compute expected soldier performance in terms of speed and accuracy by designating intersects of capabilities and tasks.

Capturing Emotions in Synthetic Agents

According to Simon (1967) and Sloman (1997), affect is incorporated into agent behavior as soon as the agent is assigned a goal or objective; obstacles to achievement of the objective are the specific instigators of affective responses (Aube and Sentini, 1996). The precise form of the affective state and its influence on behavior are additionally contingent on the intrinsic attributes of the agent (Dyer, 1987).

Several investigators in artificial intelligence and computer science have begun work on incorporating affective variables such as personality and emotion into synthetic agents (Brehmer and Dorner, 1993; Nass et al., 1997). This work is at a fairly primitive level and, as noted above, has been pursued primarily to provide a vehicle for further study of human behavior or to develop more realistic actors in synthetic entertainment environments. For example, Rao (1996) uses the concept of the BDI agent, that is, one that possesses beliefs, desires, and intentions. These attributes are assigned to agents by means of Horn-clause logic and PROLOG. Likewise, Aube and Sentini (1996) have incorporated factors such as loyalty, patriotism, pride, and anger in their synthetic agents. They refer to these concepts as "regulators." Cesta et al. (1996) assign one of four dominant attitudes to synthetic agents: solitariness, parasiticalness, selfishness, and socialness (e.g., giving and receiving help). Distinctive behavior emerges from these agents when they engage in a bargaining game, even when the agent's characters are one-dimensional.

Giunchiglia and Giunchiglia (1996) suggest that agents can be programmed to hold beliefs about the views of other agents. In context, this would be equivalent to a condition in which the OPFOR (opposing force) commander possessed ideas about the BLUFOR (blue force) commander's ideas about the OPFOR commander's behavior.

Moffat (1997) proposes a simplified model of emotion that can be used to provide synthetic agents with "personalities." In this work, personality is defined as follows (p. 133):

> Given an agent with certain functions and capabilities, in a world with certain functionally relevant opportunities and constraints, the agent's mental reactions (behavior, thought, and feeling) will be only partially constrained by the situation it finds itself in. The freedom it has in which to act forces any action to reveal choice or bias in the agent, that may or may not be shown in other similar situations. Personality is the name we give to those reaction tendencies that are consistent over situations and time.

Moffat's model is named "Will" after William of Occam (for parsimony). Some of the true emotions built into Will include happiness, anger, and pride. Each of these is characterized by valence (whether it is pleasant or unpleasant), by agency (who is responsible for the event—the world, the user, or the self), and by morality (whether the action or event is considered morally good or bad).

Among the most recent implementations are those of Padgham and Taylor (1997), who emphasize simplicity of the program structure, and Muller (1997), who emphasizes the richness of interactions between cognitive and affective elements. However, none of these implementations is directly germane to the simulation of military engagements.

Approaches to Modeling the Effects of Moderator Variables on Perception and Situation Awareness[5]

In this section, we describe how individual difference variables could be used to moderate perception and situation awareness within the context of the

[5]Borrows heavily from Hudlicka and Zacharias (1997).

types of process models described in Chapter 7. In particular, we describe how moderator parameters could modulate a perception model built on modern estimation theory, and how such parameters could also be used to modulate situation awareness models built around, respectively, the technologies of expert systems, case-based reasoning, and belief networks.[6]

Modern Estimation Models of Perception

The discussion earlier in this chapter of perceptual models that might feed into a situation awareness model outlines a hybrid model, consisting of a state estimator and an event detector, that (1) generates a current state estimate of any continuous dynamic variables and (2) identifies current task-relevant events and features (see Chapter 7 for a full discussion). The event detector component of this model can be thought of as a *cue extraction mechanism* that combines both data-directed and goal-directed processing to identify salient perceptual features and important events in the incoming data stream (i.e., the event detector module will be more efficient at extracting features for which it has been trained). The state estimator can be implemented as a Kalman filter (Kleinman et al., 1971) containing an internal model that captures the dynamic properties of each object in the environment. The event detector can incorporate a rule-based event classification mechanism with a fuzzy-logic front end, allowing for manipulations of the certainty of the detected events and cues. Stressor/moderator parameters of this model could include the following:

- **Content parameters**—sophistication and accuracy of the internal estimator model, number of events that can be detected, preference for particular events
- **Process parameters**—number of variables that can be processed simultaneously, speed and accuracy of state prediction and event detection

The primary individual differences that could be represented in the state estimator and event detector include training, skill level, and individual history, which affect both the number and type of state variables predicted and events detected (i.e., sophistication of the commander's internal model) and the efficiency of prediction and detection (e.g., well-trained events are detected more readily, requiring lower signal strength or fewer triggering cues). This approach could also be used to represent the documented effect of anxiety on interpretive functions (MacLeod, 1990) by biasing the interpretation of certain neutral stimuli as threatening.

---

[6]We recognize that there are many other submodel architectures and technologies that could be used in these examples. Those cited here are used to illustrate how moderator effects could be modeled within a "process" context, and not simply as an add-on "performance" moderator.

Models Based on Expert Systems

There are a number of ways to introduce stressor/moderator effects into situation awareness models based on expert systems, such as the following:

- **Content parameters**—number of rules, type/content/complexity of rules
- **Processing parameters**—speed and accuracy of matching, type of conflict resolution, speed of conflict resolution

All categories of individual differences can be represented in a situation awareness model based on expert systems. Varying levels of skill and training could be represented by varying the number and content of individual rules—these include both the triggering antecedent conditions and the consequents for identifying specific situations. The degree of certainty associated with the assessment, which is a function of both skill level and affective personality factors, could be modeled by varying the degree of matching accuracy required for rule triggering. Differences in situation assessment style could be represented by specific assessments associated with particular triggering conditions. Thus a model of an overly cautious commander could have a preponderance of situations characterized by high blue losses, while a more aggressive and risk-tolerant model could include a higher number of red loss consequents. Other style preferences, such as speed of assessment or specific requirements for certainty, could be modeled in terms of processing parameters, controlling the speed of rule matching and execution and the certainty of rule activation strength required to trigger action.

Additional discussion of how moderators can be introduced into production rule decision-making models, specifically Soar and adaptive control of thought (ACT-R), is provided in Chapter 6. That discussion is directly pertinent as well to a situation awareness model based on expert systems.

Models Based on Case-Based Reasoning

An approach similar to that used for models based on expert systems can be used to incorporate stressor/moderator effects into situation awareness models based on case-based reasoning. For example, the following parameters could be included in such a model:

- **Content parameters**—number of cases, elaboration of cases, type and number of index features
- **Processing parameters**—speed and accuracy of matching index cases, sophistication of adaptation rules

As with the expert system-based models, all categories of individual differences can be represented through suitable incorporation of moderator variables into episodic or case-based reasoning models (see Chapter 5 for more discussion).

Models Based on Belief Networks

With respect to a belief network approach to constructing situation awareness models, recall from Chapter 7 that the situation assessor could be implemented as a set of belief nets (Pearl, 1986) that represent knowledge as nodes and links. Nodes represent particular features and events, intermediate results, and final overall situation types. Links represent causal and correlational relations between events and situations represented by the nodes and are associated with conditional probability tables. Stressor/moderator parameters of this model could include the following:

- **Content parameters**—content of belief network nodes, number of nodes, network topology, belief network conditional probability tables, number of distinct belief networks
- **Process parameters**—speed of propagation of probabilities along belief network links, speed of node belief updating, affective valence associated with particular nodes (e.g., a value associated with a particular situation or feature indicating whether it is desirable or undesirable)

All categories of individual difference variables identified earlier in this chapter can be represented in a belief network-based situation assessor. For example, varying levels of skill and training could be represented by varying numbers and types of belief network nodes and more finely tuned link probabilities, with higher skill levels being characterized by more refined nodes allowing greater accuracy, speed, and completeness of situation assessment. Individual history could be represented by weighting certain links more heavily, thus increasing the likelihood of their activation during processing and the consequent derivation of the corresponding specific situations. Personality/affective factors could be represented by associating a specific affective value (positive or negative) with previous situations. Depending on the current affective state of the agent, certain situations might then be recalled preferentially as a result of mood-congruent recall (Bower, 1981; Blaney, 1986), increasing the likelihood of their recognition, while others might be avoided, decreasing the likelihood that they would be selected during situation assessment. Finally, differences in cognitive and decision-making styles could be represented by the setting of processing biases to favor activation of nodes with certain properties (e.g., a recency bias being reflected by the most recently activated node or the ability to favor goal-directed versus data-directed reasoning) and by node number and content (e.g., a larger number of more refined nodes representing an accommodating preference and a smaller number of more generic nodes representing an assimilating preference, or the node level of abstraction representing case-based vs. first-principles situation assessment).

An Approach to Modeling the Impact of Affective States on Command Decision Making[7]

In this section, we describe how affective variables could be used to modify command decision making within the general context of existing cognitive architectures, such as ACT-R and Soar (see Chapter 3 for further discussion of these and other such models). The basic approach focuses on the use of symbolic, modular cognitive architectures that support parameter-controlled processing, where the parameter vectors represent distinct individual differences across the range of variables discussed earlier. Figure 9.4 illustrates the general notion of individual difference parameters selectively and differentially moderating different processes and resources in a given cognitive architecture, thus leading to observable changes in the behavioral output.

Below is an outline, developed by Hudlicka, of the impact of selected affective states (anxiety, obsessiveness, and depression) on the decision-making process of a platoon leader planning an ambush.

Mission, Enemy, Terrain, Troops, and Time (METT-T) Analysis

- Depressed state induced by general combat stress level can induce recall of previous failures associated with similar situations (e.g., similar weather, enemy unit involved, friendly task elements, terrain).
- Lack of time available for planning can increase anxiety and fatigue in an obsessive platoon leader, who requires more time for decision making.
- Platoon's interim isolated position can increase leader's anxiety.

Tentative Plan Construction

- Anxious platoon leader can overestimate danger potential of situation and may commit a larger-than-necessary number of troops and resources.

Situation Estimate and Course-of-Action Development

- Anxious and obsessive leader can be concerned that enemy can bypass ambush too easily and may be tempted to spread personnel out too much to cover more ground.
- Depressed leader may overestimate likelihood of losing critical equipment (e.g., helicopters).
- Obsessive leader may micromanage the process, thereby creating additional bottlenecks (e.g., insist on speaking with helicopter pilots).
- Anxious leader may overestimate danger potential and choose an inappropriate, overly conservative course of action.

[7]This section borrows heavily from Hudlicka (1997).

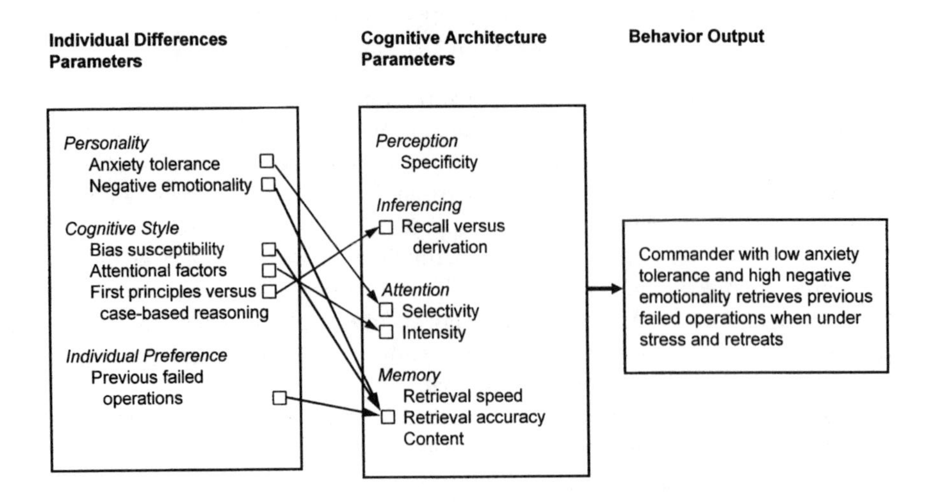

FIGURE 9.4 Summary of modeling intrinsic behavior moderators for command decision making.

- Obsessive leader may rehearse more thoroughly, thereby increasing chances of success.

Plan Monitoring and Execution

- Anxious leader may interpret a signal as approaching enemy and fire too early, thus betraying platoon's position to enemy.
- Obsessive leader may not trust incoming data about approaching enemy and wait too long before initiating action.

The above behaviors can be modeled with a cognitive architecture in two ways:

- By modifying the model content to reflect the desired characteristics, that is, by constructing the model knowledge base in such a way that the desired behavior can be generated
- By setting the model processing parameters to reflect a particular emotional orientation or emotional state, thereby biasing the model processing in a particular direction

Both of these approaches are illustrated below by examples showing how some of the specific conditions and behaviors could be represented in a model of human performance that assumes the architectural features of Soar and ACT-R.

Example 1 *Depressed State Induces Recall of Previous Failures*

To model this phenomenon successfully, a model would require the following features:

- A collection of past failure situations encoded in memory
- An affective tag associated with each memory element, indicating its affective valence (positive, negative, or neutral)
- A parameter indicating the current affective state of the commander
- An associative search process biased toward the recall of information matching the current affective state
- A perceptual process capable of encoding the current situation in terms of features characterizing encoded memory at the appropriate level of abstraction (i.e., particular weather, unit type, maneuver type)

Example 2 *Lack of Time Increases Anxiety in an Obsessive Leader*

To model this phenomenon successfully, a model would require the following features:

- A parameter indicating commander traits, set to indicate high obsessiveness
- A link between this parameter value and a set of planning preference parameters, such as time required to make a decision and number of decision iterations
- A link between the planning preference parameters and the current anxiety state of commander, allowing an increase in anxiety if the time available is less than the time required or if the number of iterations possible is less than the number of iterations required

Example 3 *Anxious Leader Overestimates Potential Danger of Situation and Commits a Larger-than-Necessary Number of Troops and Resources*

To model this phenomenon successfully, a model would require the following features:

- A parameter indicating the current affective state of the commander
- A representation of alternative plans in terms of specific features, such as number of units committed, amount of resources allocated to each unit, placement of units relative to each other, and maneuver type
- An indication of how the value of each of the above features varies with different affective states (e.g., increased anxiety increases resource requirements)
- A mapping between the above current affective state and feature values, and selection of a plan that matches the values of the current affective state

Implementing the suggestions in the above examples involves modeling and manipulating the agent's internal behavior at the information processing level.

This work is in contrast to the efforts on IMPRINT, in which the focus is on behavior modification at the task level.

CONCLUSIONS AND GOALS

Although data exist on the effects of extrinsic moderator variables on human performance, most of these data provide only cutoff points or limits, rather than performance degradation functions. As a result, these variables cannot be encoded directly into models of human behavior. However, as shown by the work on IMPRINT, it is possible to develop and test hypotheses about the effects on performance time and accuracy of specific levels of such variables as temperature, noise, and sleepless hours.

With regard to internal moderators of behavior, several theories have been developed that categorize the variables related to personality, emotion, and cultural values. In some cases, the theories associate the variables with variations in performance. Yet there is a great deal of overlap among the variables classified under personality, emotion, attitude, and cultural values, and most of the empirical data in this area are subjective and qualitative. Some preliminary success has been achieved with the introduction of internal behavior moderators into models of human behavior. However, these models include extremely simplistic representations of the relationships between specified levels of one or more of these variables and performance.

Short-Term Goals

- Apply existing knowledge about both extrinsic and internal behavior moderators to establish value settings for various parameters of human behavior. That is, estimate how specified levels of sleep loss or fatigue, for example, might affect attention, multitasking, or decision making, and observe the effects of the use of such estimates in a sample of simulated engagements.
- Study the effects of introducing emotion into models of human behavior in the form of relatively simple algorithms. For example, set up demonstrations to show how different levels and types of emotion affect decision-making outcomes in the models. Use subjective assessments by experienced operational personnel to determine the value of these additions for increasing the realism of event sequences.
- Formulate a research strategy for developing sufficient knowledge about selected behavior moderators to provide a basis for encoding these variables into computational models.

Long-Term Goals

- Develop computational models of behavior moderators that are based solidly on empirical research findings.

10

Modeling of Behavior at the Unit Level

INTRODUCTION

Within seven minutes, the crew of the American Warship Vincennes, faced with a dynamic and hostile environment, had to decide whether an approaching aircraft was hostile (Cooper, 1988; Rochlin, 1991). Whether the decision made—which resulted in shooting down a nonhostile aircraft—was right or wrong is not the issue. The decision certainly was critical: failure to respond appropriately could have resulted in the Vincennes being attacked and possibly sunk. Such decisions are the result of a variety of factors at both the individual and unit levels (Roberts and Dotterway, 1995). Among the factors at the unit level are the number of opportunities for the organization to review its decision within the abbreviated decision period (Cohen, 1988), the type and appropriateness of the training provided to the decision makers (Duffy et al., 1988), the quality and accuracy of the information available (U.S. Congress, 1989), the hierarchical command structure, and inherent bias in the operating procedures involved (Watson et al., 1988). As Admiral William J. Crowe Jr., then Chairman of the Joint Chiefs of Staff, commented: "The rules of engagement are not neutral. They're biased in favor of saving American lives."

Are there changes that could have been made to the Vincennes' command, control, and communications (C^3) structure that would have altered the course of action taken? Many have argued that changes in the C^3 structure could in fact have led to a different outcome. Research on organizations and teams suggests that a less hierarchical structure, more opportunities for review, and alternative training scenarios could all have altered the outcome. The Vincennes incident illustrates how the C^3 architecture of the organizational unit can influence its performance. Reasoning about the unit requires understanding not just how the individual agents perform their roles, but also how the distribution of resources and tasks across personnel and the specific C^3 architecture constrain individual action, alter the flow of information, and ultimately influence the unit's performance. If appropriate computational models of organizational units and their C^3 structures were available, one could begin to reason about questions regarding the design of C^3 architectures such as that of the Vincennes. One could also reason about the behavior of other structures, such as coalition C^3 and the overall behavior of opposing forces.

The purpose of this chapter is to describe current computational approaches to modeling of C^3 architectures and the tools and techniques needed to reason about C^3. The focus is on unit-level models in which each of the actors (commander and subordinates) is modeled, as well as some of the factors and procedures that link these actors together. Thus, a unit-level model of a fixed air wing might model each team member, along with the lines of communication and types of messages, reporting functions, norms about contacting or not contacting others, social knowledge about how to operate as a team, intelligence about what to do when one or more team members become disabled, the tasks, the authority structure, the procedures, differences in actors due to physical position, available resources, training both individually and as a team, and military level. We might characterize such C^3 models as social agent models in which there are multiple agents connected by one or more networks.

Such computational models of organizational units have a large number of potential uses in military settings. For example, imagine a man-in-the-loop simulation used for training a brigade commander and staff. Unit-level models could be used to simulate the behavior of units under the commander's direction. Realistic models of the subordinate units would enable the commander to use more realistic simulations of battle engagements to explore the impact of communication losses, personnel losses, manpower reduction, misunderstanding of orders, resource attrition, and so on. Having such models would also reduce training costs as not all units would have to be present at the same time. Moreover, units could take part in joint task force operations without the complete unit being sent. As another example, imagine a computational model of information flow in a C^3 architecture. Realistic unit-level models could be used by commanders to perform a series of "what if" analyses that would help determine the relative merits of putting in place various types of information and communications equipment. Models of the organizational unit could also be used for examining resource reallocation strategies, determining the fragility or flexibility of an organizational unit under stress, and exploring viable adaptation strategies. Computational models of the unit's C^3 structure may also be an important part of planning modules. Finally, C^3 models could be useful in determining areas of vulnerability in opposing forces.

Currently within the Defense Modeling and Simulation Office (DMSO), and indeed across the services, there are very few organizational unit-level models of any sort, and even fewer of the multiagent or network type that are discussed in this chapter. Those models that do exist within DMSO and the services are often inflexible and cannot be adapted to different environments. Further, the current military models of opposing forces tend to be static or to represent the unit as an aggregate, and so sidestep issues of command and control. Unit-level learning models, such as those described in this chapter, in which the collection of agents and the networks connecting them are modeled, have the potential to change this situation. Such models could be used to examine the potential impact of changes in C^3 structures. They could also be used to examine the relative effectiveness of a given C^3 structure in conflicts with opposing forces having a different C^3 structure. Unit-level learning models may be particularly useful for examining issues of unit effectiveness and flexibility in a dynamic and volatile environment. Finally, general improvements in computational models that involve multiagent learning may have general application as optimization techniques for determining the optimal C^3 structure when the relationship between performance and the various characteristics of individuals and the unit is extremely complex.

In the military context, models that attempt to speak to unit-level issues typically take one of two forms: (1) the big-picture simulator and (2) expanded individual models. The big-picture simulators are simulation models in which the unit is typically characterized by a set of very general or aggregate features, such as size and number of resources. Such models have been used to calculate general levels of attrition or overall troop movement. They are typically instantiated as a series of equations reflecting average behavior. Details on individual cognition and differences in individual behavior are typically ignored. Team- and group-level issues, such as the impact of the existing knowledge network (who knows what, and who knows who knows what), the underlying communication network (who talks to whom), effects due to the specific authority structure (which is particularly crucial in joint operations), and so on are also completely ignored. The big-picture simulators that have been used for modeling blue forces, red forces, joint forces, or coalition forces often have only a minimal, if any, model of the C^3 architecture. Such models are particularly useful for making high-level predictions—for example, about attrition—regarding very large groups when the statistical properties of the group are more important than individual differences. These models are not sufficient, however, to address most C^3 concerns. For example, a typical way of representing C^3 issues is as an error rate around some distribution of response. However, since the C^3 architecture itself can affect performance in complex and nonlinear ways that vary by task, having a computational module that alters organizational unit behavior based on the C^3 architecture could increase the validity of these models.

The expanded individual models are simulation models of individual agents in which each agent has, as part of its knowledge base, some organizational unit knowledge. An example of this approach is the Soar-intelligent forces (IFOR) model that was deployed in synthetic theater of war (STOW)-97. These models are particularly useful for modeling very small units (3 to 12 members) that are engaged in tasks requiring a high level of cognitive processing. Such knowledge might include rules about who takes over if the commander is incapacitated, rules about what communication can and should be sent to whom and in what format, and expectations about the behavior of other members of the organizational unit. A key characteristic of these models is that each agent is modeled separately, with a high level of attention being paid to the modeling of individual cognitive capabilities. Sometimes team models are built by combining a small number (fewer than 12) of these individual agent models. Organizational unit behavior is expected to arise from the ongoing interactions among the agents or to be stored as default knowledge about what to do when certain preplanned contingencies (such as attrition) occur. Details on the social organization of the unit and measures of unit-level behavior are typically not included as an integral part of these models. Norms, roles, and general social knowledge are treated as exogenous.

Team models built by expanding individual models are rare. In the few existing models of this kind, the organizational unit-level knowledge of the team member agents is often based on a characterization of standard operating procedures, expert opinion, or doctrine. Thus, these models are organizationally static because the structure is fixed by these opinions or procedures. Moreover, since the mechanisms for acquiring and using social knowledge are often major determinants of organizational unit behavior, having socially realistic agents could increase the validity and use of these models.

The vast majority of the work on organizational unit-level models has occurred outside of DMSO and the military context. Within this larger context, a third type of organizational unit-level model also exists—the social agent model. These are simulation models of organizational units as collections of multiple agents engaged in performing a distributed task. These models are particularly useful for modeling small to moderate-sized units (3 to 100 members) when the focus is on communication, reorganization, reengineering, or planning. A key characteristic of these models is that each agent is modeled separately, with a high level of attention being paid to modeling of the connections among these agents as determined by the C^3 architecture and the mechanisms for acquiring and using social knowledge. In many of these models, the agents are adaptive. Organizational unit behavior is expected to arise from the ongoing interactions among these agents as situated in the specific C^3 architecture. Details on individual cognition are often minimal.

Clearly, models of unit-level behavior in the military context are currently lacking. The next section examines the rationale for developing such models. This is followed by a brief review of prior work in unit-level modeling that has been done in nonmilitary settings. The various potential applications of such models and some of the additional techniques necessary to make them useful in the military context are then discussed relative to three areas of application—design and evaluation of C^3 architectures, modeling of opposing forces, and unit-level learning and adaptation. Next, four overarching issues that arise during unit-level modeling are reviewed. This is followed by a synopsis of the current state of unit-level modeling languages and frameworks. The final section presents conclusions and goals in the area of unit-level modeling.

WHY MODEL THE ORGANIZATIONAL UNIT?

Models at different levels are useful in different contexts. In physics, for example, Newtonian mechanics is sufficient for modeling and predicting the behavior of large masses, whereas quantum mechanics is needed for modeling and predicting the behavior of specific particles. In general, one does not need to resort to modeling the behavior of each individual particle to model large masses. Indeed, it is impossible to specify the behavior of large masses by specifying the quantum mechanics of all particles because of the large number of complex nonlinear interactions involved. Similarly, models at both the unit (Newtonian) and individual (quantum) levels play an important role in the social world. Different tasks require different models. Sometimes, particularly for man-in-the-loop simulations, both individual-level models (of, for example, key actors) and unit-level models (of, for example, secondary actors), may need to be included.

There are a variety of reasons for pursuing models at both the individual and unit levels. First, even if it were possible to build and run organizational unit models by modeling all the constituent agents, doing so might be too computationally intensive. Second, the behavior of organizational units appears to be more than a simple sum of the parts. Rather, the behavior of the organizational unit is often strongly determined by the ongoing pattern of interactions among the agents, the norms and procedures governing group interaction, and legal and institutional constraints beyond the control of any one agent. This point is discussed further in a later section. Third, at the present time, even if the computational resources needed to build and run organizational unit-level models composed of large numbers of cognitively accurate individual agents were available, it is unlikely that such models would generate organizational behavior. The reason is that to generate realistic social and organizational behavior, each agent would have to possess huge quantities of social knowledge, including knowledge about others, how to act in a group, when to follow a norm, and so on (Carley and Newell, 1994). There is no compendium of such social knowledge that is universally agreed upon and in a machine-readable form or easily translatable into such a form. Thus, the computer scientist looking for a list of the 300+ rules that govern social behavior will not find a set that will be acceptable to social scientists. Although there is a large amount of information, theory, and data that can be drawn upon, accessing such information will require either working with social scientists or making a substantial effort to understand a new field or fields.

Even if such a compendium existed, it would still not enable agents to generate all social behavior as not all behavior emerges from ongoing interactions; rather, some behavior is a result of social or institutional decisions made at earlier points in time or by agents not in the current system. Further, such a social knowledge base might well be far larger than a task knowledge base. Finally, there is a disjunction between the current state of knowledge of unit-level and individual-level behavior. A fair amount is known about how organizational units influence the information to which individual agents have access. A fair amount is also known about how individual agents in isolation learn to do tasks. However, little is known about how individual agents as members of groups perceive, interpret, and use information about the C^3 architecture in deciding how to perform tasks and partition tasks across other agents.

PRIOR WORK IN UNIT-LEVEL MODELING

Research on organizational unit-level models has a fundamentally interdisciplinary intellectual history, drawing on work in distributed artificial intelligence, multiagent systems, adaptive agents, organizational theory, communication theory, social networks, sociology, and information diffusion. One of the earliest such models is presented by Cyert and March (1992 [1963]). Theirs is a relatively simple information processing model. They use this model to explore the impact of cognitive and task limits on organizational economic performance. This particular model, and the results garnered from it, are not directly relevant in the military context, but the newer models are relevant.

Current organizational unit-level models take an information processing approach (for a review of these computational models see Carley, 1995b). The information processing approach is characterized by attention to the cognitive and social limits on the individual's ability to process and access information (Simon, 1947; March and Simon, 1958; Thompson, 1967; Galbraith, 1973; Cyert and March, 1992 [1963]). Today's models are much more realistic than the firm model of Cyert and March and the garbage can model of organizational decision making under ambiguity of Cohen et al. (1972). Today's models also draw on research in various traditions of organization science, including social information processing (Salancik and Pfeffer, 1978), resource dependency (Pfeffer and Salancik, 1978), institutionalism (Powell and DiMaggio, 1994), and population ecology (Hannan and Freeman, 1989).

These models vary dramatically in the level of abstraction and level of detail with which the agent, task, C^3 structure, and technology are modeled. Some are simple abstract models of generic decision-making behavior; examples are the garbage can model (Cohen et al., 1972) and the Computational ORganizational Performance model (CORP) (Lin and Carley, 1997a). Others are very detailed, specific models of organizational decisions or decision-making processes; examples are the virtual design team (VDT) (Cohen, 1992; Levitt et al., 1994; Jin and Levitt, 1996) and Hi-TOP (Majchrzak and Gasser, 1992). In general, increasing the level of detail in these models increases the precision of their predictions and decreases their generalizability.

The more abstract models are often referred to as *intellective models*. Intellective models are relatively small models intended to show proof of concept for some theoretical construct. Such models enable the user to make general predictions about the relative benefit of changes in generic C^3 structures. The more detailed the model, the more specific their predictions will be. Intellective models provide general guidance and an understanding of the basic principles of organizing. To run these models, one specifies a few parameters and then runs one or more virtual experiments. The results from these experiments show the range of behavior expected by organizational units with these types of characteristics. These models are particularly useful for demonstrating the necessity, inadequacy, or insufficiency of specific claims about how organizational units behave and the relative impact of different types of policies or technologies on that behavior. Additionally, these models are useful for contrasting the relative behavior of different organizational units.

The very detailed models are often referred to as *emulation models*. Emulation models are large models intended to emulate a particular organization in order to identify specific features and limitations of that unit's structure. Such models enable the user to make specific predictions about a particular organization or technology. Emulation models are difficult and time-consuming to build; however, they provide policy predictions, though typically only on a case-by-case basis. These models typically have an extremely large number of parameters or rules and may require vast quantities of input data so they can be adapted to fit many different cases. To run the emulation, the user often needs to specify these parameters or rules, using data on the particular organizational unit being studied. The model is then adapted to fit the past behavior of this unit. Once this adaptation has been done, the model can be used to explore different aspects of the organization being studied and to engage in "what if" analysis. One of the main uses of emulation models is getting humans to think about their unit and to recognize what data they have and have not been collecting. These models are particularly useful for engineering the design of a specific unit. Validated emulation models can also replace humans or groups in some training and planning exercises. An example of this was the use of Soar-IFOR and Systems Analysis International Corporation (SAIC)-IFOR in STOW-97.

APPLICATION AREAS FOR ORGANIZATIONAL UNIT-LEVEL MODELS

As previously noted, there are numerous possible applications of organizational unit-level models in the military context. Rather than trying to describe all existing models of organizational unit-level performance and their potential application in the military context, we focus here on three application areas: C^3, opposing forces, and organizational unit-level learning and adaptation. For each of these application areas, the current state of modeling within the military is discussed. Where possible, relevant findings, models, and modeling issues that have arisen in other contexts are examined as well.

Application Area 1: Command, Control, and Communications

As previously noted, C^3 issues are rarely addressed in military simulations. However, under the auspices of DMSO, many training and assessment simulations are beginning to be developed to the point where such issues need to be addressed. Some inroads have been made in the representation of communications that are either commands or required reports and in the division of knowledge in knowledge-based systems such as Soar into that which is general and that which is specific to a particular role in the command structure. Further, some of the DMSO-sponsored models have progressed in terms of command generators, parsers, and interpreters. Exploratory models have been used to look at general battlespace planning, though none of the generic planning models have moved beyond the basic research phase. Even within the existing models, a variety of details remain to be worked out, such as (1) development of standard protocols for representing and communicating commands at varying levels of detail, (2) definition of the task-based knowledge associated with different roles in the authority structure, and (3) modeling of the process whereby command relationships are restructured as various organizational units become incapacitated.

The above steps are necessary for the development of a fully functioning model of C^3, but they are hardly sufficient. The current approach is limited in large part because it focuses on C^3 at the individual level (e.g., representation of officers or the task-based knowledge of specific officers). Current models omit and tend not to consider the full ramifications of structural and cultural factors. An example of a structural factor is the distribution of resources across organizational units or the degree of hierarchy in the command structure. An example of a cultural factor is the degree to which subordinates do or do not question the authority of their commanders. Information and processes critical to C^3 are ignored in many of the models currently employed by DMSO. Such critical information and processes include allocation and reallocation of resources, task assignment and retasking, coordination of multinational forces, coordination of joint task forces, information about who knows what, the information criticality of particular positions in the C^3 structure, and the management of information flow. Current models are often built assuming either that (1) structural and cultural effects will emerge out of the aggregated behavior of individual organizational units, (2) structural and cultural features are relatively simple additions once accurate models of individual soldiers or commanders have been developed, or (3) the effects of structural and cultural factors are second order and so need not be attended to at this point.

Much of the research in the area of teams and organizations suggests that these assumptions are false. Structural and cultural factors are often imposed rather than emergent. Because of the complexity and nonlinearity of the behavior of these factors, as well as their interaction, structural factors should not be represented by simply "adding on" unit-level "rules." Moreover, much of the work on organizations suggests that structural and cultural factors with cognitive and task factors are a dominant constraint on human behavior.

Rather than being emergent, organizational units are often put in place deliberately to achieve certain goals (Etzioni, 1964); for example, joint task forces are assembled, almost on the fly, to meet specific military objectives such as securing a particular port or rescuing a group of civilians. However, the organizational unit must face many uncertainties along the way. For example, the mission plan may be vague, the weather may be changing, various intelligence (such as maps) may be unavailable, personnel may become incapacitated, or communications technology may break down. Organizational units try to maintain control by structuring themselves to reduce uncertainty (Thompson, 1967), constraining individual action (Blau, 1955; Blau and Meyer, 1956), or interacting with the environment in certain preprogrammed ways (Scott, 1981). Much of this structuring is at the C^3 level, and some C^3 forms are much less prone to the vagaries of uncertainty than are others (March and Weissinger-Baylon, 1986; Lin and Carley, 1997a).

Within organizational units, the effects of structure and culture may be so strong that effects due to individual cognition or limited rationality may not be the primary determinants of unit-level behavior. Indeed, in the organizations arena, results from many current models relevant to C^3 issues demonstrate that if the command structure, the flow of communication, and task knowledge specific to certain roles or levels in the C^3 structure are represented, a great deal of the variation in organizational unit-level behavior can be explained without considering individual differences or affect (Levitt et al., 1994; Carley, 1991b; Lant, 1994; Lant and Mezias, 1990, 1992).

In general, there is a strong interaction among C^3 structure, task, and agent cognition in affecting organizational unit-level performance. This interaction is so intricate that there may be multiple combinations of C^3 structure, task decomposition, and agent cognitive ability that will generate the same level of performance. One reason for this is that the C^3 structure can substitute for direct communication (Aldrich, 1979; Cyert and March, 1963), absorb uncertainty (Downs, 1967; March and Simon, 1958), and increase the need for support staff or reports (Fox, 1982). Moreover, the C^3 structure is a primary determinant of what individuals can and do learn (March and Simon, 1958). Similarly, the way a task is structured may affect what individuals learn (Tajfel, 1982), the flexibility of the organizational unit (Carley, 1991b), and the need for communication

(Thompson, 1967). Finally, the agent's cognitive and physical capabilities affect what actions the agent can and needs to take, whereas the social or organizational context affects which of the possible actions the agent chooses to take (Carley and Newell, 1994).

Much of the work in the area of computational organization theory has involved the development of computational models for examining the relative strengths and weaknesses of various C^3 architectures, their susceptibility to performance degradation under different conditions (such as information overload, erroneous feedback, missing information, and missing personnel), and the impact of communications technology on information flow through the organizational unit. Those who have designed, built, and evaluated these computational models of organizational units have faced a variety of modeling and measurement issues to which partial answers are now available: (1) how to measure and monitor changes in the C^3 architecture, (2) how to compare C^3 architectures, and (3) how to determine whether a C^3 architecture is effective. The designer of any C^3 model in which all personnel, resources, tasks, and so on are modeled will face these issues.

Measuring and Monitoring Changes in C^3 Architectures

As noted by Cushman (1985:18), "The nature of theater forces command and control systems is a web of interconnected subsystems supporting the entire spectrum of functionality of the operation from top to bottom. . . ." Good computational models of C^3 architectures will require an effective method for representing this entire web of subsystems and the degree to which they change in response to various threats or the utilization of new telecommunications technology. For example, suppose one wants to evaluate the impact of electronic conferencing or workdesks on a unit's performance. One approach would be to examine changes in the tasks in which personnel engage. However, when new technologies are put in place, organizational units typically restructure themselves by, for example, reassigning tasks, altering the reporting structure, or changing personnel. Consequently, differences in the workload of individual operators may not capture the full impact of the new technology. What is needed are measures of the C^3 architecture as a whole and how it changes. The ability to measure, and thus monitor, changes in the C^3 architecture makes it possible to validate and use models to conduct "what if" analyses about the impact of new policies, procedures, and technologies.

Three common approaches are used to represent the C^3 architecture in models that examine the behavior and decision making of organizational units: the rule-based approach, the network approach, and the petri net approach. Each of these is discussed below.

The Rule-Based Approach One approach to representing the C^3 architecture is to use a rule-based representation for procedures. For example, communication protocols could be represented as a series of rules for when to communicate what to whom and how to structure the communication. This approach is complementary to the network approach described below. Representing C^3 procedures as rules facilitates the linking of organizational unit-level models with models of individual humans, since the unit-level rules can be added to the individual-level models as further constraints on behavior. This is the approach taken, for example, in the team (or multiagent) Soar work (Tambe, 1996b; Tambe, 1997; Carley et al., 1992) and the AAIS work (Masuch and LaPotin, 1989).

There are several difficulties with the rule-based approach. First, much of the research on the impact of C^3 structure may not be directly, or efficiently, interpretable as rules. Second, it is difficult to measure and monitor coordination, communication, and other organizational unit-level activities when they are represented as rules. Third, procedures for segregating agent- and group-level knowledge in a rule-based system need more research attention.

The Network Approach The most common approach to representing the C^3 architecture is to use one or more networks (or equivalently matrices) representing the organizational unit as a set of interconnected networks (hence a set of matrices). For example, a communication structure might be represented as a matrix such that each cell contains a 1 if the row person can communicate with the column person and a 0 otherwise. This approach could be expanded to the full C^3 architecture by specifying the set of matrices needed to represent C^3. Other networks needed for specifying the full C^3 architecture include the skill matrix (the linkages between people and knowledge), the command network, the communication network, the resource access network, and the requirements matrix (the linkages between knowledge and task requirements).

This network approach has many desirable features. First, the representation facilitates the measurement of many factors that influence organizational unit behavior, such as throughput, breaks in communication, structural redundancy, and workload assignment errors (see, for example, Krackhardt, 1994, and Pete et al., 1993, 1994). Second, having commanders or team members specify the structure often helps them identify specific problems in the C^3 architecture. Given a set of matrices that represent the C^3 architecture mismatches, it becomes possible to calculate information flow mismatches and redundancy.

This basic approach of using sets of matrices of relationships to represent the organizational structure is widely used (Cohen et al., 1972; Jin and Levitt, 1993; Carley, 1992). Such a representation is valuable in part because it has face validity. Further, it provides a ready tool for acquiring information from commanders and organizational unit members and for providing the organizational unit with information that will help in altering the C^3 architecture to improve performance. Reconfiguration is readily measured under this scheme as the number of changed linkages. Using this approach, the developers of VDT have been able to capture the structure of various organizations engaged in routine tasks and have identified barriers to effective performance (Cohen, 1992; Levitt et al., 1994). This approach is also being implemented in a series of models for examining strategic C^3 adaptation (Carley and Svoboda, 1996). The network-based approach has spawned a huge number of measures of C^3 design (Carley, 1986a; Malone, 1987; Krackhardt, 1994). One of its major difficulties, however, is that the current visualization tools for displaying networks are extremely poor, particularly when the networks are dynamic or composed of multiple types of nodes and/or links.

The Petri Net Approach A third type of representation is based on petri nets. Petri nets can be used to represent C^3 structures by focusing on the set of allowable relations among agents, resources, and tasks. Given this information, it is possible to generate all C^3 structures that satisfy both general structural constraints (e.g., one cannot have more than five actors) and the specific designer's requirements (e.g., one wants task A to be done before task B) (Remy and Levis, 1988). Lu and Levis (1992) developed a mathematical framework based on colored petri nets for representing flexible organizational structures. This technique allows several key issues in designing flexible C^3 structures to be addressed. For example, it can be used to define the optimal communication and authority structure given a specific task structure. Petri net models can also be used to examine the relative impact of different constraints, such as task scheduling, on the optimal C^3 structure. An alternative approach to matching structure to task is the VDT framework (Levitt et al., 1994), which uses a combined PERT network and organizational chart technique to look at information flow and bottlenecks. It may be noted that the petri net and network-based approaches are not incompatible; indeed, research on integrating the two approaches is called for.

Comparing C^3 Architectures

As noted by Cushman (1985:24), "no two forces are exactly alike." This statement is true whether one is talking about small teams, a joint task force, a battalion, or an integrated battle group. In all cases, the way the unit is coordinated, the way communication occurs, the way authority is handled, and the way skills and resources are distributed vary, even for two units of the same size and nature, such as two wings. The issue is not whether two units are different, but how similar or different they are and how meaningful the differences are. Thus to analyze the relative impact of various C^3 architectures, it is necessary to determine when two C^3 architectures with apparent differences are meaningfully different.

Most statistical tools for determining whether two things are statistically different are applicable only to variable data for which one can easily calculate the mean and standard deviation. For organizational units, the data involved are often network level. Little is known about how to place statistical distributions about networks and so determine when their differences are statistically significant. Until such a capability is developed, it will be difficult to validate simulations, field studies, and experiments that suggest the effectiveness of certain changes in C^3 architecture.

Examining the potential impact of changes in C^3 architectures and being able to contrast different C^3 structures, particularly for very large organizational units, may require being able to subdivide the networks that represent the organizational unit into a series of subgroups, based on the ability of the subgroups to perform some function or meet certain fitness criteria. This capability may be especially critical in the modeling of theater forces, where the force as a whole is better modeled as a collection of parts. For example, when an organizational unit fails or succeeds, one issue is how to determine which part of the C^3 structure is responsible. To answer this and many other questions about C^3 architectures, it is often necessary to be able to subdivide organizational units into substructures. However, subdividing the networks that represent the organizational unit into a series of subgroups based on the subgroups' ability to perform some function or meet certain fitness criteria is a nontrivial problem and represents a methodological challenge.

Optimization and pattern-matching routines emerging from work in artificial intelligence—such as classifier systems and simulated annealing—can be used for partitioning graphs (binary networks) that represent various aspects of the C^3 structure and for comparing alternative C^3 architectures. The area of graph partitioning has received a great deal of attention from both social scientists (Breiger et al., 1975; White et al., 1976; Batagelj et al., 1992) and computer scientists (Bodlaender and Jansen, 1991). Recent work has used genetic algorithms (Freeman, 1993) to partition members of an organizational unit into subgroups, given various criteria. In a related vein, Krackplot (Krackhardt et al., 1994) uses a simulated annealing technique to draw networks with minimal line overlap based on a routine developed by Eades and Harel (1989). If this work is to have value in partitioning C^3 architectures at the organizational unit level, several additional factors will need to be considered, including partitioning of weighted networks, partitioning of networks with colored nodes (multiple types of nodes), and simultaneous partitioning of multiple networks. Such formal routines would help in locating portions of the C^3 architecture that served as bottlenecks or low-performance sectors. Moreover, work on network partitioning is potentially useful in many other contexts, such as circuit layout.

Determining the Effectiveness of C^3 Architectures

Research at the organizational unit level conducted by organization researchers has demonstrated that there is no one right C^3 architecture for all tasks (Lawrence and Lorsch, 1967) and that specific architectures vary in their effectiveness depending on the environment and training of the personnel within the organizational unit (Lin and Carley, 1997b). The organizational unit's C^3 structure can be ineffective for a number of reasons. For example, there may be a mismatch between individuals' knowledge or skills and the task requirements, as happens when personnel trained for one situation are employed in another—a classic problem in situations other than war. Or there may be a mismatch between individuals' access to resources and the resources needed to perform an assigned task. This can happen when, during the course of an engagement, the commander must make do with resources and personnel available, even if they are not ideal.

Methodologically, these findings pose several challenges for the building of appropriate models of C^3 architectures. One such challenge is how to model the C^3 architecture so that both the sources of effectiveness and the organizational unit's flexibility (the degree to which it reconfigures itself) can be measured. A second challenge is how to determine the robustness of specific architectures—the extent to which a particular architecture will remain a top performer even as conditions change (e.g., as the environment changes, personnel gain experience, and personnel are transferred). A third challenge is the building of models of actual organizational units that can adapt their architectures on the basis of what is possible rather than what is optimal. One strategy in the modeling of organizations that has met with success in identifying and accounting for the impact of mismatches between the unit's structure and its needs is the network approach discussed earlier.

Application Area 2: Opposing Forces

Many training and assessment processes require agents acting as a unified force. The discussion here focuses on opposing forces, although the factors examined are relevant in modeling any force. There are several minimal requirements on agents of opposing forces:

- Opposing force agents should act in accordance with the norms and doctrine of their country. Thus, these agents should shoot, stand, march, and so on as such forces would be trained to do.
- Opposing force agents should alter their behavior in response to the situation. There is a debate about whether it is necessary for changes in behavior to reflect accurately what the average person would do, or whether it is sufficient that the behavior simply be nonpredictable.
- It should be possible to reconfigure high-level simulation systems rapidly (say, within 24 hours) so that the opposing forces being modeled reflect those of concern. This would suggest the need for modeling these opposing forces as "variations on a theme." Ideally the opposing forces would also exhibit adaptive behavior, but that may not be as important for many simulation systems, at least initially, as is responsiveness.

It is often assumed that one should be able to isolate a small set of parameters that distinguish usefully among different forces. The research in the field of organizations suggests that factors affecting organizational performance include training, type of training, C^3 architecture, and level of stress. Decades of research indicate that these factors interact in complex and nonlinear ways in affecting organizational unit-level performance. The performance of opposing forces will vary if their agents are trained in different ways, exist within different C^3 structures, or are subject to different levels of stress. Such effects can, within certain limits, be modeled using models of C^3 architectures. For such factors (other than stress), research findings have been encapsulated as a series of rules in an expert system referred to as "the organizational consultant" by Baligh et al. (1987, 1990, 1994). This approach has been useful in helping private-sector organizations determine how to redesign themselves to meet the demands of the task they are facing. This approach, and these rules, could potentially be adapted to the military context. Such an expert system could then be used by scenario generators to determine how to set the parameters for the factors mentioned for a hypothetical opposing force given a particular task.

Whether other cultural factors, such as willingness to countermand orders or to act independently, have organizational unit-level consequences is the subject of current inquiry by organization and team theorists. Anecdotal evidence indicates that such factors may be crucial. For example, organizations in which workers believe their lives are controlled by fate are less prone to be concerned with safety measures. A limited amount of modeling work has focused on at the impact of cultural factors, such as norms, on behavior; the most notable is the Harrison and Carrol (1991) model of performance.

A possible approach to generating models of opposing forces is a parameter-based approach. One set of parameters would be used for individual-level agents and one for group- or organization-level agents. Then given a base set of models (e.g., of agents of different types, for different organizations), an opposing force could be rapidly generated by taking the appropriate set of base models and altering their parameters to reflect the real force of interest. Cultural indicators would in this sense act no differently than structural indicators, such as size of opposing force and number of different types of weapons. Whether such a parameter-based approach is feasible depends, at least in part, on whether opposing forces actually do differ on a series of identifiable and important parameters, such as degree of hierarchy, density of communication, specificity of commands, and ability to question orders. Making this determination might require a detailed statistical analysis of possibly classified data.

Although not impossible, it is often difficult to link the parameter-based and rule-based approaches. There is a need for more cost-effective and efficient systems combining multiple modeling approaches. Consequently, many researchers are working to develop meta-languages that will enable two or more approaches to be combined (e.g., genetic algorithms and neural networks, or rule-based and equation-based languages).

Key issues of concern with regard to the parameter-based approach include the following:

- Is the approach feasible?
- What is the minimum set of parameters needed for representing differences in forces?
- Given the parameters, how are different forces characterized?
- How can these parameters be integrated into existing models (or types of models) to alter behavior? For example, for an individual agent modeled as a subject matter expert, how would these parameters act to alter behavior? Should they be used as probabilities for altering the likelihood of rule firing or as sets of predefined alternative rules?
- Does the parameter-based approach reflect reality?
- Would adaptive agent models be better than a parameter-based approach, or should some combination of the two be used?

Application Area 3: Unit-Level Learning and Adaptation

Plans and C^3 architectures rarely last past contact with the enemy (Beaumont, 1986:41). Over the course of a battle, the rules of engagement may change, troops may die, resources may get destroyed, weather may affect deployment, and communications links may become inoperable. Further, changes in technology and legislation mean that C^3 architectures are never complete and are continually undergoing change. As noted by Alberts (1996:44), "the ability to maintain mission capability while upgrading or integrating systems remains crucial." For operations other than war, there may be little historical knowledge to act on, so plans may need to be developed as the operation is carried out. Since military organizational units can and do change, there is a need to know how to design them so that when the need arises, they can adapt (i.e., alter their C^3 architecture so as to maintain or improve performance). Moreover, for planning and analysis purposes, there is a need to be able to forecast how organizational units are likely to change given various triggering events (such as resource destruction). Computational models of adaptive units can help meet these needs. However, as noted earlier, the C^3 architectures in most computational models used in military settings at the organizational unit level for both training and planning assume a fixed C^3 architecture.

There is currently a great deal of interest in learning and/or adaptation at the organizational unit level (see also Chapter 5). DMSO, the various services, and nonmilitary funding agencies have been supporting a wide range of effort in the area of organizational unit-level adaptation. Research on organizations (Lant, 1994), on artificial life (Epstein and Axtell, 1997), and on distributed artificial intelligence (Bond and Gasser, 1988) has begun to explore the implications of agent and organizational unit adaptation for organizational unit-level behavior. The work on unit-level learning and change has been done under various rubrics, including change, learning, adaptation, evolution, and coevolution. Much of this work has been carried out using computational models composed of multiple individual agents that interact and have the ability to learn. Computational techniques that show strong promise in this area are genetic programming, neural networks, and simulated annealing.

Before continuing with a discussion of what has been done in this area, it is worth noting that computer learning techniques have been used to address many different issues of unit-level adaptation from a variety of vantage points. First, these learning approaches are used to suggest a new C^3 architecture given a task or a set of constraints. Thus they are being used not to model the specific adaptation of an organizational unit, but to model the kind of architecture that should or is likely to emerge. In this way, these models can be used to establish "confidence regions" around a set of possible structures. Second, these learning approaches are being used to develop a basic understanding of the value, cost, and benefits of changing in certain ways. For example, organizational units faced with stress often cope by becoming more rigid, which in some cases means becoming more hierarchical. These learning models can be used to estimate when such rigidity is likely to pay off and what factors in the C^3 architecture might inhibit that response. Third, real organizational units undergo both one-shot modification and (often in association with that big modification) many other, small adaptations. These learning models have typically been used to capture learning in terms of these smaller, more prevalent adaptations. Recent work using hybrid models and simulated annealing has been used to look at both sequences of big (one-shot) modifications and many small adaptations.

Many unit-level models attempt to identify the optimal C^3 architecture given certain performance criteria. Typically, this research has demonstrated that the ability of the algorithm to identify the optimal structure depends on the complexity of the fitness function (and roughness of the fitness surface) and the constraints on adaptation. The more complex the function, the more constraints there will be on what actions the organizational unit can take, and the greater the difficulty the algorithm will have in identifying the optimal form.

An important issue here is whether routines that take approaches other than optimization will be valuable. First, the environment and the constraints on the organizational unit may be sufficiently complex that it is impossible to find the optimal solution in a reasonable amount of time. Second, human organizations are rarely engaged in a process of optimization. Locating the optimal structure depends on knowing the task to be performed. Given a particular task and its functional decomposition, alignment procedures can be used to locate the optimal C^3 structure for that task (Remy and Levis, 1988). In many contexts, however, the task changes dynamically, and the organization must be able to respond rapidly to this change. This may preclude, purely on the basis of time, the luxury of locating the optimal design. Moreover, organizational units rarely locate the optimal structure; rather, they locate a satisfactory structure and use it. Thus, the fact that many of the computational algorithms have difficulty identifying the optimal structure may not be such an important concern. Rather, the value of these learning algorithms in modeling organizational unit-level behavior may lie in the extent to which the procedure they use to alter the organizational unit's C^3 structure matches the procedure used by chief executive officers and the extent to which they can be adjusted to capture the constraints on changing the C^3 structure that are present in actual military organizational units. At the organizational unit level, then, basic research on constraint-based adaptation procedures is needed. If the goal is to build models that are representative of human organizational units, more work is needed on satisficing, rather than optimizing, routines.

Research in this area has yielded a number of findings with regard to the way units learn, adapt, and evolve. This set of findings poses a challenge to the modeler, as these are precisely the kinds of behaviors that must be taken into account or should emerge from a model of organizational unit-level behavior.

- For many of the tasks faced by organizational units, there is little or no feedback, or the feedback is severely delayed, or the task never recurs so that feedback is largely irrelevant (Cohen and March, 1974; March and Weissinger-Baylon, 1986; March, 1994). Consequently, feedback-based models of learning are often insufficient or inappropriate for capturing organizational unit-level behavior. Researchers may want to focus instead on expectation-based learning, that is, learning based on expected rather than real feedback.
- The rate of learning and the quality of what is learned are affected by the task the organizational unit is performing, the unit's C^3 structure, and the constraints placed on individual agents (Carley, 1992; Lin and Carley, 1997b). The C^3 structure that works best when the organizational unit is faced with a new task or novel situation appears to be different from that which is most effective for performing routine tasks or working in known environments (Blau and Scott, 1962; Burns and Stalker, 1966). Research findings indicate that units exhibiting the best performance in novel situations tend to be those in which personnel are organized in a more collegial, less structured, and hierarchical fashion; have a broad range of experience; and are allowed to act on the basis of that experience, rather than having to follow standard operating procedures (Roberts, 1989; Lin and Carley, 1997a). More highly structured and hierarchical organizational units whose members follow standard operating procedures appear to be better performers for more routine tasks (Blau and Scott, 1962; Burns and Stalker, 1966).
- Not all learning is useful (Levitt and March, 1988; Huber, 1996; Cohen, 1996). Indeed, there can be mislearning. Organizational units develop performance histories that affect future behavior. Thus the lessons learned when the unit was formed will influence both the rate at which it learns and what it learns

in the future. Initial mislearning on the part of the organizational unit can have long-term deleterious consequences. Organizational units can also engage in superstitious learning, whereby the unit learns to attach outcomes to the wrong causes. How to prevent mislearning and superstitious learning and how to detect such errors early in the organizational unit's life cycle are areas not well understood.

- Organizational units can learn to do certain tasks so well that they become trapped by their competency and so are not flexible enough to respond to novel situations. Organizational units need to balance exploitation of known capabilities against exploration for new capabilities (March, 1996). Exploitation allows the unit to take advantage of learning by doing and fine-tuning; exploration allows the unit to identify new opportunities. Research is needed on how to ensure both competency and flexibility. Research conducted to date suggests that organizational units can prepare themselves to learn novel ideas and ways of doing work by acquiring the right mix of personnel, engaging in research and development, and being flexible enough to reassign personnel and reengineer tasks.
- In many cases, organizational units can substitute real-time coordination for an established C^3 structure (Hutchins, 1990). However, such real-time coordination can increase the time needed to make decisions.
- Improvements in the unit's performance can be achieved through coordination (Durfee and Montgomery, 1991; Tambe, 1996a) or communication (Carley et al., 1992). However, the relationship among coordination, communication, and performance is complex and depends on the cognitive capabilities of the agents (Carley, 1996a).

Techniques for Modeling Unit-Level Learning and Adaptation

Unit-level learning and adaptation can be modeled using the techniques of (1) neural networks (Rumelhart and McClelland, 1986; McClelland and Rumelhart, 1986; Wasserman, 1989, 1993; Karayiannis and Venetsanopoulos, 1993; Kontopoulos, 1993; see also Chapter 5), (2) genetic algorithms and classifier systems (Holland, 1975, 1992; Holland et al., 1986; Macy, 1991a, 1991b; Koza, 1990; Kinnear, 1994), and (3) simulated annealing (Kirkpatrick et al., 1983; Rutenbar, 1989; Carley and Svoboda, 1997; Carley, forthcoming). In a neural network, information is stored in the connections between nodes, which are typically arranged in sequential layers (often three layers) such that there are connections between nodes in contiguous layers but not within a layer. Of the three techniques, neural networks best capture the experiential learning behavior of individual humans. Genetic algorithms draw from biological theories of evolution. A genetic algorithm simulates the evolutionary process by allowing a population of entities to adapt over time through mutation and/or reproduction (crossover) in an environment in which only the fittest survive. Of the three techniques, genetic algorithms may be the least suited to military simulations that are intended to be computational analogs of the behavior of individual combatants or units since these algorithms require an evolutionary process in which there are multiple agents at the same level competing for survival through a long-term fitness function. Simulated annealers are computational analogs of the process of metal or chemical annealing. Simulated annealing is a heuristic-based search technique that attempts to find the best solution by first proposing an alternative, determining whether its fit is better than that of the current system, adopting it if it is better, and adopting even a "bad" or "risky" move with some probability. The latter probability typically decreases over time as the "temperature" of the system cools. Of the three techniques, simulated annealing best captures the expectation-based satisficing behavior engaged in by CEOs and executive teams (Eccles and Crane, 1988).

All three of these techniques can be thought of as optimization techniques for classifying objects or locating solutions; that is, given a specific environment (often referred to as a landscape) the algorithm employs some form of learning (e.g., strategic, experiential) to search out the best solution (highest or lowest point in the landscape). Each technique employs only one type of learning. However, in actual organizational units, many different types of learning are employed simultaneously. Most of the research using these techniques has assumed that the environment (the landscape) is unchanging. One research issue is whether these models would be more robust, locate better solutions, and locate solutions more quickly if they employed multiple types of learning simultaneously. Another issue is whether any of the current techniques are useful when the environment is changing.

Multiagent Models of Organizational Unit Adaptation

Most of the work on computational modeling of unit-level learning and adaptation employs multiagent models. These range from the more symbolic distributed artificial intelligence models to models using various complex adaptive agent techniques, such as those discussed in the previous section, chunking (Tambe, 1996a), and other stochastic learning techniques (Carley and Lin, 1997, forthcoming; Lin and Carley, 1997a, 1997b; Glance and Huberman, 1993, 1994). Some of this work is based on or related to mathematical models of distributed teams (Pete et al., 1993, 1994; Tang et al., 1991) and social psychological experimental work on teams (Hollenbeck et al., 1995a, 1995b). Most of these models perform specific stylized tasks; many assume a particular C^3 structure.

The artificial agents in these models are not perfect analogs of human agents (Moses and Tennenholtz, 1990). A basic research issue is how accurately the agents in these models need to behave so that an organizational unit composed of many of these agents will act like a unit of humans. Some research suggests that complete veridicality at the individual level may not be needed for reasonable veridicality at the organizational unit level (Castelfranchi and Werner, 1992; Carley, 1996a). Other work demonstrates that the cognitive capabilities of the individual agents, as well as their level and type of training, interact with the C^3 structure and the task the agents are performing to such an extent that different types of agent models may be sufficient for modeling organizational unit-level response to different types of tasks (Carley and Newell, 1994; Carley and Prietula, 1994).

All of the multiagent computational techniques in this area assume that the agents act concurrently. Concurrent interaction among low (possibly zero)-intelligence agents is capable of producing complex and detailed organizational unit-level behavior. For example, many researchers in this area have focused on the emergence of conventions for the evolution of cooperation. In this domain, it has been demonstrated repeatedly that seemingly simple rules, such as trying to attain the highest cumulative award, often give rise to interesting and nontrivial social dynamics (Shoham and Tennenholtz, 1994; Macy, 1990; Horgan, 1994). Synchronization of multiple agents is arguably the basis for both evolution and the development of hierarchy (Manthey, 1990).

Social dynamics, both equilibrium and nonequilibrium, depend on the rate of agent learning (adaptation revolution) (e.g., de Oliveira, 1992; Collins, 1992; Carley, 1991a). Various constraints on agent actions can enable faster and more robust organizational unit-level learning (Collins, 1992; Carley, 1992). Social chaos is reduced by having intelligent adaptive agents determine their actions using strategies based on observations and beliefs about others (Kephart et al., 1992) or by having their actions constrained by their position in the organizational unit. This suggests that cognition and C^3 structure play a dual defining role in emergent phenomena (Carley, 1992). Further, it has been demonstrated that environmental and institutional factors such as payoffs, population dynamics, and population structure influence the evolution of cooperation in a discontinuous fashion (Axelrod and Dion, 1988).

There have been a few models of organizational unit-level learning and adaptation that do not involve multiple models (for a review see Lant, 1994). These models typically employ either autonomous agents acting as the organizational unit or various search procedures. This work has been useful in demonstrating various points about organizational learning at the organizational unit level (Lant and Mezias, 1990, 1992). However, none of these models have been used to examine the behavior of the organizational unit in performing specific tasks. Thus, their potential usefulness in military simulations cannot be determined at this time.

OVERARCHING ISSUES

There are several overarching issues related to unit-level modeling: unit-level task scalability, task analysis and performance, ease of modeling units, and rediscovery.

Scalability

The work on organizational unit-level modeling raises an important substantive question: What changes as the size of the organizational unit increases? It is commonly assumed that the behavior, problems, and solutions involved in unit-level behavior differ only in scale, not in kind, as the size of the organizational unit increases. Thus it is assumed that units of hundreds of individuals will make decisions in the same way and face the same types of problems as units of 3 to 10 individuals. Carried to its extreme, this line of reasoning suggests that organizational units and individuals act in the same way, and the behavior of an organizational unit is just an aggregate of individual-level actions and reactions. Numerous studies indicate, however, that such is not the case. Even in a less extreme form, the scalability assumption may not hold in all cases. For example, research in crisis management suggests that disasters are not scalable (Carley and Harrald, 1997). The lessons learned in responding to a small disaster, such as a minor hurricane that damages a few houses, are not applicable to large disasters, such as a major hurricane that completely destroys all the housing in several communities. The types of response entailed are different—say, ensuring that insurance claims are accurate versus providing shelter, food, and medicine to large numbers of people. Research in many areas other than crisis management also suggests that the scalability assumption is false under various conditions.

When scalability can be assumed, the same model of organizational unit behavior can be used regardless of the size of the unit; when scalability cannot be assumed, different models are needed for units of different sizes. To date there has been little research on which types of problems, processes, behaviors, and so forth are scalable and which are not. Determining what factors are scalable is an important step in deciding when new models are or are not needed.

Further, regardless of whether problems, processes, and behaviors are scalable, different metrics may be needed for measuring these factors at different levels. For example, consider the measurement of various aspects of the command structure for small six-person teams and for theater-level forces. In both cases, one might be interested in the span of control, the number of levels, and the degree of decision decentralization. There are two different problems involved. First, in both cases these factors can be measured if one knows the complete mapping of who reports to whom. However, obtaining this mapping is more time-consuming in the case of the theater-level force than in the case of the six-person team. A second and more difficult problem is that the range of variation on certain metrics, such as span of control, depends on the size of the organizational unit. In very small organizational units, this range may be so small for some metrics that those metrics have no value in predicting performance outcomes, whereas for large organizational units, these metrics may be the critical ones for predicting performance. Consequently, different measures of C^3 structure may be needed at different scales.

Organizational Unit-Level Task Analysis and Performance

At the organizational unit level, task analysis involves specifying the nature of the task and C^3 structure in terms of such factors as assets, resources, knowledge, access, and timing. The basic idea is that the task and C^3 structure affect organizational unit-level performance. Task analysis at the organizational unit level does not involve examining the motor actions an individual must perform or the cognitive processing in which an individual must engage. Rather, it involves specifying the set of tasks the organizational unit as a whole must perform to achieve some goal, the order in which those tasks must be accomplished, the resources needed to accomplish them, the individuals or subunits that have the necessary resources, and so on. (For example, see the task analysis done for the VDT by Levitt et al., 1994, and Cohen, 1992, or applications involving petri nets in Remy and Levis, 1988).

There has been and continues to be a great deal of research in sociology, organizational theory, and management science on how to do a task analysis at the organizational unit level. For tasks, the focus has been on developing and extending project analysis techniques such as PERT charts (Levitt et al., 1994) and dependency graphs building on the dependency forms identified by Thompson (1967). For the C^3 structure, early work focused on general features such as centralization, hierarchy, and span of control (see, for example, Blau, 1960). Recently, however, network techniques have been used to measure and distinguish the formal reporting structure from the communication structure (see, e.g., Krackhardt, 1994). These various approaches have led to a series of survey instruments and analysis tools. This work involves a variety of unresolved issues, including how to measure differences in the structures and how to represent change.

Additionally, a great deal of research has been done on how the task and the C^3 structure influence performance (see Carley, 1995b, for a review of modeling work in this area). In particular, research in the past three decades has repeatedly demonstrated that there is no one right C^3 structure or design for an organizational unit (see, e.g., Lawrence and Lorsch, 1967; Pennings, 1975). Rather, the way the organizational unit should be organized depends on the specific tasks to be performed, the volatility of the environment; the extent to which personnel move into and out of various jobs in the organizational unit; the amount, quality, and accuracy of information available for making decisions; and the level and type of training or experience acquired by the participants (Malone, 1987; Roberts, 1990; Carley and Lin, forthcoming). These findings follow from empirical work in both the field and the laboratory. The theory has found expression in the form of verbal descriptions of organizing and decision-making processes, as well as computational and mathematical theories.

At the organizational unit level, task analysis is an operationalization of the information processing approach to the study of organizational units. The following factors are often addressed:

- The way the structure of the organizational unit and the nature of the task limit what information is available when to which personnel in the organizational unit, and the way such limits influence organizational unit performance
- The way the interaction between these structures and the cognitive limits of human personnel influences organizational unit performance
- The way these structures interact with the complexity, quality, quantity, and accuracy of information to determine performance
- The set of processes the organizational unit carries out and the order in which they can occur

Increasingly during the past decade, organizational unit-level theories and models with an information processing focus have found expression as computer programs (see Carley, 1995b, for a review). By calibrating these programs to actual organizational units, predictions about organizational unit behavior can be generated. The precision of these predictions varies with the level of detail used in representing the task and the C^3 structure (Carley and Prietula, 1994). In fact, simply involving a team member or CEO in specifying the structural information needed for these models helps him/her see barriers to performance in the unit. Similar models are being developed by researchers in the distributed artificial intelligence community (see, for example, the work of Durfee, 1988; Tambe, 1996b, 1996c; Gasser et al., 1993; and Gasser and Majchrak, 1992). Currently, even the most detailed of these models (e.g., the VDT model) are sufficient only for small classes of very routine design tasks. However, these models are potentially useful for evaluating alternative task assignments, organizing schemes, or new technologies because they allow the researcher or CEO to engage in "what if" analyses.

Ease of Modeling Organizational Units

One of the main problems in rapidly building organizational unit-level models is that the available languages do not have a set of primitives geared to the organizational unit level. Thus, each researcher implementing an organizational unit-level model creates his/her own tools for modeling a variety of organizational unit-level behaviors, such as moving personnel between divisions, promoting personnel, measuring performance, combining decisions, and communicating commands. Moreover, for organizational unit-level models in which the unit comprises a set of agents (multiagent models, discussed earlier) the researcher must model both the agent (or types of agents) and the organizational unit. Consequently, organizational unit-level models can be much more time-intensive to develop than individual agent models. Another difficulty with organizational unit-level models is that they are often quite computer intensive, particularly when each agent is modeled as a separate entity.

Rediscovery

One of the major difficulties in this rapidly developing, multidisciplinary area is that researchers are continually rediscovering well-known results from other areas. For example, researchers in distributed artificial intelligence rediscovered the well-known result from organizational theory that there is no one right organizational design for all tasks. Likewise, social scientists rediscovered the well-known result from computational analysis that performance improvements in neural net-like models require a huge number of training trials. Thus it is imperative in this area that artificial intelligence modelers and computer science system builders collaborate with organizational or social scientists.

ORGANIZATIONAL UNIT-LEVEL MODELING LANGUAGES AND FRAMEWORKS

Researchers have begun to develop organizational unit-level modeling languages or frameworks. None of these tools dominates, and each has some failings as a comprehensive approach for implementing organizational unit-level models. Any organizational unit-level meta-language must enable rapid modeling of agents, task, and C^3 structure. The existing systems vary in their adequacy on these dimensions, the severity of the constraints they place on the types of objects that can be modeled, and their flexibility for modeling objects. For the most part, these systems are documented only minimally, and many are still under development.

MACE

MACE (Gasser et al., 1987a, 1987b) was one of the first general (domain-independent) testbeds for modeling multiagent systems. It introduced the idea of using agents for every aspect of model construction and development (including user interaction and experiment management). MACE is one of the few distributed object systems that is truly concurrent. It includes explicit concepts drawn from social theory, such as recursive agent composition, or the idea that a group of agents (an organizational unit) can itself be treated as an agent with distributed internal structure. Further, the notion of social worlds developed by symbolic interactionists is operationalized in MACE as knowledge-based agent boundaries. Each agent has a set of potential interaction partners, and that agent's knowledge about others, rather than explicit rules, constraints, or programming structures, defines the boundaries of communication and interaction (and so the C^3 communication structure). MACE includes specific facilities for modeling a number of features of other agents, such as goals, roles, and skills. It is this type of knowledge of others that is used to define the pattern of interaction over time. Today, the idea of modeling the agent's knowledge of others is commonplace within computer science and artificial intelligence.

Constructuralism

The constructural model (Carley, 1990, 1991a; Kaufer and Carley, 1993) focuses on agent interaction. Agents are modeled as collections of knowledge and propensities for interaction. Propensities for interaction change dynamically over time as the agents interact and exchange information. Such changes can be constrained, however, by the imposition of a formal C^3 structure that imposes demands for certain interactions. Within the constructural model, a communications technology is available to agents that affects the number of agents with which they can communicate simultaneously, whether a communicated message must be the same to all receivers, how much information the agent can retain, and so on. The communications technology itself may act as an agent with particular information processing and communicative capabilities.

Virtual Design Team (VDT)

In the VDT model (Levitt et al., 1994; Jin and Levitt, 1996), the organizational unit is modeled as a set of agents linked by task and C^3 structure. The agent is modeled as an in-box, an out-box, a set of preferences for information handling, a set of skills, and so on. The VDT uses very detailed modeling of specific organizational tasks, focusing on the dependencies among subtasks (much as one would in PERT or GANT charts), but leaves the content of what is done in each subtask unspecified. Agents have a suite of communications technologies available to them, such as telephone, face-to-face meetings, and e-mail. However, the use of technology is affected by predefined preferences for the use of particular tools for certain types of messages. Performance measures focus on task completion and rework.

Strictly Declarative Modeling Language (SDML)

SDML (Moss and Edmonds, 1997; Edmonds et al., 1996; Moss and Kuznetsova, 1996) is a multiagent object-oriented language for modeling organizational units. It has been used to model both team (flat) and hierarchical C^3 structures. SDML does not contain a built-in model of the cognitive agent. Rather, it is sufficiently flexible to represent both simple agents (such as those in the constructural model) and more sophisticated agents (such as Soar agents). One feature of SDML that facilitates studying the impact of the agent is that it includes libraries for alternative architectures, such as genetic programming and

Soar. These libraries facilitate exploring the interaction between agent cognition and organizational design. The pattern of ties among agents can also be represented in SDML. The ease of representing both agents and linkages means that the C^3 structure can readily be modeled as a set of predefined linkages, roles, and knowledge bases. Within SDML, the structure of the multiagent system is represented as a container hierarchy, such that agents may be contained within small organizational units that are contained within larger organizational units. Containers and their associated agents are linked by an inheritance hierarchy.

Multiagent Soar

There are a number of multiagent or team Soar models. In these models, the individual agents are built into the Soar architecture (Laird et al., 1987). Thus the agents are goal directed, although the goals need not be articulable. Knowledge is organized in problem spaces, and actions are directed by operators and preferences. Preferences can be used to represent shared norms or cultural choices. However, it is difficult to change preferences dynamically. In multiagent Soar, each agent has a model of other agents, and chooses interactions and dynamically forms the C^3 structure by reasoning about those others (Tambe, 1996a, 1996b; Carley et al., 1992). Such knowledge may include expectations about the other agents' goals, preferences, and actions. Agents also have mental models about the social world, which can include information about the nondynamic aspects of the C^3 structure. Communication involves the passing of messages among agents. Agents may have various communication-related problem spaces, such as for determining when to communicate what to whom or how to compose and parse messages.

SWARM

SWARM is a multiagent simulation language for modeling collections of concurrently interacting agents in a dynamic environment (Stites, 1994; Minar et al., 1996; Axelrod, 1997). It was designed for artificial-life simulations and thus is best used to explore complex systems comprising large numbers of relatively simple agents. Within SWARM, agents can dynamically restructure themselves to accommodate changes in incoming data or the objective function. Over time, systems of SWARM agents come to exhibit collective intelligence beyond the simple aggregation of agent knowledge.

Task Analysis, Environment Modeling, and Simulation (TAEMS)

TAEMS is a framework for modeling a computationally intensive task environment (Decker and Lesser, 1993; Decker, 1995, 1996). For the most part, it is simply a way of representing tasks. The TAEMS framework is compatible with agent-centered approaches to team modeling. Within TAEMS, however, an agent is simply a locus of belief and action. Agents can thus, at least to a limited extent, communicate, gather information, and execute actions. Tasks can be described at three levels of abstraction: objective, subjective, and generative. At the objective level, tasks are described in terms of subtasks and their interrelations, such as enables, precedes, facilitates, and hinders. At the subjective level, the view of each agent is characterized. And at the generative level, the range of alternatives, the distributions used, and the generative processes needed to specify differences in specific instantiations of tasks are specified.

ORGAHEAD

ORGAHEAD is a framework for examining the impact of the C^3 authority structure on performance for distributed choice and classification tasks (Carley and Svoboda, 1996; Carley, forthcoming). Within ORGAHEAD, organizational units have the capability of adapting their C^3 structure dynamically over time in response to environmental changes. Organizational unit-level action results from actions of multiple agents, the C^3 structure connecting them, and the distribution of knowledge or resources across agents. Individually, agents either learn through experience or follow standard operating procedures. The organizational unit also learns structurally by altering procedures and linkages among the agents, such as who reports to whom and who does what. This latter types of learning is implemented as a simulated annealing algorithm. Within ORGAHEAD, the user can vary the authority structure, the degree of training received by the agents, the amount of information the agents recall, the rate of structural change, and the ways in which the C^3 structure can change.

CONCLUSIONS AND GOALS

Computational models of organizational unit-level behavior are becoming increasingly sophisticated. Work in this area holds great promise for the military context. At this point, however, none of the existing models can simply be plugged directly into a current DMSO platform as the model of an organizational unit. In all cases, important extensions or modifications are needed. In part, this is because existing organizational unit-level models have either too limited a repertoire of C^3 structures or too limited a representation of tasks. Likewise, existing DMSO models are not set up to capture organizational unit-level performance measures. However, several of the models could be adapted or used in concert with other models to examine aspects of organizational unit-level behavior. Further, the network-based representation of C^3 structures and the various network measures of structure could be incorporated into some current DMSO programs.

The current set of organizational unit-level models demonstrates the viability and necessity of using simulation to examine organizational unit-level behavior. To enhance the applicability of organizational unit-level computational models to the military context, several steps need to be taken, including data collection, development of measures, improvement and extension of existing models, development of visualization techniques, and basic research on models of unit-level adaptation.

Short-Term Goals

- Extend existing models so they are more directly applicable to military settings. Three types of activities should be considered: (1) extending current models to new tasks; (2) creating teams of military and civilian research personnel to extend existing models to a C^3 model that could be used in some specific military setting, such as synthetic theater of war-Europe (STOW-E); and (3) developing a list of contexts in which C^3 models are most needed.
- Develop databases against which to test and/or validate existing models. To this end, network data for key C^3 features for particular engagements, training exercises, or war games should be collected. Such data include, for example, the command structure for specific units, the associated communication structure, what kind of information is communicated by what technology, information on resource and platform assignment, and performance measures. Such data are often collected, particularly in war games, but are rarely made accessible in network form (who is talking to whom about what). Analysis of such network data would provide information on the range of existing structures, the strengths and weaknesses of those structures, and the types of factors that need to be included in the C^3 model. Data should also be collected to examine the relationship between C^3 structure and performance for a series of specific tasks.
- Locate and develop measures of C^3 architectures that are meaningful in the military context. These measures should then be integrated into new and existing models. Data on these measures could be gathered for analyzing and validating the resulting models.
- Examine and address the general principles of command—such as unity of command. Is there any empirical evidence for when these principles do and do not work? These principles, along with associated empirical constraints, should be made available as a unit-level analog of conceptual model of the mission space (CMMS). These principles could also be used to inform and validate models of C^3 architectures, thus increasing the realism of these models in the military context.
- Develop a C^3 analysis tool kit. This tool kit should include procedures for determining when C^3 architectures are meaningfully different; procedures for segmenting networks based on various performance criteria; and techniques for visualizing networks, particularly C^3 architectures.

• Begin research on constrained-adaptation and multiagent learning models. Current multiagent learning algorithms need to be extended to be more robust, learn more quickly, and find better solutions in complex and rapidly changing environments. There are many near-term needs in this area. First, finding algorithms that will generate better solutions in dynamic environments will involve extending and examining new optimization techniques (possibly hybrid techniques), as well as developing new techniques that will enable rapid and accurate learning under various constraints, such as little feedback or changes in available options. Second, the various organizational unit-level learning algorithms need to be compared. Third, organizational unit-level researchers, such as organizational theorists, should be brought together with computational modelers to develop valid organizational unit-level models. Validated unit-level models would be valuable decision aids for commanders trying to reason about the impact of novel strategies and new technologies on the forces under their command. Finally, work is needed on expanding existing models and validating them against data on actual organizational units. To represent organizational unit behavior accurately, models of unit-level learning should not only learn to make the correct decision, but also mislearn under certain circumstances to better represent real human behavior.

Intermediate-Term Goals

• Continue research on constraint-based adaptation, and try to incorporate the models of adaptation developed in the short term into models of C^3 structure. This work would be particularly useful in making forces behave more realistically when facing changes in orders or casualties.

• Develop a multiagent system that combines both a network interface for examining the impact of changes in the C^3 architecture and a set of highly intelligent agents engaged in various tasks, possibly a force-level task. The use of hybrid models that combine cognitive agent models (see Chapter 3) with social network models is a critical step in being able to address C^3 issues.

• Develop a better understanding of organizational unit-level adaptation. This can be accomplished by working in at least five areas:

—Develop measures of adaptation. Can adaptive behavior be recognized? At the organizational unit level, basic research is needed on what aspects of the C^3 architecture facilitate adaptation, given the constraints of the military situation (e.g., command hierarchy, rules for transferring personnel among organizational units, existing resources and skills).

—Determine the scalability of these measures. Are the factors that inhibit or promote adaptability at the small group level the same as those that inhibit or promote adaptability at the organizational unit level? Which metrics are valuable for which sizes of forces?

—Investigate what types of changes in C^3 structure are most likely to be adaptive, and address the way C^3 architectures can be set up so that they can adapt to changing task environments. This may involve developing optimization procedures for environments in which the performance surface changes over time.

—Begin to gather field data for evaluating dynamic C^3 systems. Models for C^3 adaptation at the organizational unit level often must take into account institutional, technological, and political factors that cannot be covered adequately by laboratory experiments in a limited time frame. A better understanding is needed of what kinds of data are required and how those data can be gathered.

—Develop organizational unit-level models with C^3 architectures that adapt dynamically in response to triggering events, such as depletion of resources or alterations in the rules of engagement.

- Develop a framework or meta-language for describing and implementing organizational unit-level models. Progress on individual-level models was facilitated by the development of platforms, such as Soar, that integrate various aspects of human cognition and/or physiology. At the unit level, one of the most pressing issues is that development of unit-level models is extremely time-consuming, and each modeler spends part of his or her time reinventing basic procedures, such as communication protocols and algorithms for traversing the command structure. Research in this area indicates the need for a meta-language that will facilitate rapid linking of different types of agent models with models of both task and organizational structure. Such a language should have built-in default procedures for measuring performance and aspects of the task and C^3 structures. No current language is sufficient for this purpose.

Long-Term Goals

- Gather information on the conditions under which organizational unit-level behavior is or is not the simple aggregate of individual-level behaviors.
- Develop unit-level models in which the organizational unit is a combined force. In combined and coalition forces, additional issues such as cultural clashes, language barriers, and technological differences combine to complicate the C^3 process (Maurer, 1996). Thus the development of coalition organizational unit-level models requires attention to details not relevant for forces from a single country.
- Examine interactions among different types of learning and the implications of such interactions for unit-level performance. Organizational unit-level learning is not simply the aggregation of individual learning. That is, it is possible for the organizational unit to learn and adapt even when all the individual agents are acting according to standard operating procedures. C^3 architectures may learn or adapt by reassigning resources or tasks or by developing new procedures or information systems. The same basic standard operating procedures may apply for individual agents, but the number of agents or the time to

complete the task may differ. At the organizational unit level, adaptation may take the form of emergent structures rather than individual learning. However, learning at the unit level may interfere with or be aided by learning at the individual level. For example, unit-level learning in corporations is frequently embedded in the connections among personnel and the roles the personnel play, but the value of such learning is often negated when personnel are replaced or the organization downsizes. There is a need to assess whether this is the case in military settings and what impact such interference would have on unit-level performance.

—In the course of developing models of planning, take unit-level issues into account.

—Explore how the output of unit-level models is turned into plans such as those that might be generated by a commander or by staff personnel for a commander. Currently, the messages passed within many of the unit-level models have minimal content (e.g., they may contain implicit decisions and belief or trust in those decisions). Research is needed to link this output to messages whose content reflects C^3 issues.

11

Information Warfare: A Structural Perspective

INTRODUCTION

What would be the impact of a leaflet dropped by a psychological operations unit into enemy territory? Would the number who surrender increase? As the amount of information available via satellite increases, will regimental staff personnel be more or less likely to notice critical changes in the tactical situation? Questions such as these are central to the information warfare context. As the amount and quality of information increase and the ability to deliver specific information to specific others improves, such questions will become more important. A better understanding of how individuals and units cope with changes in the quantity, quality, and type of information is crucial if U.S. forces are to maintain information dominance.

Information warfare, sometimes referred to as information operations, can be thought of as any type of conflict that involves the manipulation, degradation, denial, or destruction of information. Libicki (1995:7) identifies seven types of information warfare. Many aspects of information warfare identified by Libicki and others are concerned with technology issues, such as how one can identify technically that the integrity of databases has been tampered with and how one can keep networks linking computers from becoming unstable. While all these aspects of vulnerability are important, they are not the focus here. In this chapter, information warfare is examined more narrowly—from the perspective of human behavior and human processing of information. Thus the ideas discussed here bear on three aspects of information warfare identified by Libicki: (1) command and control warfare, (2) intelligence-based warfare, and (3) psychological warfare. From the perspective of this chapter, information warfare has to do with those processes that affect the commander's, staff's, opposing commander's, or civilians' decisions, ability to make decisions, and confidence in decisions by affecting what information is available and when.

This chapter focuses, then, on the human or social side of information warfare. Factors that affect the way people gather, disseminate, and process information are central to this discussion. These processes are so general that they apply regardless of whether one is concerned with command and control warfare, intelligence-based warfare, or psychological warfare. Thus rather than describing the various types of information warfare (which is done well by Libicki) this chapter addresses the social and psychological processes that cut across those types. Clearly, many of the factors discussed in earlier chapters, such as attention, memory, learning, and decision making, must be considered when modeling the impact of information warfare on individuals and units. In addition, several other aspects of human behavior—the structure of the unit (the way people are connected together), information diffusion, and belief formation—are relevant, and the discussion here concentrates on these three aspects. It is these aspects that will need to be integrated into existing models of information warfare and existing estimates of vulnerability if these models and estimates are to better reflect human behavior.

Consider the role played by these three aspects (structure, diffusion, and belief formation) in the area of information dominance. Information dominance has to do with maintaining superiority in the ability to collect, process, use, interpret, and disseminate information. Essentially, the way the commander acts in a particular situation, or indeed the way any individual acts, depends on the information characteristics of the situation, including what other individuals are present, what they know, what they believe, and how they interact. For example, in joint operations, commanders may have staffs with whom they have had only limited interactions. Vice Admiral Metcalf (1986) addresses this issue in discussing the Grenada Rescue Operation. He refers repeatedly to the importance of the Nelsonian principle—"know your commanders"—and of having a staff that has been trained to work together. He stresses the importance of understanding what others know and developing a sense of trust. In general, individuals are more likely to trust and/or act on information (including orders) if it comes from the "right" source. For example, Admiral Train notes that when he became the commander of the Sixth Fleet, he asked the incumbent, who believed the Beirut evacuation plans were inadequate, why he had not spoken up sooner. The incumbent replied that since he thought the evacuation order was from a valid authority, he was reluctant to object. Additionally, during missions, as crises come to the fore, the amount of information involved can escalate. The resulting information overload leads people to find ways of curtailing the information to which they attend. One means of doing so is to use one's interaction network (the web of trusted advisors) to decide what information is most important. All of these factors point to the essential role of social structure, the connections among personnel, in the diffusion of information and the development of an individual's ideas, attitudes, and beliefs.

Being able to measure, monitor, and model these information networks is important in evaluating weaknesses and strengths in a personnel system from an information warfare perspective. Understanding how information diffuses within and between groups, how individuals acquire information, and how their beliefs and decisions change as a function of the information available to them and those with whom they interact, as well as the decisions and beliefs of friends and foes, is critical in a variety of military contexts, including information warfare, coalition management, and intelligence. Models of information warfare that can be used for examining alternative strategies (for defense and offense) can build upon models of information diffusion and belief formation. Similarly, models of information diffusion and belief formation can be used to help address questions such as who should get what information, how much information is needed, and who should filter the information. In this chapter, formal models of information diffusion and belief formation are discussed.

Individuals' decisions, perceptions, attitudes, and beliefs are a function of the information they know. Individuals learn the vast majority of what they know during interactions with others (i.e., vicariously), rather than through direct experience. Processes of communication, information diffusion, and learning are key to understanding individual and group behavior. These means of gathering information include both direct interactions with a specific person, such as talking with someone, and indirect interactions, such as reading a memorandum or monitoring a radar screen. Factors that affect these interactions in turn affect what information individuals learn, what attitudes and beliefs they hold, and what decisions they make. These factors fall roughly into two categories: structural and technological. Structural factors are those that influence who interacts with whom. Technological factors are those that influence the mode of interaction—face-to-face, one-to-many, e-mail, phone, and so forth. This chapter focuses primarily on the role of structural factors and existing models of the way they influence information diffusion and belief formation. We note in passing that there is also a great deal of work on the technological factors associated with information warfare; however, such concerns are not the focus here.

Collectively, the information diffusion and belief formation models and the empirical studies on which they are based suggest that an individual or group can control the attitudes, beliefs, and decisions of others by controlling the order in which the others receive information and the individuals from whom they receive it and by fostering the idea that others share the same beliefs. Social control does not require that information be hidden or that people be prevented from accessing it. It may be sufficient simply to make certain topics taboo so that people think they know each other's beliefs and spend little time discussing them. Or to turn this argument on its head, educational programs designed to provide individuals with all the facts are not guaranteed to change people's beliefs. Research in this area suggests that in general, once an individual learns others' beliefs and concludes that they are the object of widespread social agreement, he/she will continue to hold those beliefs, despite repeated education to the contrary, until his/her perception of others' beliefs changes.

The next two sections examine the modeling of information diffusion and belief formation, respectively. Next is a discussion of the role of communications technology in the diffusion of information and the formation of beliefs. The final section presents conclusions and goals in the area of information warfare.

MODELS OF INFORMATION DIFFUSION

Branscomb (1994:1) argues that "in virtually all societies, control of and access to information became instruments of power, so much so that information came to be bought, sold, and bartered by those who recognized its value." However, the ability of individuals to access information, to recognize its value, and to buy, sell, or barter it depends on their position in their social network. This simple fact—that the underlying social or organizational structure, the underlying networks, in which individuals are embedded is a major determinant of information diffusion—has long been recognized (Rogers, 1983). Numerous empirical studies point to the importance of structure in providing, or preventing, access to specific information. Nonetheless, most models of information diffusion (and of belief formation) do not take structure into account.

Social or organizational structure can be thought of as the pattern of relationships in social networks. There are several ways of characterizing this pattern. One common approach is to think in terms of the set of groups of which an individual is currently or was recently a member (e.g., teams, clubs, project groups). The individual has access to information through each group.

A second common approach is to think in terms of the individual's set of relationships or ties to specific others, such as who gives advice to whom, who reports to whom, and who is friends with whom. A familiar organizational structure is the command and control (C^2) or command, control, communications, and intelligence (C^3I) architecture of a unit (see also Chapter 10). In most social or organizational structures, multiple types of ties link individuals (see, e.g., Sampson, 1968, and Roethlisberger and Dickson, 1939). This phenomenon is referred to as multiplexity (White et al., 1976). For example, in the C^3I architecture of a unit, there are many structures, including the command structure, the communication structure, the resource access structure, and the task structure. Different types of ties within the organization can be used to access different types of information.

A common way of representing social structure is as a network. For example, Figure 11.1 shows two illustrative structures. The structure on the left is known as a formal structure, or organization chart; it is characterized by a pattern

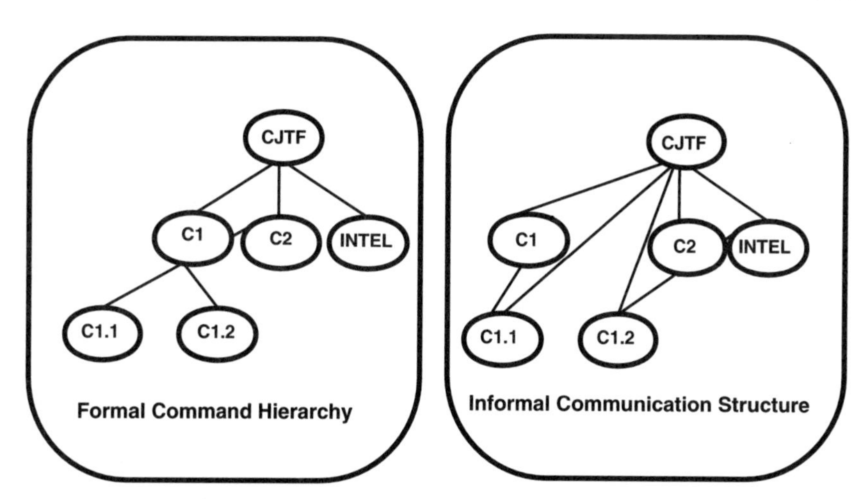

FIGURE 11.1 Illustrative structures as networks.

of command or reporting ties. The structure on the right is known as an informal structure; it is characterized by a pattern of ties among individuals based on who communicates with whom. When social or organizational structures are represented as a network, the nodes can be groups, teams, individuals, organizations, subunits, resources, tasks, events, or some combination of these. The ties can be based on any type of linkage, such as economic, advice, friendship, command, control, access, trust, or allied.

The study of social networks is a scientific discipline focused on examining social structure as a network and understanding its impact on individual, group, organizational, and community behavior (Wellman and Berkowitz, 1988). Most models combine the formal and informal structures, and in many empirical studies, the informal subsumes the formal. A large number of measures of structure and of the individual's position within the structure have been developed (Scott, 1981; Wellman, 1991; Wasserman and Faust, 1994). Empirical studies of information diffusion have demonstrated the utility of this approach and the value of many of these measures for understanding the diffusion process (Coleman et al., 1966; Burt, 1973, 1980; Valente, 1995; Morris, 1994; Carley and Wendt, 1991). These studies suggest that what information individuals have, what decisions they make, what beliefs they hold, and how strongly they hold those beliefs are all affected by their position (peripheral or central) in the network, the number of other individuals with whom they communicate, and the strength of the relationships with those other individuals. Some empirical research suggests that each individual has an internal threshold for accepting or acting on new information that depends on the type of information and possibly on individual psychological traits, such as the need for acceptance (Valente, 1995). This threshold can be interpreted as the number of surrounding others who need to accept or act on a piece of information before the individual in question does so. Moreover, much of the work on thresholds is more descriptive than prescriptive.

There is almost no work, either empirical or modeling, that directly links network and psychological factors. A first empirical attempt at making such a link is the work of Kilduff and Krackhardt (1994) and Hollenbeck et al. (1995b). In terms of modeling, Carley's (1990, 1991a) constructural model links cognition (in terms of information) and structure, but does not consider other aspects of human psychology.

While many researchers acknowledge the importance of social structure to information diffusion (Katz, 1961; Rapoport, 1953), there are relatively few models of how the social structure affects the diffusion process (Rogers, 1983:25). Empirical studies have shown that differential levels of ties within and between groups, the strength of ties among individuals, and the level of integration all impact information diffusion (for example, see, Burt, 1973, 1980; Coleman et al., 1966; Granovetter, 1973, 1974; and Lin and Burt, 1975).

There are several computational models of information diffusion. Most such models are limited in one or more of the following ways: structural effects are completely ignored, the social or organizational structure is treated as static and unchanging, only a single piece of information diffuses at a time, or the role of communications technology is not considered. These limitations certainly decrease the overall realism of these models. However, it is not known whether these limitations are critical with regard to using these models in an information warfare context.

Early diffusion models were based on a contagion model, in which everyone who came into contact with new information learned it. Following are brief descriptions of three current models: (1) the controversial information model, (2) the constructural model, and (3) the dynamic social impact model. These models are less limited than those based on the contagion model. Each is also capable of being combined with belief formation models such as those discussed in the next section.

The Controversial Information Model

The controversial information model (Krackhardt, forthcoming) is based on a sociobiological mathematical model of information diffusion developed by Boorman and Levitt (1980). This model of information diffusion applies only to the diffusion of "controversial" innovations or beliefs. A controversial innovation or belief is one whose value (and subsequent adoption) is socially and not rationally determined; that is, there is no exogenous superior or inferior quality to the innovation or belief that determines its eventual adoption. Thus, whether an individual adopts the innovation or belief depends not just on what he/she knows, but also on whether others also adopt it.

In this model, there are two competing innovations or beliefs. Each individual can adopt or hold only one of the two at a time. Individuals are subdivided into a set of groups, each of which has a prespecified pattern of connection with other groups. Individuals and the innovations or beliefs to which they subscribe can move between groups, but only along the prespecified connections. Whether individuals remain loyal to an innovation or belief depends on whether they encounter other individuals who subscribe to it. The particular pattern in which groups are connected, the rate of movement of individuals between groups, the size of the groups, the likelihood of individuals changing the innovations or beliefs to which they subscribe as they encounter others, and the initial distribution of innovations and beliefs across groups all influence the rate at which a particular innovation or belief will be adopted and which innovation or belief will dominate.

Using a computational version of this model, Krackhardt (forthcoming) examines the conditions under which an innovation may come to dominate an organization. The results from Krackhardt's model parallel the theoretical results obtained by Boorman and Levitt (1980):

- In a large undifferentiated organization, no controversial innovation can survive unless a large proportion of the organization's members adopt it at the outset.
- There are structured conditions under which even a very small minority of innovators can take over a large organization.
- Once an innovation has taken hold across the organization, it is virtually impossible for the preinnovation state to recover dominance, even if it begins with the same structural conditions the innovators enjoyed.

Contrary to intuition and some of the literature on innovation, the conditions that lead to unexpected adoption across an entire organization are enhanced by the following factors:

- A high level of viscosity (that is, a low degree of free movement of and interaction among people throughout the organization)
- The initial group of innovators being located at the periphery of the organization, not at the center

The controversial information model has several key features. First, the impact of various structures on information diffusion and the resultant distribution of beliefs within groups can be examined. For example, the model could be used to explore different C^3I structures. Second, two competing messages or beliefs are diffusing simultaneously; they are treated as being in opposition, so that an individual cannot hold both simultaneously. A limitation of this model, however, is that the structure itself (the pattern of groups) is predefined and does not change dynamically over time. Thus this model is most applicable to examining information diffusion in the short run.

The Constructural Model

According to the constructural model, both the individual cognitive world and the sociocultural world are continuously constructed and reconstructed as individuals move concurrently through a cycle of action, adaptation, and motivation. During this process, not only does the sociocultural environment change, but social structure and culture coevolve in synchrony.

Carley (1991a) defines the following primary assumptions in describing the constructural model:

(1) individuals are continuously engaged in acquiring and communicating information
(2) what individuals know influences their choices of interaction partners
(3) an individual's behavior is a function of his or her current knowledge

According to the basic formulation (Carley, 1990, 1991a, 1995a; Kaufer and Carley, 1993), individuals engage in a fundamental interaction-shared knowledge cycle. During the action phase of the cycle, individuals can communicate to and receive information from any source. During the adaptation phase, individuals can learn, augment their knowledge, alter their beliefs, and as a result cognitively reposition themselves in the sociocultural environment. During the motivation phase, individuals select a potential interaction partner or artifact.

The constructural model suggests that the concurrent actions of individuals necessarily lead to the coevolution of social structure and culture. Further, the impact of any individual on the sociocultural environment is affected by that individual's integration into the society, both structurally and culturally, and the individual in turn impacts the sociocultural environment by engaging others in exchanges of information. Thus, for example, the rapidity with which a new idea diffuses is a function of the innovator's position in the sociocultural environment, and that position is in turn affected by the diffusion of the new idea. A consequence is that different sociocultural configurations may facilitate or hinder information diffusion and consensus formation. Because communications technologies affect the properties of the actor and hence the way the actor can engage others in the exchange of information, they influence which sociocultural configurations best facilitate information diffusion and consensus formation.

In the constructural model, beliefs and attitudes mediate one's interpersonal relationships through a process of "social adjustment" (Smith et al., 1956; Smith, 1973), and social structure (the initial social organization that dictates who will interact with whom) affects what attitudes and beliefs the individual holds (Heider, 1946), as well as other behavior (White et al., 1976; Burt, 1982). It follows from this model that if those with whom the individual interacts hold an erroneous belief, the individual can become convinced of that belief despite factual evidence to the contrary and will in turn persuade others of its validity.

The constructural model has several key features. First, the diffusion of novel information through different social and organizational structures can be examined. These structures can, though they need not, change dynamically over time. Second, multiple messages (many more than two) can diffuse simultaneously. In this case, the messages compete for the individual's attention. These messages are not valued and so do not reflect beliefs. Thus, in contrast with the controversial information model, the individuals in the constructural model can hold all messages simultaneously. Beliefs are calculated on the basis of what messages are held. Third, some types of communications technologies can be represented within this model as actors with special communicative properties or as alterations in the way individuals interact. A limitation of this model, however, is that the messages or pieces of information are undifferentiated. Thus, content or type of information cannot influence action, nor can individuals choose to communicate a specific piece of information in preference to another.

Dynamic Social Impact Theory

Latane's dynamic social impact theory (Latane, 1996; Huguet and Latane, 1996) suggests that individuals who interact with and influence each other can produce organized patterns at the group or unit level that serve as a communicable representation identifiable by others. A key feature of this model is that individuals tend to be most influenced by those who are physically nearby. Thus spatial factors that influence who interacts with whom can result in locally consistent patterns of shared attitudes, meaning, and beliefs. Empirical evidence supports the idea that through interaction over time, group members become more alike, and their attitudes and beliefs become correlated (Latane and Bourgeois, 1996). Further, empirical evidence confirms that the physically closer in space individuals are, the more frequent are the interactions they recall. Results suggest that the relationship between distance and interaction frequency may possibly be described by an inverse power law with a slope of –1 (Latane et al., 1995). Latane (1996) uses an evolutionary approach to suggest how communication can lead to changes in attitude as individuals develop cognitive bundles of information that then become distributed through the social space.

A simplified version of the associated simulation model can be characterized as a modified game of artificial life. Individuals are laid out spatially on a grid and can interact with those nearest (e.g., those to the north, south, west, and east). They begin with one of two competing beliefs (or messages or attitudes) that diffuse simultaneously and are treated as being in opposition so that an individual cannot hold both simultaneously. Initially, beliefs may be distributed randomly. Over time, however, individuals come to hold beliefs similar to those held by others near them. A limitation of this model is that the social structure is typically defined as the set of physically proximal neighbors in some type of city-block pattern. Therefore, all individuals typically have the same number of others with whom they can interact. An approach similar to that of Latane is taken in the work on A-Life by Epstein and Axtell (1997).

MODELS OF BELIEF FORMATION

Models and theories of belief (and attitude) formation range from those that focus on social processes to those that focus almost exclusively on psychological processes. Four lines of research in this area are discussed below: mathematical models of belief formation in response to message passing, belief networks, social information processing models, and transactive memory.

Mathematical Models Focusing on Information Passing

Much of the social psychology work on belief formation focuses on how message, message content, and sender attributes affect beliefs. Two important theoretical traditions in this respect are reinforcement theory (Fishbein and Ajzen, 1975; Hunter et al., 1984) and information processing theory (Hovland and Pritzker, 1957; Anderson and Hovland, 1957; Anderson, 1959; Anderson, 1964; Anderson, 1971; Hunter et al., 1984). In contrast with these social-psychological models, social network models of belief formation focus on how the individual's position in the social network and the beliefs of other group members influence the individual's beliefs. These social network models are often referred to as models of social influence.

Empirical and theoretical research on belief formation leads to a large number of findings or predictions, some of which conflict. For example, one perspective provides some empirical evidence suggesting that individuals hold beliefs that meet particular psychological (often emotional) needs (Katz, 1960; Herek, 1987); hence erroneous beliefs might be held because they reduce stress or increase feelings of self-esteem. Stability of beliefs would be attributable, at least in part, to emotional stability. In contrast, the symbolic interactionist perspective suggests that the stability of social structures promotes stability of self-image and hence a resistance to change in beliefs, attitudes, and behaviors (Stryker, 1980; Stryker and Serpe, 1982; Serpe, 1987; Serpe, 1988). There is some empirical evidence to support this perspective as well. An implication of this perspective that can be garnered from the models is that erroneous beliefs would be held if the social structure were such that interactions reinforced those beliefs. To be sure, much of the work in this area has also looked at the content of the message and whether the message coming from the social structure was positive or negative. However, the point here is that a great deal of the literature also focuses almost exclusively on the structure (e.g., social context, who talks to whom) and its impact on attitudes independent of the content of specific messages. Attitude reinforcement theory suggests that erroneous beliefs persist only if they are extreme (Fishbein and Ajzen, 1975; Ajzen and Fishbein, 1980; Hunter et al., 1984). Information processing theories suggest that erroneous beliefs persist only if the most recent information supports an erroneous conclusion (Hovland and Pritzker, 1957; Anderson and Hovland, 1957; Anderson, 1959; Anderson, 1964).

Both reinforcement theory and information processing theory model the individual as receiving a sequence of messages about a topic. In these models, the individual adjusts his or her belief about the topic on the basis of each new message. A message is modeled as the information the individual learns during an interaction (i.e., either a fact or someone's belief). These theories differ, however, in the way the adjustment process is modeled. Reinforcement theorists and information processing theorists generally do not distinguish among types of messages, nor do they postulate that different types of messages will affect the individual's attitude in different ways.

Reinforcement theorists argue that the change in belief caused by a message is in the direction of the message (Hunter et al., 1984). Hunter et al. (1984) formulate Fishbein's model as a model of belief change by essentially stipulating that individuals' beliefs about an object change as they receive new messages, each of which alters an underlying belief about the presence of an attribute for that object. In this model, there is a gradual change in these underlying beliefs as new messages are received. Thus positive information leads to a more positive belief, negative information leads to a more negative belief, and neutral information has no effect. The individual's belief at a particular time is simply a weighted sum of the messages he or she has received. Various models of this ilk differ in whether the weights are a function of the individual, the message, the particular attribute, or some combination of these.

In contrast to reinforcement models, information processing models argue that the individual receives a sequence of messages and adjusts his or her belief in the direction of the discrepancy between the message and the current belief (Hunter et al., 1984). The various information processing models differ in whether the weights that determine the extent to which individuals shift their beliefs are a linear or nonlinear function of the individual, the message, or some combination of the two. In contrast with reinforcement theory, however, these weights are not a function of the individual's belief or the underlying beliefs (see for example, Hovland and Pritzker, 1957; Anderson, 1959; Anderson, 1964; Anderson, 1971). Thus in these models, positive information may lead to a less positive belief if the message is less positive than the current belief.

Numerous empirical studies have suggested that more established beliefs are more difficult to change (Cantril, 1946; Anderson and Hovland, 1957; Danes et al., 1984). In addition, Danes et al. (1984:216), in an empirical study controlling for both the amount of information already known by the individual and the extremity of the belief, found that "beliefs based on a large amount of information are more resistant to change" regardless of their level of extremity, and that extreme beliefs based on little information are less resistant to change than extreme beliefs based on much information. Thus extreme beliefs will be resistant to change to the extent that they are those for which the individual has the most information. A large number of empirical studies (see Hunter et al., 1984) have found support for the discrepancy hypothesis of information processing theory (Whittaker, 1967; Insko, 1967; Kiesler et al., 1969). A detailed analysis of these empirical results, however, leads to the following conclusions: (1) extreme beliefs, unless associated with more information, are generally more affected by contradictory information; (2) neutral messages may or may not lead to a belief change, but if they do, the change is typically that predicted by a discrepancy model; and (3) belief shifts are in the direction of the message for non-neutral messages. Thus these results provide basic support for the idea that there is a negative correlation between belief shift and current belief regardless of message content as predicted by information processing theory, but not for the idea that messages supporting an extreme belief will evoke a belief shift in the opposite direction.

Both the reinforcement models and the information processing models are astructural; that is, they do not consider the individual's position in the underlying social or organizational structure. This is not the case with social network models of belief. Certainly, the idea that beliefs are a function of both individuals' social positions and their mental models of the world is not new, and indeed reflects ideas of researchers such as James (1890), Cooley (1902), Mead (1934), and Festinger (1954). Numerous studies have provided ample empirical evidence that social pressure, in terms of what the individual thinks others believe and normative considerations, affects an individual's attitudes (e.g., Molm, 1978; Humphrey et al., 1988). However, formal mathematical treatments of this idea are rare. In a sense, the information diffusion models discussed above fall in the category of social network models. There also exists another class of models—referred to as social influence models—that ignore the diffusion of information and focus only on the construction of beliefs after information has diffused.

Belief Networks

One of the most promising approaches for understanding issues of information warfare and unit-level behavior is hybrid models employing both models of agency and models of the social or communication network. By way of example, consider the formation of a unit-level belief.

Within units, beliefs and decision-making power are distributed across the members, who are linked together through the C^3I architecture. One way of modeling belief formation and its impact on decision making at the unit level is to treat the entire unit as an extremely large belief network in which the different individuals act, in part, as nodes or a collection of nodes in the larger network. This image, however, is not accurate. Viewing the entire unit as a single belief network assumes that the connections within and between individuals are essentially the same. Moreover, it does not allow for the reduction in data that occurs when individuals interact. Such reductions can occur when, for example, one individual paraphrases, forgets, interprets, or delegates as unimportant comments made by others. Further, this approach does not allow for extrapolations of data that can occur when individuals interact. For example, individuals can decide to promote their own ideas rather than the group consensus.

An alternative approach is to model each subunit or group member as a belief network and to model the connections among these members as a communication or social network. In this approach, each subunit or individual takes in and processes information, alters its belief, and generates some decision or belief as output. These subunits or individuals as nodes in the social network then communicate these pieces of reduced and processed information, these decisions or beliefs, to others. This is a hybrid model that combines organizational and cognitive models. In this type of hybrid model, information is reduced and transformed within nodes and delayed through communication constraints between nodes. Such a model can combine the power of social network analysis to evaluate aspects of communication and hierarchy with the power of belief networks to evaluate aspects of individual information processing. Such models use the social network to make the belief networks dynamic.

Social Information Processing

Numerous researchers have discussed and examined empirically the social processes that impact individuals' attitudes and behaviors in organizational units. This work has led to a number of theories about the way individuals process and use social information, including social comparison theory (Festinger, 1954), social learning theory (Bandura, 1977), social information processing theory (Salancik and Pfeffer, 1978), and social influence theory (Freidkin and Johnson, 1990). The earliest and most influential of these was social comparison theory. Festinger (1954) argues that individuals observe and compare themselves with others, and if they see themselves as similar to others, will then alter their attitudes accordingly. Salancik and Pfeffer (1978) argue, on the basis of both theory and empirical evidence that individuals acting for their organizational units will make decisions on the basis of not only what information is available about the problem, but also what information they have about other people and how these others see the problem. This point has been extended by numerous researchers; for example, social network theorists argue that part of this social information is the set of relevant others as determined by the social network in which the individual is embedded (Krackhardt and Brass, 1994). Freidkin and Johnson (1990) extend these views as part of social influence theory and argue that individuals change their attitudes based on the attitudes of those with whom they are in direct contact. Their work includes a formal mathematical model of this process.

Currently, measures and ideas drawn from the social network literature are increasingly being used as the basis for operationalizing models of structural processes that influence belief and attitude formation (e.g., Rice and Aydin, 1991; Fulk, 1993; Shah, 1995). With the exception of the models discussed in the previous section, the processes by which individuals are influenced by others are generally not specified completely at both the structural and cognitive levels. Current models usually posit a process by which individuals interact with a small group of others. A typical model is the following (e.g., Rice and Aydin, 1991; Fulk, 1993):

$$y = aWy + Xb + e$$

where:

y is a vector of self's and other's attitude or belief on some point
X is a matrix of exogenous factors
W is a weighting matrix denoting who interacts with or has influence on whom
a is a constant
b is a vector (individualized weights)
e is a vector (error terms)

Models differ in the way attitudes are combined to determine how the group of others affects the individual. More specifically, models differ dramatically in the way they construct the W matrix.

Transactive Memory

Transactive memory (Wegner, 1995; Wegner et al., 1991) refers to the ability of groups to have a memory system exceeding that of the individuals in the group. The idea is that knowledge is stored as much in the connections among individuals as in the individuals. Wegner's model of transactive memory is based on the metaphor of human memory as a computer system. He argues that factors such as directory updating, information allocation, and coordination of retrieval that are relevant in linking computers together are also relevant in linking the memories of individuals together into a group or unit-level memory.

Empirical research suggests that the memory of natural groups is better than the memory of assigned groups even when all individuals involved know the same things (even for groups larger than dyads; see Moreland [forthcoming]). The implication is that for a group, knowledge of who knows what is as important as knowledge of the task. Transactive knowledge can improve group performance (Moreland et al., 1996). There is even a small computational model of transactive memory; however, both the theoretical reasoning and the model have thus far been applied primarily to dyads (two people).

ROLE OF COMMUNICATIONS TECHNOLOGY

While the literature on information diffusion demonstrates the role of social structure (and to a lesser extent culture) in effecting change through information diffusion, it ignores the role of communications technology. The dominant underlying model of communication that pervades this work is one-to-one, face-to-face communication. The literature on belief formation, particularly the formal models, also ignores communications technology. On the other hand, the literature on communications technology largely ignores the role of the extant social structure and the processes of belief formation. Instead, much of the literature focuses on technological features and usage (Enos, 1990; Rice and Case, 1983; Sproull and Kiesler, 1986), the psychological and social-psychological consequences of the technology (Eisenstein, 1979; Freeman, 1984; Goody, 1968; Kiesler et al., 1984; Rice, 1984), or historical accounts of its development (Innis, 1951; de Sola Poole, 1977; Reynolds and Wilson, 1968). Admittedly, a common question asked is whether communications technology will replace or enhance existing networks or social structures (Thorngen, 1977). Further, there is an abundance of predictions and evidence regarding changes to social structure due to the technology. For example, it is posited that print made the professions possible by enabling regular and rapid contact (Bledstein, 1976) and that electronic communication increases connectedness and decreases isolation (Hiltz and Turoff, 1978). Nevertheless, research on communications technology, information diffusion, and belief formation has remained largely disassociated. One exception is the work on community structures by Wellman and colleagues (Haythornthwaite et al., 1995; Wellman et al., 1996). Another is the work by Rice and colleagues (for a review see Rice, 1994). Currently, mathematical and computational modelers are seeking to redress this gap.

Empirical research has demonstrated that various communications technologies can have profound social and even psychological consequences (see, for example, Price, 1965; Rice, 1984; Sproull and Kiesler, 1991). Such consequences are dependent on various features of the technology. One salient feature of many technologies is that they enable mass or one-to-many communication. Another salient feature of many communications technologies, such as books and videotapes, is that they enable an individual's ideas to remain preserved over time and to be communicated without the individual being present, thus allowing communication at great geographical and temporal distances (Kaufer and Carley, 1993). In this sense, the technology enables the creation of artifacts that can themselves serve as interaction partners.

In those few formal models that have attempted to incorporate aspects of communications technology, two approaches have been considered. The first models the technology by characterizing alterations in the communication channels (e.g., rate of information flow, number of others with whom one can simultaneously interact). This approach is used in the VDT model of Levitt et al. (1994). The second approach models the artifacts or artificial agents created by the technology, such as books, telegrams, and e-mail messages, and allows individuals to interact with them. This approach is used in the constructural model of Carley (1990, 1991a).

CONCLUSIONS AND GOALS

The argument that the individual who receives a message changes his or her attitude toward the source of the message as a function of the message is made by most communication theories (for a review, see Hunter et al., 1984), including reinforcement theory (Rosenberg, 1956; Fishbein, 1965; Fishbein and Ajzen, 1974; Fishbein and Ajzen, 1975; Ajzen and Fishbein, 1980); information processing theory (Hovland and Pritzker, 1957; Anderson and Hovland, 1957; Anderson, 1959; Anderson, 1964; Anderson, 1971); social judgment theory (Sherif and Hovland, 1961; Sherif et al., 1965); and affective consistency theories, such as dissonance theory (Newcomb, 1953; Festinger, 1957), balance theory (Heider, 1946; Heider, 1958), congruity theory (Osgood and Tannenbaum, 1955; Osgood et al., 1957), and affect control theory (Heise, 1977, 1979, 1987). The typical argument is that messages have emotive content and so provide emotional support or punishment; thus individuals adjust their attitude toward the source in order to enhance the level of support or decrease the level of punishment, or because they agree with the source, or some combination of these. Regardless of the specific argument, it follows from such theories that changes in both belief and attitude toward the source (1) should be systematically related and (2) will under most conditions be either positively or negatively correlated, and that change in attitude toward the source is a function only of the message, the individual's current belief, and the individual's current attitude toward the source. In other words, there is a cycle in which beliefs and attitudes change as a function of what information is communicated to the individual by whom, and with whom the individual interacts as these beliefs and attitudes change.

The strength of the diffusion models discussed in this chapter is that they focus on the dynamic by which information or beliefs are exchanged and the resultant change in the underlying social network. By and large, however, the diffusion models are relatively weak in representing belief formation. In contrast, the strength of the belief formation and social influence models is in their ability to accurately capture changes in beliefs. However, their weakness is that they fail to represent change in the underlying social structure of who interacts with whom. Combining these approaches into a model that captures both structural and belief change is the next step.

A cautionary note is also in order. Beliefs are typically more complex than the models described herein would suggest. For example, a single affective dimension is often not sufficient to capture a true belief (Bagozzi and Burnkrant, 1979; Schlegel and DiTecco, 1982), factual evidence is not necessarily additive, some people's beliefs may be more important to the individual than others (Humphrey et al., 1988), the specific content of a message may impact its credibility, and so on. Adding such features to models such as those described here would increase the models' realism and possibly enable them to make even finer-grained predictions. An examination of these and other possible alterations to these models is thus called for.

Finally, this chapter opened with examples of questions that are typical of an information warfare context. None of the current military models can even begin to address questions such as these. The nonmilitary models discussed in this chapter provide some basic concepts that could be used to address such questions, and each has at its core a simple mechanism that could be utilized in the military context. Each, however, would require significant alteration for this purpose.

Short-Term Goals

- Augment existing models so that information about who communicates with or reports to whom and when is traceable and can be stored for use by other programs. Doing so would make it possible to add a module for tracking the impact of various information warfare strategies.
- Incorporate social network measures of the positions of communicators and their mental models. Such measures are necessary for models of information warfare and information diffusion based on structural and psychological features of personnel.
- Gather and evaluate structural information about who sends what to whom and when from war games. Given the ability to capture structural information routinely, it would be possible to begin systematically evaluating potential weak spots in existing C^3I structures from an information warfare perspective. This information would also provide a basis for validating computational models in this area.
- Develop visual aids for displaying the communication network or C^3I structure being evaluated from an information warfare perspective. Such aids might rely on intelligent agent techniques for displaying graphs. None of the current models of diffusion, network formation, or vulnerability assessment across personnel have adequate visualization capabilities. While there are a few visualization tools for network data, they are not adequate, particularly for networks

that evolve or have multiple types of nodes or relations. The capability to visualize these networks and the flow of information (both good and bad) would provide an important decision aid.

Intermediate-Term Goals

• Utilize multidisciplinary teams to develop prototype information warfare decision aids incorporating elements of both information diffusion and belief formation models. Efforts to develop such decision aids would illuminate the important military context issues that must be addressed. The development teams should include researchers in the areas of multiagent modeling, social networks, the social psychology of belief or attitude formation, and possibly graph algorithms. Such a multidisciplinary approach is necessary to avoid the reinvention of measures and methods already developed in the areas of social psychology and social networks.

• Develop and validate models that combine a social-psychological model of belief formation with a social network model of information diffusion. Explore how to model an individual decision maker's belief and decision as a function of the source of the knowledge (from where and from whom). Such an analysis would be a way of moving away from a focus on moderators to a model based on a more complete understanding of belief formation. Such a model could be used to show conditions under which the concurrent exchange of information between individuals and the consequent change in their beliefs result in the formation or persistence of erroneous beliefs and so lead to decision errors. Even a simple model of this type would be useful in the information warfare context for locating potential weak spots in different C^3 architectures. As noted, one of the reasons a model combining information diffusion and belief formation models is critical for the information warfare context is that commanders may alter their surrounding network on the basis of incoming information. For example, suppose the incoming information is ambiguous, misleading, contradictory, or simply less detailed than that desired. In the face of such uncertainty, different commanders might respond differently: some might respond by increasing their capacity to get more information; others might act like Napoleon and deal with uncertainty by "reducing the amount of information needed to perform at any given level" (Van Creveld, 1985:146).

Long-Term Goals

• Support basic research on when specific beliefs will be held and specific decisions will be made. Research is needed to move from several statements that beliefs, the rate of decision making, the confidence in decisions, and the likelihood of erroneous decisions are high or low to statements about which specific beliefs will be held or which decisions will be made. Basic research is needed toward the development and validation of models that combine detailed models of information technology, message content, belief formation, and the underlying social and organizational structure and are detailed enough to give specific predictions. In particular, emphasis should be placed on message content and its impact on the commander's mental model and resultant decisions. Selecting what information is available can affect intelligence decisions. However, without a thorough understanding of the role of cultural barriers, trust, and language barriers, detailed predictions about the impact of specific messages are not possible.

- Support basic research on converting moderator functions into models of how people cope with novel and extensive information.

12

Methodological Issues and Approaches

The purpose of this chapter is to provide general methodological guidelines for the development, instantiation, and validation of models of human behavior. We begin with a section describing the need for the tailoring of models that incorporate these representations in accordance with specific user needs. The core of the chapter is a proposed methodological framework for the development of human behavior representations.

THE NEED FOR SITUATION-SPECIFIC MODELING

At present, we are a long way from having either a general-purpose cognitive model or a general-purpose organizational unit model that can be incorporated directly into any simulation and prove useful. However, the field has developed to the point that simulations incorporating known models and results of cognition, coordination, and behavior will greatly improve present efforts by the military, if—and only if—the models are developed and precisely tailored to the demands of a given task and situation, for example, the tasks of a tank driver or a fixed-wing pilot. It is also important to note that clear measures of performance of military tasks are needed. Currently, many measures are poorly defined or lacking altogether.

Given the present state of the field at the individual level, it is probably most useful to view a human operator as the controller of a large number of programmable components, such as sensory, perceptual, motor, memory, and decision processes. The key idea is that these components are highly adaptable and may be tuned to interact properly in order to handle the demands of each specific task in a particular environment and situation. Thus, the system may be seen as a framework or architecture within which numerous choices and adaptations must be made when a given application is required. A number of such architectures have been developed and provide examples of how one might proceed, although the field is still in its infancy, and it is too early to recommend a commitment to any one architectural framework (see Chapter 3).

Given the present state of the field at the unit level, it is probably most useful to view a human as a node in a set of overlaid networks that connect humans to each other in various ways, connect humans to tasks and resources, and so forth. One key idea is that these networks (1) contain information; (2) are adaptable; and (3) can be changed by orders, technology, or actions taken by individuals. Which linkages in the network are operable and which nodes (humans, technology, tasks) are involved will need to be specified in accordance with the specific military application. Some unit-level models can be thought of as architectures in which the user, at least in principle, can describe an application by specifying the nodes and linkages. Examples include the virtual design team (Levitt et al., 1994) and ORGAHEAD (Carley and Svoboda, 1996; Carley, forthcoming).

The panel cannot overemphasize how critical it is to develop situation-specific models within whatever general architecture is adopted. The situations and tasks faced by humans in military domains are highly complex and very specific. Any effective model of human cognition and behavior must be tailored to the demands of the particular case. In effect, the tailoring of the model substitutes for the history of training and knowledge by the individual (or unit), a history that incorporates both personal training and military doctrine.

At the unit level, several computational frameworks for representing teams or groups are emerging. These frameworks at worst supply a few primitives for constructing or breaking apart groups and aggregating behavior and at best facilitate the representation of formal structure, such as the hierarchy, the resource allocation structure, the communication structure, and unit-level procedures inherited by all team members. These frameworks provide only a general language for constructing models of how human groups perform tasks and what coordination and communication are necessary for pursuing those tasks. Representing actual units requires filling in these frameworks with details for a specific team, group, or unit and for a particular task.

A METHODOLOGY FOR DEVELOPING HUMAN BEHAVIOR REPRESENTATIONS

The panel suggests that the Defense Modeling and Simulation Office (DMSO) encourage developers to employ a systematic methodology in developing human behavior representations. This methodology should include the following steps:

- Developers should employ interdisciplinary teams.

• They should review alternatives and adopt a general architecture that is most likely to be useful for the dominant demands of the specific situation of interest.

• They should review available unit-level frameworks and support the development of a comprehensive framework for representing the command, control, and communications (C^3) structure. (The cognitive framework adopted should dictate the way C^3 procedures are represented.)

• They should review available documentation and seek to understand the domain and its doctrine, procedures, and constraints in depth. They should prepare formal task analyses that describe the activities and tasks, as well as the information requirements and human skill requirements, that must be represented in the model. They should prepare unit-level task analyses that describe resource allocation, communication protocols, skills, and so forth for each subunit.

• They should use behavioral research results from the literature, procedural model analysis, ad hoc experimentation, social network analysis, unit-level task analysis, field research, and, as a last resort, expert judgment to prepare estimates of the parameters and variables to be included in the model that are unconstrained by the domain or procedural requirements.

• They should systematically test, verify, and validate the behavior and performance of the model at each stage of development. We also encourage government military representatives to work with researchers to define the incremental increase in model performance as a function of the effort required to produce that performance.

The sections that follow elaborate on the four most important of these methodological recommendations.

Employ Interdisciplinary Teams

For models of the individual combatant, development teams should include cognitive psychologists and computer scientists who are knowledgeable in the contemporary literature and modeling techniques. They should also include specialists in the military doctrine and procedures of the domain to be modeled. For team-, battalion-, and force-level models, as well as for models of command and control, teams composed of sociologists, organizational scientists, social psychologists, computer scientists, and military scientists are needed to ensure that the resultant models will make effective use of the relevant knowledge and many (partial) solutions that have emerged in cognitive psychology, artificial intelligence, and human factors for analyzing and representing individual human behavior in a computational format. Similarly, employing sociology, organizational science, and distributed artificial intelligence will ensure that the relevant knowledge and solutions for analyzing and representing unit-level behavior will be employed.

Understand the Domain in Depth, and Document the Required Activities and Tasks

The first and most critical information required to construct a model of human behavior for military simulations is information about the task to be performed by the simulated and real humans as regards the procedures, strategies, decision rules, and command and control structure involved. For example, under what conditions does a combat air patrol pilot engage an approaching enemy? What tactics are followed? How is a tank platoon deployed into defensive positions? As in the Soar-intelligent forces (IFOR) work (see Chapter 2), military experts have to supply information about the desired skilled behavior the model is to produce. The form in which this information is collected should be guided by the computational structure that will encode the tasks.

The first source of such information is military doctrine—the "fundamental principles by which military forces guide their actions in support of national objectives" (U.S. Department of the Army, 1993b). Behavioral representations need to take account of doctrine (U.S. doctrine for own forces, non-U.S. doctrine for opposing forces). On the one hand, doctrinal consistency is important. On the other hand, real forces deviate from doctrine, whether because of a lack of training or knowledge of the doctrine or for good reason, say, to confound an enemy's expectations. Moreover, since doctrine is defined at a relatively high level, there is much room for behavior to vary even while remaining consistent with doctrine. The degree of doctrinal conformity that is appropriate and the way it is captured in a given model will depend on the goals of the simulation.

Conformity to doctrine is a good place to start in developing a human behavior representation because doctrine is written down and agreed upon by organizational management. However, reliance on doctrine is not enough. First, it does not provide the task-level detail required to create a human behavior representation. Second, just as there are both official organization charts and informal units, there are both doctrine and the ways jobs really get done. There is no substitute for detailed observation and task analysis of real forces conducting real exercises.

The Army has a large-scale project to develop computer-generated representations of tactical combat behavior, such as moving, shooting, and communicating. These representations are called combat instruction sets. According to the developers (IBM/Army Integrated Development Team, 1993), each combat instruction set should be:

- Described in terms of a detailed syntax and structure layout.
- Explicit in its reflection of U.S. and opposing force tactical doctrines.
- Explicit in the way the combat instruction set will interface with the semiautomated forces simulation software.
- Traceable back to doctrine.

Information used to develop the Army combat instruction sets comes from written doctrine and from subject matter experts at the various U.S. Army Training and Doctrine Command schools who develop the performance conditions and standards for mission training plans. The effort includes battalion, company, platoon, squad, and platform/system-level behavior. At the higher levels, the mission, enemy, troops, terrain, and time available (METT-T) evaluation process is used to guide the decision-making process. The combat instruction sets, like the doctrine itself, should provide another useful input to the task definition process.

At the individual level, although the required information is not in the domain of psychology or of artificial intelligence, the process for obtaining and representing the information is. This process, called task analysis and knowledge engineering, is difficult and labor-intensive, but it is well developed and can be performed routinely by well-trained personnel.

Similarly, at the unit level, although the required information is not in the domain of sociology or organizational science, the process for obtaining and representing the information is. This process includes unit-level task analysis, social network analysis, process analysis, and content analysis. The procedures involved are difficult and labor-intensive, often requiring field research or survey efforts, but they can be performed routinely by well-trained researchers.

At the individual level, task analysis has traditionally been applied to identify and elaborate the tasks that must be performed by users when they interact with systems. Kirwan and Ainsworth (1992:1) define task analysis as:

> . . . a methodology which is supported by a number of specific techniques to help the analyst collect information, organize it, and then use it to make judgments or design decisions. The application of task analysis methods provides the user with a blueprint of human involvement in a system, building a detailed picture of that system from the human perspective. Such structured information can then be used to ensure that there is compatibility between system goals and human capabilities and organization so that the system goals will be achieved.

This definition of task analysis is conditioned by the purpose of designing systems. In this case, the human factors specialist is addressing the question of how best to design the system to support the tasks of the human operator. Both Kirwan and Ainsworth (1992) and Beevis et al. (1994) describe in detail a host of methods for performing task analysis as part of the system design process that can be equally well applied to the development of human behavior representations for military simulations.

If the human's cognitive behavior is being described, cognitive task analysis approaches that rely heavily on sophisticated methods of knowledge acquisition are employed. Many of these approaches are discussed by Essens et al. (1995). Specifically, Essens et al. report on 32 elicitation techniques, most of which rely either on interviewing experts and asking them to make judgments and categorize material, or on reviewing and analyzing documents.

Descriptions of the physical and cognitive tasks to be performed by humans in a simulation are important for guiding the realism of behavior representations. However, developing these descriptions is time-consuming and for the most part must be done manually by highly trained individuals. Although some parts of the task analysis process can be accomplished with computer programs, it appears unlikely that the knowledge acquisition stage will be automated in the near future. Consequently, sponsors will have to establish timing and funding priorities for analyzing the various aspects of human behavior that could add value to military engagement simulations.

At the unit or organizational level, task analysis involves specifying the task and the command and control structure in terms of assets, resources, knowledge, access, timing, and so forth. The basic idea is that the task and the command and control structure affect unit-level performance (see Chapter 10). Task analysis at the unit level does not involve looking at the motor actions an individual must perform or the cognitive processing in which an individual must engage. Rather, it involves laying out the set of tasks the unit as a whole must perform to achieve some goal, the order in which those tasks must be accomplished, what resources are needed, and which individuals or subunits have those resources.

A great deal of research in sociology, organizational theory, and management science has been and is being done on how to do task analysis at the unit level. For tasks, the focus has been on developing and extending project analysis techniques, such as program evaluation and review technique (PERT) charts and dependency graphs. For the command and control structure, early work focused on general features such as centralization, hierarchy, and span of control. Recently, however, network techniques have been used to measure and distinguish the formal reporting structure from the communication structure. These various approaches have led to a series of survey instruments and analysis tools. There are a variety of unresolved issues, including how to measure differences in the structures and how to represent change.

Instantiate the Model

A model of human behavior must be made complete and accurate with specific data. Ideally, the model with its parameters specified will already be incorporated into an architectural framework, along with the more general properties of human information processing mechanisms. Parameters for selected sensory and motor processes can and should be obtained from the literature. However, many human behavior representations are likely to include high-level decision-making, planning, and information-seeking components. For these components, work is still being done to define suitable underlying structures, and general models at this level will require further research. In many cases, however, the cognitive activities of interest should conform to doctrine or are highly proceduralized. In these cases, detailed task analyses provide data that will permit at least a first-order approximation of the behavior of interest.

Sometimes small-scale analytical studies or field observations can provide detailed data suitable for filling in certain aspects of a model, such as the time to carry out a sequence of actions that includes positioning, aiming, and firing a rifle or targeting and launching a missile. Some of these aspects could readily be measured, whereas others could be approximated without the need for new data collection by using approaches based on prediction methods employed for time and motion studies in the domain of industrial engineering (Antis et al., 1973; Konz, 1995), Fitts' law (Fitts and Posner, 1967), or GOMS[1] (John and Kieras, 1996; Card et al., 1983). These results could then be combined with estimates of perceptual and decision-making times to yield reasonable estimates of human reaction times for incorporation into military simulations.

Inevitably, there will be some data and parameter requirements for which neither the literature nor modeling and analysis will be sufficient and for which it would be too expensive to conduct even an ad hoc study. In those cases, the developer should rely on expert judgment. However, in conducting this study, the panel found that expert judgment is often viewed as the primary source of the necessary data; we emphasize that it should be the alternative of *last resort* because of the biases and lack of clarity or precision associated with such judgments.

Much of the modeling of human cognition that will be necessary for use in human behavior representations—particularly those aspects of cognition involving higher-level planning, information seeking, and decision making—has not yet been done and will require new research and development. At the same time, these new efforts can build productively on many recent developments in the psychological and sociological sciences, some of which are discussed in the next chapter.

Verify, Validate, and Accredit the Model

Before a model can be used with confidence, it must be verified, validated, and accredited. Verification refers here to the process of checking for errors in the programming, validation to determining how well the model represents reality, and accreditation to official certification that a model or simulation is acceptable for specific purposes. According to Bennett (1995), because models and simulations are based on only partial representations of the real world and are modified as data describing real events become available, it is necessary to conduct verification and validation on an ongoing basis. As a result, it is not possible to ensure

[1]GOMS (goals, operators, methods, and selection rules) is a relatively simple methodology for making quantitative estimates of the performance times for carrying out well-structured procedural tasks.

correct and credible use of such models for a specific application early in the process.

Verification may be accomplished by several methods. One is to develop tracings of intermediate results of the program and check them for errors using either hand calculations or manual examination of the computations and results. Verification may also be accomplished through modular programming, structured walkthroughs, and correctness proofs (Kleijnen and van Groenendaal, 1992).

Validation is a more complex matter. Indeed, depending on the characteristics of the model, its size, and its intended use, adequate demonstration of validity may not be possible. According to DMSO, validation is defined as "the process of determining the degree to which a model is an accurate representation of the real world from the perspective of the intended users of the model" (U.S. Department of Defense, 1996). The degree of precision needed for a model is guided by the types and levels of variables it represents and its intended use. For example, some large models have too many parameters for the entire model to be tested; in these cases, an intelligent testing strategy is needed. Sensitivity analysis may be used to provide guidance on how much validity is needed, as well as to examine the contributions of particular models and their associated costs. Carley (1996b) describes several types of models, including emulation and intellective models. Emulation models are built to provide specific advice, so they need to include valid representations of everything that is critical to the situation at hand. Such models are characterized by a large number of parameters, several modules, and detailed user interfaces. Intellective models are built to show proof of concept or to illustrate the impact of a basic explanatory mechanism. Simpler and smaller than emulation models, they lack detail and should not be used to make specific predictions.

Validation can be accomplished by several methods, including grounding, calibration, and statistical comparisons. Grounding involves establishing the face validity or reasonableness of the model by showing that simplifications do not detract from credibility. Grounding can be enhanced by demonstrating that other researchers have made similar assumptions in their models or by applying some form of ethnographic analysis. Grounding is appropriate for all models, and it is often the only level of validation needed for intellective models.

Calibration and statistical comparisons both involve the requirement for real-world data. Real-life input data (based on historical records) are fed into the simulation model, the model is run, and the results are compared with the real-world output. Calibration is used to tune a model to fit detailed real data. This is often an interactive process in which the model is altered so that its predictions come to fit the real data. Calibration of a model occurs at two levels: at one level, the model's predictions are compared with real data; at another, the processes and parameters within the model are compared with data about the processes and parameters that produce the behavior of concern. All of these procedures are relevant to the validation of emulation models.

Statistical or graphical comparisons between a model's results and those in the real world may be used to examine the model's predictive power. A key requirement for this analysis is the availability of real data obtained under comparable conditions. If a model is to be used to make absolute predictions, it is important that not only the means of the model and the means of the real world data be identical, but also that the means be correlated. However, if the model is to be used to make relative predictions, the requirements are less stringent: the means of the model and the real world do not have to be equal, but they should be positively correlated (Kleijnen and van Groenendaal, 1992).

Since a model's validity is determined by its assumptions, it is important to provide these assumptions in the model's documentation. Unfortunately, in many cases assumptions are not made explicit. According to Fossett et al. (1991), a model's documentation should provide an analyst not involved in the model's development with sufficient information to assess, with some level of confidence, whether the model is appropriate for the intended use specified by its developers.

It is important to point out that validation is a labor-intensive process that often requires a team of researchers and several years to accomplish. It is recommended that model developers be aided in this work by trained investigators not involved in developing the models. In the military context, the most highly validated models are physiological models and a few specific weapons models. Few individual combatant or unit-level models in the military context have been validated using statistical comparisons for prediction; in fact, many have only been grounded. Validation, clearly a critical issue, is necessary if simulations are to be used as the basis for training or policy making.

Large models cannot be validated by simply examining exhaustively the predictions of the model under all parameter settings and contrasting that behavior with experimental data. Basic research is therefore needed on how to design intelligent artificial agents for validating such models. Many of the more complex models can be validated only by examining the trends they predict. Additional research is needed on statistical techniques for locating patterns and examining trends. There is also a need for standardized validation techniques that go beyond those currently used. The development of such techniques may in part involve developing sample databases against which to validate models at each level. Sensitivity analysis may be used to distinguish between parameters of a model that influence results and those that are indirectly or loosely coupled to outcomes. Finally, it may be useful to set up a review board for ensuring that standardized validation procedures are applied to new models and that new versions of old models are docked against old versions (to ensure that the new versions still generate the same correct behavior as the old ones).

13

Conclusions and Recommendations

The area of human behavior representation is at least as important to the mission of the Defense Modeling and Simulation Office (DMSO) as the area of modeling of the battlefield environment, yet at the time of this study, human behavior representation was receiving only a fraction of the resources assigned to modeling of the environment. The panel's review has indicated that there are many areas today in which models of human behavior are needed and in which the models being used are inadequate. There are many other areas in which there is a clear need for models, but none exist. Successful development of the needed models will require the sustained application of resources both to the infrastructure supporting human behavior representation and to the development of the models themselves.

The panel believes continued advances in the science and practice of human behavior representation will require effort on many fronts from many perspectives. We recognize that DMSO is primarily a policy-making body and that, as an organization, it does not sponsor or execute specific modeling projects. In those areas in which DMSO does not participate directly, we recommend that it advocate an increased focus by the individual services on the objective of incorporating enhanced human behavior representation in the models used within the military modeling and simulation community.

The overall recommendations of the panel are organized into two major areas—a programmatic framework and infrastructure/information exchange. The next section presents the framework for a program plan through which the goal of developing models of human behavior to meet military needs can be realized. The section that follows presents recommendations in the area of infrastructure and information exchange that can help promote vigorous and informed development of such models. Additional recommendations that relate to specific topics addressed in the report are presented as goals in the respective chapters.

A FRAMEWORK FOR THE DEVELOPMENT OF MODELS OF HUMAN BEHAVIOR

The panel has formulated a general framework that we believe can guide the development of models of human behavior for use in military simulations. This framework reflects the panel's recognition that given the current state of model development and computer technology, it is not possible to create a single integrative model or architecture that can meet all the potential simulation needs of the services. Figure 13.1 presents the elements of a plan for DMSO to apply in pursuing the development of models of human behavior to meet short-, intermediate-, and long-term goals. For the short term (as shown on the left side of the figure), the panel believes it is important to collect real-world, war-game, and laboratory data in support of the development of new models and the development and application of human model accreditation procedures. For the intermediate term (as shown in the center of the figure), we believe DMSO should extend the scope of useful task analysis and encourage sustained model development in focused areas. And for the long term (as shown on the right hand side of the figure), we believe DMSO should advocate theory development and behavioral research that can lead to future generations of models of human and organizational behavior. Together, as illustrated in Figure 13.1, these initiatives constitute a program plan for human behavior representation development for many years to come.

Work on achieving these short-, intermediate-, and long-term goals should begin concurrently. The panel recommends that these efforts proceed in accordance with four themes, listed below in order of priority and discussed in more detail in the following subsections:

- Collect and disseminate human performance data.
- Create accreditation procedures for models of human behavior.
- Support sustained model development in focused domains.
- Support theory development and basic research in relevant areas.

Collect and Disseminate Human Performance Data

As indicated on the left of Figure 13.1, the panel has concluded that all levels of model development depend on the sustained collection and dissemination of human behavior data. Figure 13.2 expands on this requirement by elaborating the range of data collection needs, which extend from real-world field data to laboratory studies of basic human capacities. Examples of salient field research are

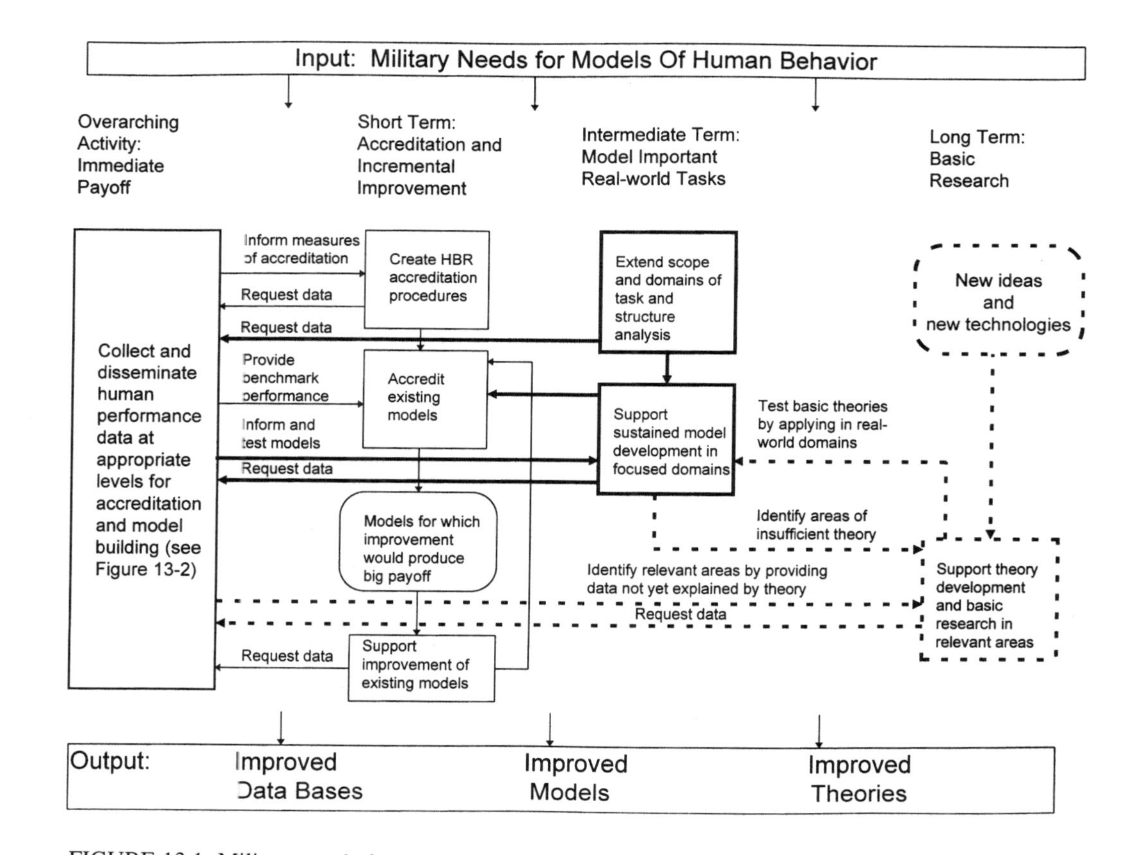

FIGURE 13.1 Military needs for models of human behavior.

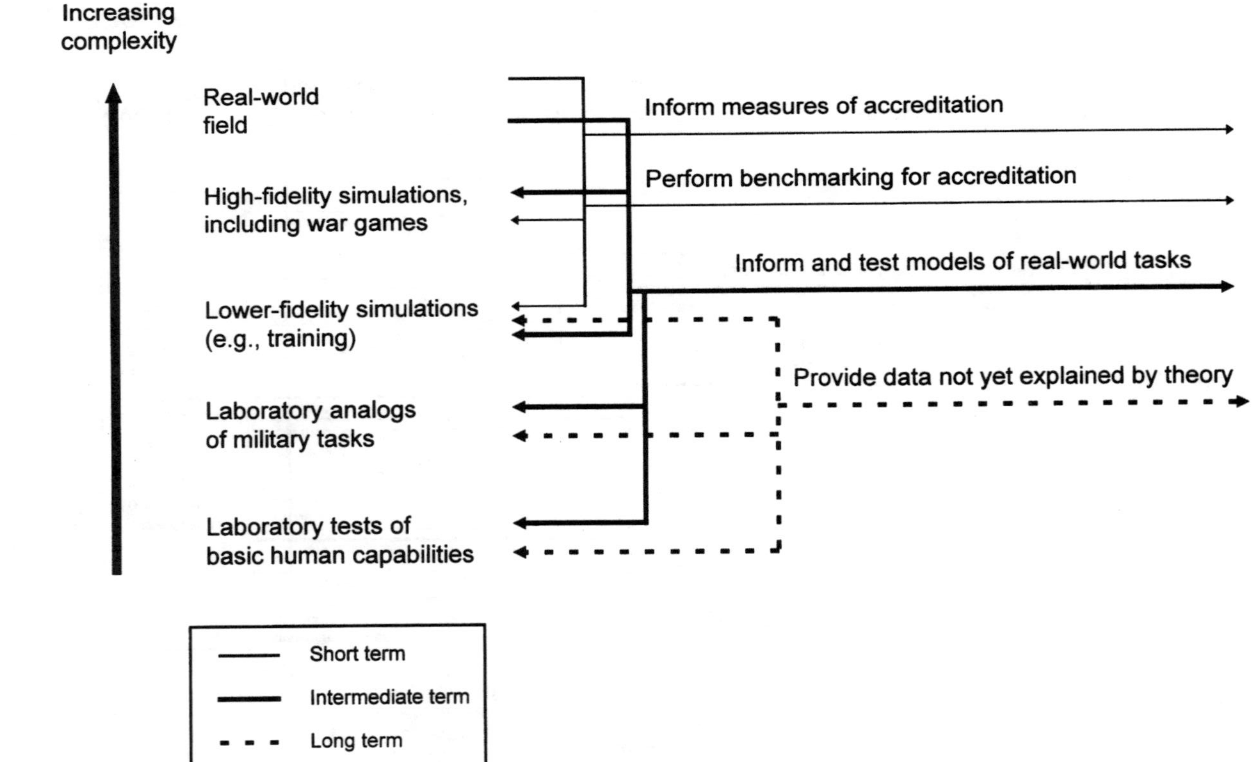

FIGURE 13.2 Collect and disseminate human performance data at appropriate levels for accreditation and model building.

studies of the effects of fatigue on troop movement speeds and the gathering of quantitative data on the communication patterns within and across echelons in typical joint task force command and control exercises. Examples of laboratory work that has importance for real-world contexts are studies of the orientation of behavior to coordinated acoustic and visual stimuli and studies of the relationship between risk taking and information uncertainty. Between these extremes there is a need for data derived from high-fidelity simulations and war games and for data from laboratory analogs to military tasks. Examples of high-fidelity data of value to the modeler are communication logs and mission scenarios.

Chapter 12 emphasizes the need for detailed task analyses, but such analyses are not sufficient for the development of realistic human behavior representations. There is also a need for the kind of real-world military data that reflect, in context, the way military forces actually behave, are coordinated, and communicate. There are some good data on how fast soldiers can walk on various kinds of terrain and how fast tanks can move, but data are sparse on such things as the accuracy of localization of gun shots in the battlefield, the time it takes to communicate a message from one echelon to the next by various media, and the flexibility of different command and control structures.

These data are needed for a variety of purposes, as indicated in Figure 13.2: to support the development of measures of accreditation, to provide benchmark performance for comparison with model outputs in validation studies, to help set the parameters of the actual models of real-world tasks and test and evaluate the efficacy of those models, and to challenge existing theory and lead to new conceptions that will provide the grist for future models. It is not enough simply to advocate the collection of these data. There also must be procedures to ensure that the data are codified and made available in a form that can be utilized by all the relevant communities—from military staffs who need to have confidence in the models to those in the academic sphere who will develop the next generation of models. Some of these data, such as communication logs from old war games, already exist; however, they need to be categorized, indexed, and made generally available. Individual model and theory builders should be able to find out what data exist and obtain access to specific data on request.

Create Accreditation Procedures for Models of Human Behavior

The panel has observed very little quality control among the models that are used in military simulations today. Just as there is a need for accreditation of constructive models that are to be used in training and doctrine development, there is a need for accreditation of models of human and organizational behavior. DMSO should develop model accreditation procedures specifically for this purpose.

One component needed to support robust accreditation procedures is quantitative measures of human performance. In addition to supporting accreditation, such measures would facilitate evaluation of the cost-effectiveness of alternative models so that resource allocation judgments could be made on the basis of data rather than opinion. The panel does not believe that the people working in the field are able to make such judgments now, but DMSO should promote the development of simulation performance metrics that could be applied equivalently to live exercises and simulations. These metrics would be used to track the relative usefulness (cost-effectiveness, staff utilization efficiency, elapsed time, training effectiveness, transfer-of-training effectiveness, range of applicability) of the available models (across different levels of human behavior fidelity and psychological validity) as compared with performance outcomes obtained from live simulations and exercises. In the initial stages, these metrics would be focused on evaluation of particular models and aspects of exercises; in the long term, however, the best of these metrics should be selected for more universal application. The goal would be to create state-of-health statistics that would provide quantitative evidence of the payoff for investments in human behavior representation. These statistics would show where the leverage is for the application of models of human behavior to new modeling needs. For this goal to be achieved, there must be sustained effort that is focused on quantitative performance metrics and that can influence evaluation across a range of modeling projects.

There are special considerations involved in human behavior representation that warrant having accreditation procedures specific to this class of behavioral models. The components of accreditation should include those described below.

Demonstration/Verification. Provide proof that the model actually runs and meets the design specifications. This level of accreditation is similar to that for any other model, except that verification must be accomplished with human models in the loop, and to the extent that such models are stochastic, will require repeated runs with similar but not identical initial conditions to verify that the behavior is as advertised.

Validation. Show that the model accurately represents behavior in the real world under at least some conditions. Validation with full generality is not possible for models of this complexity; rather, the scope and level of the required validation should be very focused and matched closely to the intended uses of each model. One approach to validation is to compare model outputs with data collected during prior live simulations conducted at various military training sites (e.g., the National Training Center, Red Flag, the Joint Readiness Training Center). Another approach is to compare model outputs with data derived from laboratory experiments or from various archival sources. Other approaches are discussed in Chapter 12.

As previously indicated, the procedures for conducting validation of models as complex as those involving human behavior are not well developed or understood; however it is important to move toward more quantitative evaluation, rather than simply relying on subject matter experts to judge that a model is "good enough." One reason these models are difficult to validate is that they are so complex that all their aspects cannot be validated within a reasonable amount of time. A second reason is that these models are often dynamic, predicting changes over time, but longitudinal data on individuals and units have rarely been collected because doing so is extremely time intensive and preempts valuable resources. Unit-level models are often particularly difficult to validate in depth because of the large amounts of data required.

Finally, to bring objectivity and specialized knowledge to the validation process, the panel suggests that the validation team include specialists in modeling and validation who have not participated in the actual model development.

Analysis. Describe the range of predictions that can be generated by the model. This information is necessary to define the scope of the model; it can also be used to link this model with others. Analysis is hampered by the complexity of these models, which makes it difficult to extract the full range of behavior covered. Thus investment in analysis tools is needed to assist in this task.

Documentation Requirements. The accreditation procedures should include standards for the documentation that explains how to run and modify the model and a plan for maintaining and upgrading the model. Models will be used only if they are easy to run and modify to meet the changing needs of the user organization. Evaluation of the documentation should include exercising specific scenarios to ensure that the documentation facilitates the performance of the specified modeling tasks. It should also be noted that models that are easy to use and modify run the danger of being used in situations where they are neither validated nor appropriate. The documentation for a model should indicate clearly the model's scope of application and the situations in which it has been validated. In the case of models that learn from simulator experience, a record of past engagement history should be included with the model's documentation. Specific limitations of the models should be listed as well. Finally, as models are modified, they should be revalidated and reaccredited.

As a high priority, the panel recommends that the above accreditation procedures be applied to military models of human behavior that are either currently in use or being prepared for use, most of which have not had the benefit of rigorous quantitative validation, and that the results of these analyses be used to identify high-payoff areas for improvement. Significant improvements may thereby be achievable relatively quickly for a small investment. The analysis results can also be used to identify the most successful models and the resources and methodologies responsible for their success, thus providing a starting point for determining the resources and methods required for successful modeling efforts. The accreditation procedures described here can lead directly to a program to make the identified improvements to existing models with an immediate payoff.

Support Sustained Model Development in Focused Domains

Several specific activities are associated with model development. They include the following.

Develop Task Analysis and Structure

It is important to continue and expand the development of detailed descriptions of military contexts—the tasks, procedures, and structures (such as common models of mission space [CMMS]; command, control, communications, and intelligence [C^3I] architectures; and military scenarios) that provide the structure for modeling human behavior at the individual, unit, and command levels. The development of combat instruction sets at the individual level, CMMS at the unit level, and joint mission essential task lists (JMETLs) at the joint service level represents a starting point for understanding the tasks to be incorporated into models and simulations—but only a starting point. For example, combat instruction sets need to be extended to all services, deeper task analyses are required for model development purposes, and existing descriptions do not cover typical requirements for modeling large units. Databases describing resources, platforms, weapon systems, and weather conditions need to be constructed. There is also a need to codify the variants in organizational structures in the form of a taxonomy. At the individual level, combat instruction sets need to move from describing lower-level physical tasks, to describing higher-level perceptual and cognitive tasks, to describing even higher-level unit tasks and the assignment of subtasks to subunits. Since task analysis is such a knowledge-intensive operation, there is a need to develop tools for knowledge assessment and knowledge acquisition for these higher-level individual- and unit-level tasks.

Further, more in-depth structural analysis of existing C^3I architectures is required as a basis for improving models of command and control. To accomplish such structural analysis, there is a need to develop a data archive describing the current architectures and those being used in various simulation systems. Likewise, there is a need for a common scheme for describing and providing data on these architectures. Since structural analysis is a data-intensive operation, it has to be supported with data collection and visualization tools. Finally, there is a need for a data archive of typical scenarios for single-service and joint-service operations in a variety of circumstances, including operations other than war. These scenarios could be used as the tasks against which to examine new models.

and should take advantage of interpretation, feedback, and revision based on previous exercises.

Promote Interoperability

In concert with model development, DMSO should evolve policy to promote interoperability among models representing human behavior. Although needs for human behavior representation are common across the services, it is simplistic to contemplate a single model of human behavior that could be used for all military simulation purposes, given the extent to which human behavior depends on both task and environment. Therefore, the following aspects of interoperability should be pursued:

- Parallels among simulation systems that offer the prospect of using the same or similar human behavior representation modules in more than one system
- The best means for assembling and sharing human behavior representation modules among simulation system developers
- Procedures for extracting components of human behavior representation modules that could be used in other modules
- Procedures for passing data among modelers
- Modules for measuring individual- and unit-level data that could be used in and calculated by more than one system

One cannot assume that integrated, interoperable models can be achieved merely by interconnecting component modules. Even differing levels of model sophistication within a specific domain may require different modeling approaches. One must generate integrative models at the level of principles and theory and adapt the implementation as needed to accommodate the range of modeling approaches.

Employ Substantial Resources

Improving the state of human behavior representation will require substantial resources. Even when properly focused, this work is at least as resource demanding as environmental representation. Further, generally useful unit-level models are unlikely to emerge simply through minor adjustments in integrative individual architectures. Integrative unit-level architectures are currently being developed, but lag behind those at the individual level both in their comprehensiveness and in their direct applicability to military issues. There are two reasons for this. First, moving from the individual to the unit level is arguably an increase in complexity. Second, far more resources and researchers have been focused on individual-level models. It is important to recognize that advances at the unit level comparable to those at the individual level will require the expenditure of comparable resources and time.

Support Theory Development and Basic Research in Relevant Areas

As illustrated in the discussion of cognition and command and control behavior in earlier chapters of this report, there are many areas in which adequate theory either is entirely lacking or has not been integrated to a level that makes it directly applicable to the needs of human behavior representation. There is a need for continued long-term support of theory development and basic research in areas such as decision making, situation awareness, learning, and organizational modeling. It would be short-sighted to focus only on the immediate payoffs of modeling. Support for future generations of models needs to be sustained as well. It might be argued that the latter is properly the role of the National Science Foundation or the National Institutes of Health. However, the kinds of theories needed to support human behavior representation for military situations are not the typical focus of these agencies. Their research tends to emphasize toy problems and predictive modeling in restricted experimental paradigms for which data collection is relatively easy. To be useful for the representation of military human behavior, the research needs to be focused on the goal of integration into larger military simulation contexts and on specific military modeling needs.

RECOMMENDATIONS FOR INFRASTRUCTURE AND INFORMATION EXCHANGE

The panel has identified a set of actions we believe are necessary to build consensus more effectively within the Department of Defense modeling and simulation community on the need for and direction of human performance representation within military simulations. The focus is on near-term actions DMSO can undertake to influence and shape modeling priorities within the services. These actions are in four areas: collaboration, conferences, interservice communication, and education/training.

Collaboration

The panel believes it is important in the near term to encourage collaboration among modelers, content experts, and behavioral and social scientists, with emphasis on unit/organizational modeling, learning, and decision making. It is recommended that specific workshops be organized in each of these key areas. We view the use of workshops to foster collaboration as an immediate and low-cost step to promote interchange within each of these areas and provide opportunities for greater interservice understanding.

Conferences

The panel recommends an increase in the number of conferences focused on the need for and issues associated with human behavior representation in military models and simulations. The panel believes the previous biennial conferences on computer-generated forces and behavioral representation have been valuable, but could be made more useful through changes in organization and structure. We recommend that external funding be provided for these and other such conferences and that papers be submitted in advance and refereed. The panel believes organized sessions and tutorials on human behavior representation with invited papers by key contributors from the various disciplines associated with the field can provide important insights and direction. Conferences can also provide a proactive stimulus for the expanded interdisciplinary cooperation the panel believes is essential for success in this arena.

Expanded Interservice Communication

There is a need to actively promote communication across the services, model developers, and researchers. DMSO can lead the way in this regard by developing a clearinghouse for human behavior representation, perhaps with a base in an Internet web site, with a focus on information exchange. This clearinghouse might include references and pointers to the following:

- Definitions
- Military task descriptions
- Data on military system performance
- Live exercise data for use in validation studies
- Specific models
- Resource and platform descriptions
- DMSO contractors and current projects
- Contractor reports
- Military technical reports

Education and Training

The panel believes that opportunities for education and training in the professional competencies required for human behavior representation at a national level are lacking. We recommend that graduate and postdoctoral fellowships in human behavior representation and modeling be provided. Institutions wishing to offer such fellowships would have to demonstrate that they could provide interdisciplinary education and training in the areas of human behavior representation, modeling, and military applications.

A FINAL THOUGHT

The modeling of cognition and action by individuals and groups is quite possibly the most difficult task humans have yet undertaken. Developments in this area are still in their infancy. Yet important progress has been and will continue to be made. Human behavior representation is critical for the military services as they expand their reliance on the outputs from models and simulations for their activities in management, decision making, and training. In this report, the panel has outlined how we believe such modeling can proceed in the short, medium, and long terms so that DMSO and the military services can reap the greatest benefit from their allocation of resources in this critical area.

References

Aasman, J.

1995 Modelling driver behavior in Soar. Unpublished Doctoral Dissertation. University of Groningen. Royal PTT Nederland NV, KPN Research.

Adams, M.J., Y.J. Tenny, and R.W. Pew

1991 *State-of-the-Art Report: Strategic Workload and the Cognitive Management of Advanced Multi-Task Systems*. SOAR CSERIAC 91-6. Crew System Ergonomics Information Analysis Center, Wright-Patterson Air Force Base, OH.

Agre, P.E., and D. Chapman

1987 PENGI: An implementation of a theory of activity. In *AIAA Proceeding of the 6th National Conference on Artificial Intelligence*. Menlo Park, CA: AIAA.

Air Force Systems Command

1980 AFSC Design Handbook Series 1-0. Human Factors Engineering, DH 1-3.

Ajzen, I.

1996 The social psychology of decision making. In *Social Psychology: Handbook of Basic Principles*, E.T. Higgins and A.W. Kruglanski, eds. New York, NY: The Guilford Press.

Ajzen, I., and M. Fishbein

1980 *Understanding Attitudes and Predicting Social Behavior*. Englewood Cliffs, NJ: Prentice-Hall.

Akyurek, A.

1992 On a computational model of human planning. Pp. 81-108 in *Soar: A Cognitive Architecture in Perspective*, J. Michon and A. Akyurek, eds. Netherlands: Kluwer Academic Publishers.

Alberts, D.

1996 *The Unintended Consequences of Information Age Technologies*. Washington, DC: U.S. Government Printing Office.

Aldrich, H.

1979 *Organizations and Environments*. Englewood Cliffs, NJ: Prentice-Hall.

Allais, M.

1953 Le comportement de l'homme rationnel devant le risque: Critique des postulates et axiomes de l'école Américaine. *Econometrica* 21:503-546.

Allender, L., L. Salvi, and D. Promisel
1997a Evaluation of human performance under diverse conditions via modeling technology. In *Proceedings of Workshop on Emerging Technologies in Human Engineering Testing and Evaluation*. Brussels, Belgium: NATO Research Study Group 24.
Allender, L., T. Kelley, S. Archer, and R. Adkins
1997b IMPRINT: The transition and further development of a soldier-system analysis tool. *MANPRINT Quarterly* 5(1):1-7.
Allender, L., J. Lockett, T.D. Kelley, L. Salvi, D. Mitchell, D.B. Headley, D. Promisel, C. Richer, and T. Feng
1995 Verification, validation, and accreditation of a soldier-system modeling tool. In *Proceedings of the Human Factors and Ergonomics Society* (39th Annual Meeting). San Diego, CA: Human Factors and Ergonomics Society.
Allport, D.A., E.A. Styles, and S. Hsieh
1994 Shifting intentional set: Exploring the dynamic control of tanks. In *Attention and Performance XV*, C. Umilta and M. Moscovitch, eds. Hillsdale, NJ: Erlbaum.
Alterman, R.
1988 Adaptive planning. *Cognitive Science* 12:393-421.
Altmann, E.M., and B.E. John
forthcoming Episodic indexing of external information. To appear in the *Cognitive Science Journal*.
Anderson, J.A.
1997 *An Introduction to Neural Networks*. Cambridge, MA: MIT Press.
Anderson, J.A., J.W. Silverstein, S.A. Ritz, and R.S. Jones
1977 Distinctive features, categorical perception, and probabililty learning: Some applications of a neural model. *Psychological Review* 84:413-451.
Anderson, J.R.
1976 *Language, Memory, and Thought*. Hillsdale, NJ: Lawrence Erlbaum.
1983 *The Architecture of Cognition*. Cambridge, MA: Harvard University Press.
1990 *The Adaptive Character of Thought*. Hillsdale, NJ: Lawrence Erlbaum.
1993 *Rules of the Mind*. Hillsdale, NJ: Lawrence Erlbaum.
Anderson, J.R., and G.H. Bower
1980 *Human Associative Memory*. Hillsdale, NJ: Erlbaum Associates.
Anderson, J.R., and C. Lebiere
1998 *Atomic Components of Thought*. Hillsdale, NJ: Lawrence Erlbaum.
Anderson, J.R., L.M. Reder, and C. Lebiere
1996 Working memory: Activation limitations on retrieval. *Cognitive Psychology* 30(3):221-256.
Anderson, N.H.
1959 Test of a model for opinion change. *Journal of Abnormal and Social Psychology* 59:371-381.
1964 Linear models for responses measured on a continuous scale. *Journal of Mathematical Psychology* 1:121-142.
1971 Integration theory and attitude change. *Psychological Review* 78:171-206.
Anderson, N.H., and C. Hovland
1957 The representation of order effects in communication research. In *The Order of Presentation in Persuasion*, C. Hovland, ed. New Haven, CT: Yale University Press.
Anno, G.H., M.A. Dore, and T.J. Roth
1996 Taxonomic Model for Performance Degradation in Combat Tasks. DNA-TR-95-115. Defense Nuclear Agency, Alexandria, VA.
Anonymous
1997 Federated Laboratory Workshop on Commander's Intent, North Carolina Agricultural and Technical State University, Greensboro, NC, October 28-29.

Antis, W., J.M. Honeycutt, and E.N. Koch
1973 *The Basic Motions of MTM.* Naples, FL: The Maynard Foundation.
Arbib, M.A., ed.
1995 *The Handbook of Brain Theory and Neural Networks.* Cambridge, MA: MIT Press.
Archer, R.D., and J.F. Lockett
1997 WinCrew—A tool for analyzing performance, mental workload and function allocation among operators. In *Proceedings of the International Conference on Allocation of Functions.* Galway, Ireland: COHSS.
Argyris, C.
1957 *Personality and Organization.* New York, NY: Harper and Row.
Arkes, H.R., and C. Blumer
1985 The psychology of sunk cost. *Organizational Behavior and Human Decision Processes* 35:124-140.
Aschenbrenner, K., D. Albert, and F. Schmalhofer
1983 Stochastic choice heuristics. *Acta Psychologica* 56:153-166.
Atkinson, R.C., and R.M. Shiffrin
1968 Human memory: A proposed system and its control processes. Pp. 89-195 in *The Psychology of Learning and Motivation: Advances in Research and Theory* (Vol. 2), K.W. Spence and J.T. Spence, eds. New York, NY: Academic Press.
Aube, M., and A. Sentini
1996 Emotions as commitments operators: A foundation for control structure in multi-agent systems. In *Agents Breaking Away*, W. Van de Velde and J.W. Perram, eds. New York, NY: Springer-Verlag.
Axelrod, R.
1997 Advancing the Art of Simulation in the Social Sciences. Working Paper 97-05-048. Santa Fe Institute.
Axelrod, R.M., and D. Dion
1988 The further evolution of cooperation. *Science* 242(4884):1385-1390.
Baddeley, A.D.
1986 *Working Memory.* Oxford: Oxford University Press.
1990 *Human Memory: Theory and Practice.* Boston: Allyn and Bacon.
Badler, N., C. Phillips, and B. Webber
1993 *Simulating Humans.* Oxford University Press.
Badre, A.N.
1978 Selecting and Representing Information Structures for Battlefield Decision Systems. ARI Technical Report 79-A20. U.S. Army Research Institute for the Behavioral and Social Sciences, Alexandria, VA.
Bagozzi, R.P., and R.E. Burnkrant
1979 Attitude organization and the attitude-behavior relationship. *Journal of Personality and Social Psychology* 37:913 919.
Baligh, H.H., R.M. Burton, and B. Obel
1987 Design of organizational structures: An expert system method. In *Economics and Artificial Intelligence*, J.L. Roos, ed. Oxford, United Kingdom: Pergamon.
1990 Devising expert systems in organization theory: The organizational consultant. In *Organization, Management, and Expert Systems*, M. Masuch, ed. Berlin, Germany: Walter De Gruyter.
1994 Validating the organizational consultant on the fly. In *Computational Organization Theory*, K.M. Carley and M.J. Prietula, eds. Hillsdale, NJ: Lawrence Erlbaum.
Banda, C., D. Bushnell, S. Chen, A. Chiu, B. Constantine, J. Murray, C. Neukom, M. Prevost, R. Shankar, and L. Staveland
1991 Army-NASA Aircrew/Aircraft Integration Program: Phase IV A^3I Man-Machine Integration Design and Analysis System (MIDAS) Software Detailed Design Document. NASA Contractor Report 177593. NASA Ames Research Center, Moffett Field, CA.

Bandura, A.
1977 *Social Learning Theory*. Englewood Cliffs, NJ: Prentice-Hall.

Banks, J.H.
1985 Lessons from the NTC: Common Battalion Task Force Training Needs (draft). U.S. Army Research Institute for the Behavioral and Social Sciences, Presidio of Monterey Field Unit, November.

Barnes, M.J., and B.G. Knapp
1997 Collaborative planning aids for dispersed decision making at the brigade level. In *Conference Proceedings of the First Annual Symposium of Advanced Displays and Interactive Displays Federated Laboratory, January 28-29*. Adelphi, MD: Army Research Laboratory.

Baron, S., G. Zacharias, R. Muralidharan, and R. Lancraft
1980 PROCRU: A model for analyzing flight crew procedures in approach to landing. In *Proceedings of the Eighth IFAC World Congress*, Tokyo, Japan.

Barrick, M.R., and M.K. Mount
1991 The Big Five personality dimensions and job performance: A meta-analysis. *Personnel Psychology* 44(1):1-26.

Bass, B.M.
1964) Business gaming for organizational research. *Management Science* 10(8):545-555.

Bass, E.J., G.D. Baxter, and F.E. Ritter
1995 Using cognitive models to control simulations of complex systems. *AISB Quarterly* 93:18-25.

Batagelj, V., P. Doreian, and A. Ferligoj
1992 An optimizational approach to regular equivalence. *Social Networks* 14(1-2):121-136.

Bauer, M.I., and B.E. John
1995 Modeling time-constrained learning in a highly-interactive task. Pp. 19-26 in *Proceedings of CHI* (Denver, CO, May 7-11). New York, NY: ACM.

BBN Systems and Technologies
1997 *OMAR User/Programmer Manual, Version 2.0*. Cambridge, MA: BBN Systems and Technologies.

Beevis, D., R. Bost, B. Döring, E. Nordø, J.-P. Papin, I.H. Schuffel, and D. Streets
1994 Analysis Techniques for Man-Machine System Design. AC/243 (Panel 8) TR/7. Defence Research Group, North Atlantic Treaty Organization.

Belenky, G.
1986 Sustaining and enhancing individual and unit performance in continuous operations. In *Proceedings of the Soldier Performance Research and Analysis Review*. Fort Belvoir, VA. April.

Belenky, G.L., G.P. Kreuger, T.J. Balking, D.B. Headley, and R.E. Solick
1987 Effect of Continuous Operations (CONOPS) on Soldier and Unit Performance: Review of the Literature and Strategies for Sustaining the Soldier in CONOPS. WRAIR Report BB-87-1. Walter Reed Army Institute of Research, Washington, DC.

Bell, D.E.
1982 Regret in decision making under uncertainty. *Operations Research* 30:961-981.

Bennett, B.W.
1995 Observations on Verification, Validation, and Accreditation (VV&A). PM-462-OSD. Acquisition and Technology Center, National Defense Research Institute, The RAND Corporation, Santa Monica, CA.

Bennett, B.W., S. Gardiner, D.B. Fox, and N.K.J. Witney
1994 *Theater Analysis and Modeling in an Era of Uncertainty: The Present and Future of Warfare*. Santa Monica, CA: The RAND Corporation.

Berry, J.W., M.H. Segall, and C. Kagitcibasi, eds.
1997 *Handbook of Cross-cultural Psychology*. Boston, MA: Allyn and Bacon.
Berry, J.W., Y.H. Poortinga, M.H. Segall, and P.R. Dasen
1992 *Cross-Cultural Psychology: Research and Applications*. Cambridge, United Kingdom: Cambridge University Press.
Binder, K.S., and R.K. Morris
1995 Eye movements and lexical ambiguity resolution: Effects of prior encounter and discourse topic. *Journal of Experimental Psychology: Learning, Memory and Cognition* 21(5):1186-1196.
Binder, J., D. Koller, S. Russell, K. Kanazawa
1995 Adaptive probabilistic networks with hidden variables. In *Proceedings of the International Joint Council on Artificial Intelligence*.
Bjork, R.A., and D. Druckman
1991 *In the Mind's Eye: Enhancing Human Performance*. Committee on Techniques for the Enhancement of Human Performance. Washington, DC: National Academy Press.
Blaney, P.H.
1986 Affect and memory. *Psychological Bulletin* 99(2):229-246.
Blau, P.M.
1955 *The Dynamics of Bureaucracy*. Chicago, IL: University of Chicago Press.
1960 Structural effects. *American Sociological Review* 25:178-193.
Blau, P.M., and M. Meyer
1956 *Bureaucracy in Modern Society*. New York, NY: Random House.
Blau, P.M., and W.R. Scott
1962 *Formal Organizations: A Comparative Approach*. San Francisco, CA: Chandler Publishing Company.
Bledstein, B.J.
1976 The Culture of Professionalism: The Middle Class and the Development of Higher Education in America. New York, NY: Norton.
Bodlaender, H.L., and K. Jansen
1991 Restrictions of Graph Partition Problems. Part 1. RUU-CS-91-44. Department of Computer Science, Utrecht University, The Netherlands.
Bond, A., and L. Gasser
1988 *Readings in Distributed Artificial Intelligence*. San Mateo, CA: Kaufmann.
Bond, B.
1996 *The Pursuit of Victory*. New York, NY: Oxford University Press.
Bonney, R.A.
1996 Human response to vibration: Principles and methods. In *Evaluation of Human Work*, J.R. Wilson and E.N. Corlett, eds. Bristol, PA: Taylor and Francis.
Boorman, S.A. and P.R. Levitt
1980 *The Genetics of Altruism*. New York, NY: Academic Press.
Bovair, S., D.E. Kieras, and P.G. Polson
1990 The acquisition and performance of text-editing skill: A cognitive complexity analysis. *Human-Computer Interaction* 5:1-48.
Bower, G.H.
1981 Mood and memory. *American Psychologist* 36:129-148.
Bradski, G., and S. Grossberg
1995 Fast learning VIEWNET architectures for recognizing 3-D objects from multiple 2-D views. *Neural Networks* 8:1053-1080.
Bradski, G., G.A. Carpenter, and S. Grossberg
1994 STORE working memory networks for storage and recall of arbitrary temporal sequences. *Biological Cybernetics* 71:469-480.

Branscomb, A.W.
1994 *Who Owns Information? From Privacy to Public Access.* New York, NY: Basic Books.

Brecke, F.H., and M.J. Young
1990 Training Tactical Decision-Making Skills: An Emerging Technology. AFHRL-TR-36. AFHRL/LRG, Wright-Patterson Air Force Base, OH.

Brehmer, B., and D. Dorner
1993 Experiments with computer simulated microworlds: Escaping both the narrow straits of the laboratory and the deep blue sea of the field study. *Computers in Human Behavior* 9(1):171-184.

Breiger, R.L., S.A. Boorman, and P. Arabie
1975 An algorithm for clustering relational data with applications to social network analysis and comparison with multidimensional scaling. *Journal of Mathematical Psychology* 12:328-383.

Broadbent, D.E.
1957 A mechanical model for human attention and immediate memory. *Psychological Review* 64:205-215.
1958 *Perception and Communication.* New York, NY: Pergamon.

Broadbent, D.E., M.H.P. Broadbent, and J.L. Jones
1986 Performance correlates of self-reported cognitive failure and obsessionality. *British Journal of Clinical Psychology* 25:285-299.

Brockner, J.
1992 The escalation of commitment to a failing course of action: Toward theoretical progress. *Academy of Management Review* 17(1):39-61.

Brown, J.
1989 *Environmental Threats: Social Science Approaches to Public Risk Assessment.* London, England: Bellhaven.

Bryan, W.L., and N. Harter
1899 Studies on the telegraphic language. The acquisition of a hierarchy of habits. *Psychological Review* 6:345-375.

Buchanan, B.G., and E.H. Shortliffe
1984 *Rule-Based Expert Systems: The MYCIN Experiments of the Stanford Heuristic Programming Project.* Reading, MA: Addison-Wesley.

Bukszar, E., and T. Connolly
1987 Hindsight and strategic choice: Some problems in learning from experience. *Academy of Management Journal* 31:628-641.

Burkett, L.N., J. Chisum, J. Pierce, and K. Pomeroy
1990 Increased dermal activity over paralyzed muscle after peak exercise. *Adapted Physical Activity Quarterly* 7(1):67-73.

Burnod, Y., P. Grandguillaume, I. Otto, S. Ferraina, P.B. Johnson, and R. Caminiti
1992 Visuomotor transformations underlying arm movements towards visual targets: A neural network model of cerebral cortical operations. *Journal of Neurosciences* 12:1435-1453.

Burns, T., and G.M. Stalker
1966 *The Management of Innovation.* London, United Kingdom: Tavistock Publications. (Originally published 1961.)

Burt, R.S.
1973 The differential impact of social integration on participation in the diffusion of innovations. *Social Science Research* 2:125-144.
1980 Innovation as a structural interest: Rethinking the impact of network position innovation adoption. *Social Networks* 4:337-355.
1982 *Toward a Structural Theory of Action.* New York, NY: Academic Press.

Busemeyer, J.R., and J.T. Townsend

1993 Decision field theory: A dynamic-cognitive approach to decision making in an uncertain environment. *Psychological Review* 100:432-459.

Bush, R.B., and F. Mosteller

1955 *Stochastic Models for Learning*. New York, NY: Wiley.

Byrn, D., and K. Kelley

1981 *An Introduction to Personality*. Englewood Cliffs, NJ: Prentice-Hall.

Byrne, M.D., and J.R. Anderson

1998 Perception and action. In *Atomic Components of Thought*, J.R. Anderson and C. Lebiere, eds. Mahwah, NJ: Lawrence Erlbaum.

Cacioppo, J.T., and L.G. Tassinary

1990 *Principles of Psychophysiology: Physical, Social and Inferential Elements*. New York, NY: Cambridge University Press.

Calder, R.B., R.L. Carreiro, J.N. Panagos, G.R. Vrablik, and B.P. Wise

1996 Architecture of a Command Forces Entity. In Proceedings of the Sixth Conference on Computer Generated Forces and Behavioral Representation. Report IST-TR-96-18. Institute for Simulation and Training, Orlando, FL.

Campbell, D.T.

1996 Unresolved issues in measurement validity: An autobiographical overview. *Psychological Assessment* 8(4):363-368.

Campbell, J.P., and L.M. Zook, eds.

1991 Improving the Selection, Classification and Utilization of Army Enlsted Personnel: Final Report on Project A. AA5-RR1597. U.S. Army Research Institute, Alexandria, VA.

Cantril, H.

1946 The intensity of an attitude. *Journal of Abnormal and Social Psychology* 41:129-135.

Carbonell, J.

1986 Derivational analogy: A theory of reconstructive problem solving and expertise acquisition. Pp. 371-392 in *Machine Learning: An Artificial Intelligence Approach 2*, R.S. Michalski, J.G. Carbonell, and T.M. Mitchell, eds. Los Altos, CA: Morgan Kaufmann.

Carbonell, J.G., C.A. Knoblock, and S. Minton

1991 Prodigy: An integrated architecture for planning and learning. Pp. 241-278 in *Architectures for Intelligence*, K. VanLehn, ed. Hillsdale, NJ: Lawrence Erlbaum.

Carbonell, J.R.

1966 A queueing model of many-instrument visual sampling. *IEEE Transactions on Human Factors in Electronics* 7(4):157-164.

Carbonell, J.R., J.L. Ward, and J.W. Senders

1968 A queueing model of visual sampling: Experimental validation. *IEEE Transactions on Man-Machine System* 9(3):82-87.

Card, S.K., T.P. Moran, and A. Newell

1983 *The Psychology of Human-Computer Interaction*. Hillsdale, NJ: Lawrence Erlbaum.

Caretta, T.R., D.C. Perry, and M.J. Ree

1994 The ubiquitous three in the prediction of situation awareness: Round up the usual suspects. Pp. 125-137 in *Situational Awareness in Complex Systems*, R.D. Gilson, D.J. Garland, and J.M. Koonce, eds. Daytona Beach, FL: Embry-Riddle Aeronautical University Press.

Carey, J.P.

1994 Multinational enterprises. In *Cross Cultural Topics in Psychology*, L.L. Adler and U.P. Gielen, eds. Westport, CT: Praeger.

Carley, K.M.

1986a Measuring efficiency in a garbage can hierarchy. Pp. 165-194 in *Ambiguity and Command*, J.G. March and R. Weissinger-Baylon, eds. New York, NY: Pitman.

1986b An approach for relating social structure to cognitive structure. *Journal of Mathematical Sociology* 12(2):137-189.
1990 Group stability: A socio-cognitive approach. Pp. 1-44 in *Advances in Group Processes, Vol. 7,* E. Lawler, B. Markovsky, C. Ridgeway, and H. Walker, eds. Greenwich, CT: JAI.
1991a A theory of group stability. *American Sociological Review* 56(3):331-354.
1991b Designing organizational structures to cope with communication breakdowns: A simulation model. *Industrial Crisis Quarterly* 5:19-57.
1992 Organizational learning and personnel turnover. *Organization Science* 3(1):20-46. [Reprinted in 1996, *Organizational Learning*, M.D. Cohen and L.S. Sproull, Thousand Oaks, CA, Sage]
1995a Communication technologies and their effect on cultural homogeneity, consensus, and the diffusion of new ideas. *Sociological Perspectives* 38(4): 547-571.
1995b Computational and mathematical organization theory: Perspective and directions. *Computational and Mathematical Organization Theory* 1(1):39-56.
1996a A comparison of artificial and human organizations. *Journal of Economic Behavior and Organization* 31:175-191.
1996b Validating Computational Models. Social and Decision Sciences Working Paper. Pittsburgh, PA: Carnegie Mellon University.
forthcoming Organizational adaptation. *Annals of Operations Research.*

Carley, K., and A. Newell
1994 The nature of the social agent. *Journal of Mathematical Sociology* 19(4):221-262.

Carley, K.M., and Z. Lin
forthcoming A theoretical study of organizational performance under information distortion. *Management Science.*

Carley, K.M., and J. Harrald
1997 Organizational learning under fire: Theory and practice. *American Behavioral Scientist* 40(3):310-332.

Carley, K.M., and M.J. Prietula
1994 ACTS theory: Extending the model of bounded rationality. Pp. 55-88 in *Computational Organization Theory*, K.M. Carley and M.J. Prietula, eds. Hillsdale, NJ: Lawrence Erlbaum.

Carley, K.M., and D.M. Svoboda
1996 Modeling organizational adaptation as a simulated annealing process. *Sociological Methods and Research* 25(1):138-168.

Carley, K.M., and K. Wendt
1991 Electronic mail and scientific communication: A study of the Soar Extended Research Group. *Knowledge: Creation, Diffusion, Utilization* 12(4):406-440.

Carley, K., J. Kjaer-Hansen, M. Prietula, and A. Newell
1992 Plural-Soar: A prolegomenon to artificial agents and organizational behavior. Pp. 87-118 in *Artificial Intelligence in Organization and Management Theory*, M. Masuch and M. Warglien, eds. Amsterdam, Netherlands: Elsevier Science Publishers.

Carneiro, R.L.
1994 War and peace: Alternating realities in human history. In *Studying War: Anthropological Perspectives*, S.P. Reyna and R.E. Downs, eds. Langhorne, PA: Gordon and Breach.

Carpenter, G.A., and S. Grossberg
1987a A massively parallel architecture for a self-organizing neural pattern recognition machine. *Computer Vision, Graphics, and Image Processing* 37:54-115.
1987b ART 2: Stable self-organization of pattern recognition codes for analog input patterns. *Applied Optics* 26:4919-4930.

1990 ART 3: Hierarchical search using chemical transmitters in self-organizing pattern recognition architectures. *Neural Networks* 3:129-152.

Carpenter, G.A., and S. Grossberg, eds.
1991 *Pattern Recognition by Self-Organizing Neural Networks*. Cambridge, MA: MIT Press.

Carpenter, G.A., and W.D. Ross
1995 ART-EMAP: A neural network architecture for object recognition by evidence accumulation. *IEEE Transactions on Neural Networks* 6:805-818.

Carpenter, G.A., S. Grossberg, N. Markuzon, J.H. Reynolds, and D.B. Rosen
1992 Fuzzy ARTMAP: A neural network architecture for incremental supervised learning of analog multidimensional maps. *IEEE Transactions on Neural Networks* 3:698-713.

Castelfranchi, C., and E. Werner, eds.
1992 Artificial Social Systems: 4th European Workshop on Modeling Autonomous Agents in a Multi-Agent World. MAAMAW 1992, S. Martinoal Cimino, Italy, July 29-31.

Ceranowicz, A.
1994 Modular Semi-Automated Forces. http://www/mystech.com/~smithr/elecsim94/modsaf/modsaf.txt.

Cesta, A., M. Miceli, and P. Rizzo
1996 Effects of different interaction attitudes on a multi-agent system performance. In *Agents Breaking Away*, W. Van de Velde and J.W. Perram, eds. New York, NY: Springer-Verlag.

Chase, W.G., and H.A. Simon
1973 The mind's eye in chess. Pp. 215-281 in *Visual Information Processing*, W.G. Chase, ed. New York, NY: Academic Press.

Cho, B., P.S. Rosenbloom, and C.P. Dolan
1991 Neuro-Soar: A neural-network architecture for goal-oriented behavior. In *Proceedings of the 13h Annual Conference of the Cognitive Science Society*, August. Chicago, IL: Cognitive Science Society.

Chong, R.S., and J.E. Laird
1997 Towards learning dual-task executive process knowledge using EPIC-Soar. In Proceedings of the Nineteenth Annual Conference of the Cognitive Science Society.

Chou, C.D., D. Madhavan, and K. Funk
1996 Studies of cockpit task management errors. *International Journal of Aviation Psychology* 6(4):307-320.

Christensen-Szalanski, J.J.J., and C.S. Fobian
1991 The hindsight bias: A meta-analysis. *Organizational Behavior and Human Decision Processes* 48:147-168.

Chu, Y.-Y., and W.B. Rouse
1979 Adaptive allocation of decisionmaking responsibility between human and computer in multitask situations. *IEEE Transactions on Systems, Man, and Cybernetics* 9(12):769-777.

Cleermans, A., and J.L. McClelland
1991 Learning the structure of event sequences. *Journal of Experimental Psychology: General* 120(3):235-253.

Clemen, R.T.
1996 *Making Hard Decisions*. Belmont, CA: Wadsworth Publishing Company/Duxbury Press.

Cohen, G.P.
1992 The Virtual Design Team: An Information Processing Model of the Design Team Management. Ph.D. Thesis. Department of Civil Engineering, Stanford University.

Cohen, J.D., K. Dunbar, and J.M. McClelland
1990 On the control of automatic processes: A parallel distributed processing account of the Stroop effect. *Psychological Review* 97:332-361.

Cohen, M.D.

1996 Individual learning and organizational routine. In *Organizational Learning*, M.D. Cohen and L.S. Sproull, eds. Thousand Oaks, CA: Sage.

Cohen, M.D., and J.G. March

1974 *Leadership and Ambiguity*. New York, NY: McGraw Hill.

Cohen, M.D., J.G. March, and J.P. Olsen

1972 A garbage can model of organizational choice. *Administrative Sciences Quarterly* 17(1):1-25.

Cohen, M.S., and B.B. Thompson

1995 A Hybrid Architecture for Metacognitive Learning. Presented at the ONR Hybrid Learning Kick-Off Meeting, Washington, DC, August 3-4.

Cohen, R.

1988 Blaming men, not machines. *Time*(August 15):19.

Coleman, J.S., E. Katz, and H. Menzel

1966 *Medical Innovation: A Diffusion Study*. New York, NY: Bobbs-Merrill Company, Inc.

Collins, R.J.

1992 Studies in Artificial Evolution. CSD-920037. Computer Science Department. University of California, Los Angeles.

Coltheart, M., B. Curtis, P. Atkins, and M. Haller

1993 Models of reading aloud: Dual-route and parallel-distributed-processing approaches. *Psychological Review* 100:589-608.

Connolly, T., and D. Deane

1997 Decomposed versus holistic estimates of effort required for software writing tasks. *Management Science* 43:1029-1045.

Connolly, T., H. Arkes, and K.R. Hammond, eds.

forthcoming *Judgment and Decision Making: An Interdisciplinary Reader* (second edition). Cambridge, England: Cambridge University Press.

Connolly, T., L.D. Ordonez, and R. Coughlan

1997 Regret and responsibility on the evaluation of decision outcomes. *Organizational Behavior and Human Decision Processes* 70:73-85.

Conroy, J., and S. Masterson

1991 Development of Environmental Stressors for PER-SEVAL. U.S. Army Research Institute, Alexandria, VA.

Cooley, C.

1902 *Human Nature and Social Order*. New York, NY: Scribner.

Coombs, C., and G.S. Avrunin

1977 Single-peaked functions and the theory of preference. *Psychological Review* 84:216-230.

Cooper, N.

1988 Seven minutes to death. *Newsweek* (July 18):18-23.

Cowan, N.

1993 Activation, attention and short-term memory. *Memory and Cognition* 21:162-167.

Crowston, K.

forthcoming An approach to evolving novel organizational forms. *Computational and Mathematical Organization Theory* 26(1):29-48.

Cushman, J.H.

1985 *Command and Control of Theater Forces: Adequacy*. Washington, DC: AFCEA International Press.

Cyert, R., and J.G. March

1992 *A Behavioral Theory of the Firm. 2nd Edition*. Cambridge, MA: Blackwell Publishers. [1963]

Damos, D.L.

forth-coming Using interruptions to identify task prioritization in part 121 air carrier operations. Department of Aerospace Engineering, Applied Mechanics, and Aviation. In *Proceedings of the Ninth International Symposium on Aviation Psychology*. Columbus, OH: The Ohio State University.

Danes, J.E., J.E. Hunter, and J. Woelfel

1984 Belief, change, and accumulated information. In *Mathematical Models of Attitude Change: Change in Single Attitudes and Cognitive Structure,* J. Hunter, J. Danes, and S. Cohen, eds. Orlando, FL: Academic Press.

Davis, P.K.

1989 Modeling of Soft Factors in the RAND Strategy Assessment Center. P-7538. Santa Monica, CA: The RAND Corporation.

Dawes, R.M.

1997 Behavioral decision making and judgment. In *Handbook of Social Psychology*, S. Fiske and G. Lindzey, eds. New York, NY: McGraw-Hill.

Dawes, R.M., and M. Mulford

1996 The false consensus effect and overconfidence: Flaws in judgment or flaws in how we study judgment? *Organizational Behavior and Human Decision Making Process* 65:201-211.

Decker, K.S., and V.R. Lesser

1993 Quantitative modeling of complex computational task environments. Pp. 217-224 in *Proceedings of the Eleventh National Conference on Artificial Intelligence*, July, Washington, DC.

Decker, K.

1995 A framework for modeling task environment. Chapter 5 in Environment-Centered Analysis and Design of Coordination Mechanisms. Ph.D. Dissertation, University of Massachusetts.

1996 TAEMS: A framework for environment centered analysis and design of coordination mechanisms. In *Foundations of Distributed Artificial Intelligence*, G.M.P. O'Hare and N.R. Jennings, eds. New York, NY: John Wiley and Sons.

Deckert, J.C., E.B. Entin, E.E. Entin, J. MacMillan, and D. Serfaty

1994 Military Command Decisionmaking Expertise. Report 631. Alphatech, Burlington, MA.

Defense Advanced Research Project Agency

1996 DARPA Proposer's Information Package. Washington, D.C.: U.S. Government Printing Office.

DeGroot, M.H.

1970 *Optimal Statistical Decisions*. New York: McGraw-Hill.

de Oliveira, P.P.B.

1992 Enact: An Artificial-life World in a Family of Cellular Automata. Cognitive Science Research Papers, CSRP 248. School of Cognitive and Computing Sciences, University of Sussex, Brighton [East Sussex], England.

de Sola Poole, I., ed.

1977 *The Social Impact of the Telephone*. Cambridge, MA: MIT Press.

De Soete, G., H. Feger, and K.C. Klauer

1989 *New Developments in Psychological Choice Modeling*. New York: Elsevier-North Holland.

Detweiler, M., and W. Schneider

1991 Modeling the acquisition of dual-task skill in a connectionist/control architecture. In *Multiple Task Performance*, D.L. Damos, ed. London, United Kingdom: Taylor and Francis Ltd.

Deutsch, J.A., and D. Deutsch

1963 Attention: Some theoretical consideration. *Psychological Review* 70:80-90.

Deutsch, S.E., and M. Adams
1995 The operator-model architecture and its psychological framework. In *Proceedings of the Sixth IFAC Symposium on Man-Machine Systems*. Cambridge, MA: Massachusetts Institute of Technology.

Deutsch, S.E., J. Macmillan, N.L. Cramer, and S. Chopra
1997 Operator Model Architecture (OMAR) Final Report. BBN Report No. 8179. BBN Corporation, Cambridge, MA.

Deutsch, S.E., M.J. Adams, G.A. Abrett, N.L. Cramer, and C.E. Feehrer
1993 Research, Development, Training, and Evaluation: Operator Model Architecture (OMAR) Software Functional Specification. AL/HR-TP-1993-0027. Air Force Material Command, Wright-Patterson Air Force Base, OH.

Diederich, A.
1997 Dynamic stochastic models for decision making under time pressure. *Journal of Mathematical Psychology* 41:260-274.

Dietterich, T.G., and N.S. Flann
1997 Explanation-based learning and reinforcement learning: A unified view. *Machine Learning* 28(2):**??.**

Dixon, N.
1976 *Military Incompetence*. New York, NY: Basic Books.

Doll, T.H., S.W. McWhorter, D.E. Schmieder, M.C. Hertzler, J.M. Stewart. A.A. Wasilewski, W.R. Owens, A.D. Sheffer, G.L. Galloway, and S.D. Herbert
1997 Biologically-based vision simulation for target-background discrimination and camouflage/lo design. Paper No. 3062-29 in *Targets and Backgrounds: Proceedings of the International Society of Photo-optical Instrumentation Engineers*, W.R. Watkins and D. Clemens, eds. Orlando, FL: Society of Photo-optical Instrumentation Engineers.

Dominguez, C.
1994 Can SA be defined? In *Situation Awareness: Papers and Annotated Bibliography (U)*, M. Vidulich, C. Dominguez, E. Vogel, and G. McMillan, eds. Vol. AL/CF-TR-1994-0085. Armstrong Laboratory, Wright-Patterson Air Force Base, OH.

Downes-Martin, S.
1995 A Survey of Human Behavior Representation Activities for Distributed Interactive Simulation: Final Report. Prepared for SAIC Defense Modeling and Simulation Office Support Office, Alexandria, VA. November.

Downey, J.E., and J.E. Anderson
1915 Automatic writing. *The American Journal of Psychology* 26:161-195.

Downs, A.
1967 *Inside Bureaucracy*. Boston, MA: Little, Brown and Company.

Drummond, M., J. Bresina, and S. Kendar
1991 The entropy reduction engine: Integrating planning, scheduling, and control. *SIGART Bulletin* 2:48-52.

Duffy, B., R. Kaylor, and P. Cary
1988 How good is this Navy, anyway? *U.S. News and World Report* July 18:18-19.

Dupuy, T.N.
1979 The Effects of Combat Losses and Fatigue on Combat Performance, Historical Evaluation and Research Organization. AD-B038029L, U.S. Government Limited. Training and Doctrine Command, Fort Monroe, VA. January.

Durfee, E.H.
1988 *Coordination of Distributed Problem Solvers*. Boston, MA: Kluwer Academic Publishers.

Durfee, E.H., and T.A. Montgomery
1991 Coordination as distributed search in a hierarchical behavior space. *IEEE Transactions on Systems, Man, and Cybernetics* 21(6):1363-1378.

Dyer, M.G.
1987 Emotions and their computation: Three computer models. *Cognition and Emotion* 3:323-347.

Eades, P., and D. Harel
1989 Drawing Graphs Nicely Using Simulated Annealing. Technical Report CS89-13. Department of Applied Mathematics Computer Science, The Weizmann Institute of Science, Rehovot, Israel.

Eccles, R.G., and D.B. Crane
1988 *Doing Deals: Investment Banks at Work.* Boston, MA: Harvard Business School Press.

Eden, C., S. Jones and D. Sims
1979 *Thinking in Organizations.* London, England: MacMillan Press.

Edgell, S.E.
1980 A set-theoretical random utility model of choice behavior. *Journal of Mathematical Psychology* 21:265-278.

Edmonds, B., M. Scott, and S. Wallis
1996 Logic, reasoning and a programming language for simulating economic and business processes with artificially intelligent agents. Pp. 221-230 in *Artificial Intelligence in Economics and Management*, P. Ein-Dor, ed. Boston, MA: Kluwer Academic Publishers.

Eisenstein, E.L.
1979 *The Printing Press as an Agent of Change, Communications, and Cultural Transformations in Early Modern Europe; 2 Vols.* London, England: Cambridge University Press.

Elkind, J.I., S.K.Card, J. Hochberg, and B.M. Huey
1990 *Human Performance Models for Computer-Aided Engineering.* San Diego, CA: Academic Press, Inc.

Elliott, C.
1994 Components of two-way emotion communication between humans and computers using a broad, rudimentary, model of affect and personality. *Cognitive Studies: Bulletin of the Japanese Cognitive Science Society, Volume 2.* (Special Issue on Emotion, in Japanese, to appear.)

Endsley, M.
1987 SAGAT: A Methodology for the Measurement of Situation Awareness. NOR DC 87-83. Northrop Corporation, Los Angeles, CA.
1988 Design and evaluation of situation awareness enhancement. Pp. 97-101 in *Proceedings of the Human Factors Society 32nd Annual Meeting.* Santa Monica, CA: Human Factors and Ergonomics Society.
1989 A methodology for the objective measurement of pilot situation awareness. Pp. 1-1 to 1-9 in *AGARD Conference Proceedings No. 178 Situational Awareness in Aerospace Operations.* Neuilly-sur-Seine, France: AGARD.
1990 Predictive utility of an objective measure of situation awareness. Pp. 41-45 in *Proceedings of the Human Factors Society 34th Annual Meeting.* Santa Monica, CA: Human Factors and Ergonomics Society.
1993 A survey of situation awareness requirements in air-to-air combat fighters. *International Journal of Aviation Psychology* 3:157-168.
1995 Toward a theory of situation awareness in dynamic systems. *Human Factors* 37(1):32-64.

Enos, R.L., ed.
1990 *Oral and Written Communication: Historical Approaches.* Newbury Park, CA: Sage.

Epstein, J., and R. Axtell
1997 *Growing Artificial Societies.* Boston, MA: MIT Press.

Ericsson, K.A., ed.
1996 *The Road to Excellence.* Mahwah, NJ: Erlbaum.

Ericsson, R.A., and J. Smith, eds.
1991 *Toward a General Theory of Expertise: Prospects and Limits*. Cambridge, England: Cambridge University Press.

Ericsson, K.A., R.T. Krampe, and C. Tesch-Romer
1993 The role of deliberate practice in the acquisition of expert performance. *Psychological Review* 100:363-406.

Erman, L.D., F. Hayes-Roth, V.R. Lesser, and D.R. Reddy
1980 The Hearsay-II Speech Understanding System: Integrating knowledge to resolve uncertainty. *Computing Surveys* (12):213-253.

Essens, P., J. Fallesen, C. McCann, J. Cannon-Bowers, and G. Dorfel
1995 COADE—A Framework for Cognitive Analysis, Design, and Evaluation. AC/243 (Panel 8) TR/17. Defence Research Group, North Atlantic Treaty Organization.

Estes, W.K.
1950 Toward a statistical theory of learning. *Psychological Review* 57:94-107.

Fallesen, J.J.
1993 Overview of Army Tactical Planning Performance Research. Technical Report 984. U.S. Army Research Institute for Behavioral and Social Sciences, Alexandria, VA.

Fallesen, J.J., and R.R. Michel
1991 Observation on Command and Staff Performance During CGSC Warrior '91. Working Paper LVN-91-04. U.S. Army Research Institute for Behavioral and Social Sciences, Alexandria, VA.

Fallesen, J.J., C.F. Carter, M.S. Perkins, R.R. Michel, J.P. Flanagan, and P.E. McKeown
1992 The Effects of Procedural Structure and Computer Support Upon Selecting a Tactical Course of Action. Technical Report 960 (AD-A257 254). U.S. Army Research Institute for Behavioral and Social Sciences, Alexandria, VA.

Fasciano, M.
1996 Everyday-World Plan Use. TR-96-07. Computer Science Department, The University of Chicago, IL.

Feldman, J.A., and D.H. Ballard
1982 Connectionist models and their properties. *Cognitive Science* 6:205-254.

Festinger, L.
1954 A Theory of social comparison processes. *Human Relations* 7:114-140.
1957 *A Theory of Cognitive Dissonance*. Evanston, IL: Row, Peterson.

Fiebig, C.B., J. Schlabach, and C.C. Hayes
1997 A battlefield reasoning system. In *Conference Proceedings of the First Annual Symposium of Advanced Displays and Interactive Displays Federated Laboratory, January 28-29*. Adelphi, MD: Army Research Laboratory.

Fikes, R.E., and N.J. Nilsson
1971 STRIPS: A new approach to the application of theorem proving to problem solving. *Artificial Intelligence* 2:189-208.

Fineberg, M.L., G.E. McClellan, and S. Peters
1996 Sensitizing synthetic forces to suppression on the virtual battlefield. Pp. 470-489 in *Proceedings of the Sixth Conference on Computer Generated Forces and Behavioral Representation*, Orlando, FL, July 23-25.

Fiol, C.M.
1994 Consensus, diversity, and learning in organizations. *Organizational Science* 5(3):403-420.

Fischhoff, B.
1975 Hindsight = foresight: The effect of outcome knowledge on judgment under uncertainty. *Journal of Experimental Psychology: Human Perception and Performance* 1:288-299.

Fischhoff, B.
1982 Debiasing. In *Judgment Under Uncertainty: Heuristics and Biases*, D. Kahneman, P. Slovic, and A. Tversky, eds. Cambridge, England: Cambridge University Press.
Fischhoff, B., and R. Beyth
1975 "I knew it would happen": Remembered probabilities of once-future things. *Organizational Behavior and Human Performance* 13:1-6.
Fishbein, M.
1965 A consideration of beliefs, attitudes, and their relationships. In *Current Studies in Social Psychology: Current Studies in Social Psychology*, I. Steiner and M. Fishbein, eds. New York, NY: Holt, Rinehart, and Winston.
Fishbein, M., and I. Ajzen
1974 Attitudes towards objects as predictors of single and multiple behavioral criteria. *Psychological Review* 1981:50-74.
1975 *Belief, Attitude, Intention and Behavior.* Reading, MA: Addison-Wesley.
Fisk, M.D.
1997 Marine Corps Modeling and Simulation Management Office. In *Conference Proceedings of the Sixth Annual Modeling and Simulation Briefing to Government and Industry, May 22-23.* Alexandria, VA: Defense Modeling and Simulation Office.
Fitts, P.M., and M.I. Posner
1967 *Human Performance.* Westport, CT: Greenwood Press.
Flach, J.M.
1995 Situation awareness: Proceed with caution. *Human Factors* 37(1):149-157.
Fleishman, E.A., and M.K. Quaintance
1984 *Taxonomies of Human Performance: The Description of Human Tasks.* Orlando, FL: Academic Press.
Forgas, J.P.
1995 Mood and judgment: The affect infusion model (AIM). *Psychological Bulletin* 117(1):39-66.
Forgas, J.P., and M.H. Bond
1985 Cultural influences on the perception of interaction episodes. *Personality and Social Psychology Bulletin* 11(1):75-88.
Fossett, C.A., D. Harrison, H. Weintrob, and S.I. Gass
1991 An assessment procedure for simulation models: A case study. *Operations Research* 39(5):710-723.
Fox, J.M.
1982 *Software and its Development.* Englewood Cliffs, NJ: Prentice-Hall.
Fracker, M.L.
1988 A theory of situation assessment: Implications for measuring situation awareness. In *Proceedings of the Human Factors Society*, 32nd Annual Meeting. Santa Monica, CA: Human Factors and Ergonomics Society.
1990 Attention gradients in situation awareness. Pp. 6/1 - 6/10 in *Situational Awareness in Aerospace Operations AGARD-CP-478.* Neuilly-sur-Seine, France: Advisory Group for Aerospace Research and Development.
1991a Measures of Situation Awareness: An Experimental Evaluation. Technical Report AL-TR-1991-0127. Armstrong Laboratory, Wright-Patterson Air Force Base, OH.
1991b Measures of Situation Awareness: Review and Future Directions. Technical Report AL-TR-1991-0128. Armstrong Laboratory, Wright-Patterson Air Force Base, OH.
Freeman, L.C.
1984 Impact of computer-based communication on the social structure of an emerging scientific specialty. *Social Networks* 6:201-221.

1993 Finding groups with a simple genetic algorithm. *Journal of Mathematical Sociology* 17:227-241.

Freidkin, N.E., and E.C. Johnson
1990 Social influence and opinions. *Journal of Mathematical Sociology* 15:193-205.

Frijda, N.H., and J. Swagerman
1987 Can computers feel? Theory and design of an emotional system. *Cognition and Emotion* 3:235-247.

Fulco, C.S., and A. Cymerman
1988 Human performance and acute hypoxia. In *Human Performance Physiology and Environmental Medicine at Terrestrial Extremes*, K.B. Pandolf, M.N. Sawka, and R.S. Gonzalez, eds. Indianapolis, IN: Benchmark Press, Inc.

Fulk, J.
1993 Social construction of communication technology. *Academy of Management Journal* 36(5):921-950.

Funk, K.H.
1991 Cockpit task management: Preliminary definitions, normative theory, error taxonomy, and design recommendations. *International Journal of Aviation Psychology* I(4):271-285.

Funk, K., and B. McCoy
1996 A functional model of flightdeck agenda management. Pp. 254-258 in *Proceedings of the Human Factors and Ergonomics Society 40th Annual Meeting*. Santa Monica, CA: Human Factors and Ergonomics Society.

Gaba, D.M., S.K. Howard, and S.D. Small
1995 Situation awareness in anesthesiology. *Human Factors* 37(1).

Galbraith, J.R.
1973 *Designing Complex Organizations*. Addison-Wesley Publishing Company.

Garland, H.
1990 Throwing good money after bad: The effect of sunk costs on the decision to escalate commitment to an ongoing project. *Journal of Applied Psychology* 75(6):728-731.

Gasser, L., C. Braganza, and N. Herman
1987a MACE: A flexible testbed for distributed AI research. Pp. 119-152 in *Distributed Artificial Intelligence*, M.N. Huhns, ed. Lanham, MD: Pitman Publishers.
1987b Implementing distributed AI systems using MACE. Pp. 315-320 in *Proceedings of the 3rd IEEE Conference on Artificial Intelligence Applications*, Orlando, FL, February.

Gasser, L., and A. Majchrzak
1992 HITOP-A: Coordination, infrastructure, and enterprise integration. Pp. 373-378 in *Proceedings of the First International Conference on Enterprise Integration*. Hilton Head, SC: MIT Press.

Gasser, L., I. Hulthage, B. Leverich, J. Lieb, and A. Majchrzak
1993 Organizations as complex, dynamic design problems. P. 727 in *Progress in Artificial Intelligence*, M. Filgueiras and L. Damas, eds. New York, NY: Springer-Verlag.

Gat, E.
1991 Integrating planning and reacting in a heterogeneous asynchronous architecture for mobile robots. *SIGART Bulletin* 2:70-74.

Gentner, D., and A.L. Stevens
1983 *Mental Models*. Hillsdale, NJ: Lawrence Erlbaum.

Georgeff, M.P.
1987 Planning. *Annual Review of Computer Science* 2:359-400.

Georgeff, M.P., and A.L. Lansky
1990 Reactive reasoning and planning. Pp. 729-734 in *Readings in Planning*, J. Allen, J. Hendler, and A. Tate, eds. San Mateo, CA: Morgan Kaufmann Publishers, Inc.

Gerhart, G., T. Meitzler, E. Sohn, G. Witus, G. Lindquist, and J.R. Freeling

1995 Early vision model for target detection. In *Proceedings of the 9th SPIE AeroSense*, Orlando, FL.

Geva, N.

1988 Executive Decision-Making Processes During Combat: A Pretest of the Content Analysis Schema. 030/1-02-88. Israeli Institute for Military Studies, Zikhron Ya'akovi.

Gibson, F., M. Fichman, and D.C. Plaut

1997 Learning in dynamic decision tasks: Computational model and empirical evidence. *Organizational Behavior and Human Decision Processes* 71:1-36.

Gigerenzer, G.

1994 Why the distinction between single-event probabilities and frequencies is relevant for psychology (and vice versa). In *Subjective Probability*, G. Wright and P. Ayton, eds. New York, NY: Wiley.

Gillund, G., and R.M. Shiffrin

1984 A retrieval model for both recognition and recall. *Psychological Review* 91:1-67.

Gilovich, T., and V.H. Medvec

1995 The experience of regret: What, when and why. *Psychological Review* 102(2):379-395.

Giunchiglia, F., and E. Guinchiglia

1996 Ideal and real belief about belief: Some intuitions. In *Agents Breaking Away*, W. Van de Velde and J.W. Perram, eds. New York, NY: Springer-Verlag.

Glance, N.S., and B. Huberman

1993 The outbreak of cooperation. *Journal of Mathematical Sociology* 17(4):281-302.

1994 Social dilemmas and fluid organizations. In *Computational Organization Theory*, K.M. Carley and M.J. Prietula, eds. Hillsdale, NJ: Lawrence Erlbaum.

Glenn, F.A., S.M. Schwartz, and L.V. Ross

1992 Development of a Human Operator Simulator Version V (HOS-V): Design and Implementation. U.S. Army Research Institute for the Behavioral and Social Sciences, PERI-POX, Alexandria, VA.

Gluck, M.A., and G.H. Bower

1988 Evaluating an adaptive network model of human learning. *Journal of Memory and Language* 27(2):166-195.

Golden, R.M.

1996 Mathematical methods for neural network analysis and design. Cambridge, MA: MIT Press.

Goldstein, W.M., and R.M. Hogarth, eds.

1997 *Research in Judgment and Decision Making*. Cambridge, England: Cambridge University Press.

Gonsalves, P., C. Illgen, G. Rinkus, C. Brunetti, and G. Zacharias

1997 A Hybrid System for AI Enhanced Information Processing and Situation Assessment. Report No. R95381. Charles River Analytics, Cambridge, MA.

Goody, J.

1986 *The Logic of Writing and the Organization of Society*. Cambridge, United Kingdom: Cambridge University Press.

Granovetter, M.S.

1973 The strength of weak ties. *American Journal of Sociology* 68:1360-1380.

1974 *Getting a Job: A Study of Contacts and Careers*. Cambridge, MA: Harvard University Press.

Gratch, J.

1996 Task decomposition planning for command decision making. In *Proceedings of the Sixth Conference on Computer Generated Forces and Behavioral Representations, July 23-25*. Report IST-TR-96-18. Orlando, FL: Institute for Simulation and Training.

Gratch, J., R. Hill, P. Rosenbloom, M. Tambe, and J. Laird
1996 Command and control for simulated air agents. In *Conference Proceedings of the Defense Modeling and Simulation Office Workshop on C2 Human Behavior Representation Modeling, February*. Alexandria, VA: Defense Modeling and Simulation Office.

Gray, W.D., S.S. Kirschenbaum, and B.D. Ehret
1997 The précis of Project Nemo, phase 1: Subgoaling and subschemas for submariners. In *Proceedings of the Nineteenth Annual Conference of the Cognitive Science Society*. Hillsdale, NJ: Lawrence Erlbaum.

Greenstein, J.S., and W.B. Rouse
1982 A model of human decisionmaking in multiple process monitoring situations. *IEEE Transactions on Systems, Man, and Cybernetics* 12(2):182-193.

Griffin, D., and A. Tversky
1992 The weighing of evidence and the determinants of confidence. *Cognitive Psychology* 24(3):411-435.

Grossberg, S.
1980 How does the brain build a cognitive code? *Psychological Review* 87:1-51.
1988 *Neural Networks and Natural Intelligence*. Cambridge, MA: MIT Press.

Grossberg, I.N., and D.G. Cornell
1988 Relationship between personality adjustment and high intelligence: Terman versus Hollingworth. *Exceptional Children* 55(3):266-272.

Grossberg, S.
1972 Neural expectation: Cerebellar and retinal analogs of cells fired by learnable or unlearned pattern classes. *Kybernetik* 10:49-57.
1976 Adaptive pattern classification and universal recoding, I: Parallel development and coding of neural feature detectors and II: Feedback, expectation, olfaction, and illusions. *Biological Cybernetics* 23:121-134 and 187-202.
1995 The attentive brain. *American Scientist* 83:438-449.

Grossberg, S., and J.W.L. Merrill
1996 The hippocampus and cerebellum in adaptively timed learning, recognition, and movement. *Journal of Cognitive Neuroscience* 8:257-277.

Grossberg, S., I. Boardman, and M. Cohen
1997 Human perception and performance. *Journal of Experimental Psychology* 23:481-503.

Guralnik, D.B., ed.
1986 *Webster's New World Dictionary, 2nd College Edition*. New York, NY: Simon and Schuster.

Hamilton, B.E., D. Folds, and R.R. Simmons
1982 Performance Impact of Current United States and United Kingdom Aircrew Chemical Defense Ensembles. Report No. 82-9. U.S. Army Aero Medical Research, Ft. Rucker, AL.

Hammond, K.J.
1986 CHEF: A model of case-based planning. *American Association for Artificial Intelligence* 86:261-271.

Hammond, K.R., R.M. Hamm, J. Grassia, and T. Pearson
1987 Direct comparison of the efficacy of intuitive and analytical cognition in expert judgment. *IEEE Transactions on Systems, Man and Cybernetics* 17(5):753-770.

Hancock, P.A., and M.H. Chignell
1987 Adaptive control in human-machine systems. In *Human Factors Psychology*, P.A. Hancock, ed. Amsterdam, NL: North-Holland Press.
1988 Mental workload dynamics in adaptive interface design. *IEEE Transactions on Systems, Man, and Cybernetics* 18(4):647-656.

Hancock, P.A., and M. Vercruyssen
1988 Limits of behavioral efficiency for workers in best stress. *International Journal of Industrial Ergonomics* 3:149-158.

Hancock, P.A., and J.S. Warm
1989 A dynamic model of stress and sustained attention. *Human Factors* 31:519-537.

Hancock, P.A., J.A. Duley, and S.F. Scallen
1994 *The Control of Adaptive Function Allocation.* Report HFRL, Nav-6, September.

Hannan, M.T., and J. Freeman
1989 *Organizational Ecology.* Cambridge, MA: Harvard University Press.

Harris, D.
1985 *A Degradation Methodology for Maintenance Tasks.* Alexandria, VA: U.S. Army Research Institute.

Harrison, J.R., and G.R. Carrol
1991 Keeping the faith: A model of cultural transmission in formal organizations. *Administrative Science Quarterly* 36:552-582.

Hart, S.G.
1986 Theory and measurement of human workload. In *Human Productivity Enhancement (Volume 1)*, J. Zeidner, ed. New York, Praeger.
1989 Crew workload management strategies: A critical factor in system performance. Pp. 22-27 in *Proceedings of the Fifth International Symposium on Aviation Psychology*, R.S.Jensen, ed. Department of Aviation. Columbus, OH: The Ohio State University.

Hart, S.G., and C.D. Wickens
1990 Workload assessment and prediction. In *MANPRINT*, H.R. Booher, ed. New York, NY: Van Nostrand Reinhold.

Hartman, B., and G. Secrist
1991 Situation awareness is more than exceptional vision. *Aviation, Space and Environmental Medicine* 62:1084-1091.

Hartzog, S.M., and M.R. Salisbury
1996 Command forces (CFOR) program status report. In *Proceedings of the Sixth Conference on Computer Generated Forces and Behavioral Representation, July 23-25.* Report IST-TR-96-18. Orlando, FL: Institute for Simulation and Training.

Harwood, K., B. Barnett, and C.D. Wickens
1988 Situational awareness: A conceptual and methodological framework. Pp. 316-320 in *Proceedings of the 11th Symposium of Psychology in the Department of Defense.* Colorado Springs, CO: U.S. Air Force Academy.

Haslegrave, C.M.
1996 Auditory environment and noise assessment. In *Evaluation of Human Work*, J.R. Wilson and E.N. Corlett, eds. Bristol, PA: Taylor and Francis.

Hayes, P.
1975 A representation for robot plans. Pp. 181-188 in *Proceedings of the Fourth International Joint Conference on Artificial Intelligence*, Menlo Park, CA.

Hayes, T.L.
1996 How do athletic status and disability status affect the five-factor personality model? *Human Performance* 9(2):121-140.

Hayes-Roth, B.
1985 A blackboard architecture for control. *Artificial Intelligence* 26:251-321.
1991 Making intelligent systems adaptive. Pp. 301-321 in *Architectures for Intelligence*, K. VanLehn, ed. Hillsdale, NJ: Lawrence Erlbaum.

Hayes-Roth, B., and F. Hayes-Roth
1979 A cognitive model of planning. *Cognitive Science* 3(4):275-310.
Haykin, S.
1994 *Neural Networks: A Comprehensive Foundation.* New York, NY: MacMillan.
Haythornthwaite, C., B. Wellman, and M. Mantei
1995 Work Relationships and Media Use: A Social Network Analysis. *Group Decision and Negotiation* (special issue on Distributed Groupware) 4(3):193-211.
Heath, C.
1995 Escalation and de-escalation of commitment in response to sunk costs: The role of budgeting in mental accounts. *Organizational Behavior and Human Decision Processes* 62(1):38-54.
Heider F.
1946 Attitudes and cognitive organization. *Journal of Psychology* 21:107-112.
1958 *The Psychology of Interpersonal Relations.* New York, NY: Wiley.
Heine, S.J., and D.R. Lehman
1997 Culture, dissonance and self-affirmation. *Personality and Social Psychology Bulletin* 23:389-400.
Heise, D.R.
1977 Social action as the control of affect. *Behavioral Science* 22:161-177.
1979 *Understanding Events: Affect and the Construction of Social Action.* New York, NY: Cambridge University Press.
1987 Affect control theory: Concepts and model. *Journal of Mathematical Sociology* 13:1-34.
1992 Affect control theory and impression formation. In *Encyclopedia of Sociology (Volume 1)*, E. Borgotta and M. Borgotta, eds. New York, NY: MacMillan.
Helmreich, R.L., A.C. Merritt, and P.J. Sherman
1996 Human factors and national culture. *ICAO Journal* 51(8):14-16.
Herek, G.
1987 Can functions be measured? A new perspective on the functional approach to attitudes. *Social Psychology Quarterly* 50:285-303.
Hiltz, S.R., and M. Turoff
1978 *The Network Nation: Human Communication via Computer.* Reading, MA: Addison-Wesley.
Hinrichs, T.R.
1988 Towards an architecture for open world problem solving. Pp. 182-189 in *Proceedings of the Fourth International Workshop on Machine Learning.* San Mateo, CA: Morgan Kaufmann.
Hinton, G.E., and T.J. Sejnowski
1986 Learning and relearning in Boltzmann machines. In *Parallel Distributed Processing: Explorations in the Microstructure of Cognition, Volume 1: Foundations*, D.E. Rumelhart and J.L. McClelland, eds. Cambridge, MA: MIT Press.
Hinton, G.E., and T. Shallice
1991 Lesioning an attractor network: Investigations of acquired dyslexia. *Psychological Review* 98:74-95.
Hintzman, D.L.
1986 "Schema abstraction" in a multiple-trace memory model. *Psychological Review* 93:411-428.
1988 Judgments of frequency and recognition memory in a multiple-trace memory model. *Psychological Review* 95(4):528-551.
Hoff, B.R.
1996 USMC individual combatants. In *Proceedings of the Defense Modeling and Simulation Office Individual Combatant Workshop, July 1-2.* Alexandria, VA: Defense Modeling and Simulation Office.

Hofstede, G.
1980 *Cultures Consequences: International Differences in Work Related Values.* Beverly Hills, CA: Sage Publications.
Hogan, R., and J. Hogan
1992 Hogan Personality Inventory Manual. Tulsa, OK: Hogan Assessment Systems, Inc.
Holland, J.H.
1975 *Adaptation in Natural and Artificial Systems.* Ann Arbor, MI: University of Michigan Press.
1992 Genetic algorithms. *Scientific American* 267(July):66-72.
Holland, J.K., K. Holyoak, R. Nisbett, and P. Thagard
1986 *Induction: Processes of Inference, Learning, and Discovery.* Cambridge, MA: MIT Press.
Hollenbeck, J.R., D.R. Ilgen, D.J. Sego, J. Hedlund, D.A. Major, and J. Phillips
1995a The multi-level theory of team decision making: Decision performance in teams incorporating distributed expertise. *Journal of Applied Psychology* 80:292-316.
Hollenbeck, J.R., D.R. Ilgen, D. Tuttle, and D.J. Sego
1995b Team performance on monitoring tasks: An examination of decision errors in contexts requiring sustained attention. *Journal of Applied Psychology* 80:685-696.
Hopfield, J.
1982 Neural networks and physical systems with emergent collective computational properties. *Proceedings of the National Academy of Sciences* 79:2554-2558.
Horgan, J.
1994 Sex, death and sugar. *Scientific American* 271(5):20-24.
Hornoff, A.J., and D.E. Kieras
1997 Cognitive modeling reveals menu search is both random and systematic. Pp. 107-114 in *Proceedings of CHI, 1997* (March 22-27, Atlanta, GA). New York, NY: ACM.
Hovland, C., and H. Pritzker
1957 Extent of opinion change as a function of amount of change advocated. *Journal of Abnormal and Social Psychology* 54:257-261.
Howes, A., and R.M. Young
1996 Learning consistent, interactive, and meaningful task-action mappings: A computational model. *Cognitive Science* 20:301-356.
1997 The role of cognitive architecture in modeling the user: Soar's learning mechanism. *Human-Computer Interaction* 12(4):311-344.
Huber, G.P.
1996 Organizational learning: The contributing processes and the literatures. In *Organizational Learning*, M.D. Cohen and L.S. Sproull, eds. Thousand Oaks, CA: Sage.
Hudlicka, E.
1997 Modeling Behavior Moderators in Military Performance Models. Technical Report 9716. Psychometrix Associates, Inc., Lincoln, MA
Hudlicka, E., and G. Zacharias
1997 Individual Difference Evaluation Architecture for Free Flight System Analysis. Technical Report R98141, October. Charles River Analytics, Cambridge, MA.
Huey, B.M., and C.D. Wickens, eds.
1993 *Workload Transition: Implications for Individual and Team Performance.* Panel on Workload Transition. Washington, DC: National Academy Press.
Huguet, P. and B. Latane
1996 Social representations as dynamic social impact. *Journal of Communication* 46(4): 57-63.
Hull, C.L.
1943 *Principles of Behavior.* New York, NY: Appleton-Century-Crofts.

Humphrey, R.H., P.M. O'Malley, L.D. Johnston, and J.G. Bachman
1988 Bases of power, facilitation effects, and attitudes and behavior: Direct, indirect, and interactive determinants of drug use. *Social Psychology Quarterly* 51:329-345.

Humphreys, M.S., J.D. Bain, and R. Pike
1989 Different ways to cue a coherent memory system: A theory for episodic, semantic and procedural tasks. *Psychological Review* 96(2):208-233.

Hunter, J.E., J.E. Danes, and S.H. Cohen
1984 *Mathematical Models of Attitude Change*. New York, NY: Academic Press.

Hutchins, E.
1990 The technology of team navigation. In *Intellectual Teamwork*, J. Galegher, R. Kraut, and C. Egido, eds. Hillsdale, NJ: Lawrence Erlbaum.
1991a Organizing work by adaptation. *Organizational Science* 2:14-39.
1991b The social organization of distributed cognition. Pp. 238-307 in *Perspectives on Socially Shared Cognition*, L.B. Resnick, J.M. Levine, and S.D. Teasley, eds. Washington, DC: American Psychological Association.

IBM
1993 *Close Combat Tactical Trainer (CCTT): Semi-Automated Forces (SAF) Combat Instruction Set (CIS) Development*. Orlando, FL: IBM Integrated Development Team.

Illgen, C., G.L., Zacharias, G. Rinkus, E. Hudlicka, V. O'Neil, and K. Carley
1997 A Decision Aid for Command and Control Warfare. Report No. R96421. Charles River Analytics, Inc., Cambridge, MA.

Innis, H.
1951 *The Bias of Communication*. Toronto, Canada: University of Toronto Press.

Insko, C.
1967 *Theories of Attitude Change*. New York, NY: Appleton-Century-Crofts.

Isen, A.M.
1993 Positive affect and decision making. In *Handbook of Emotions*, J.M. Haviland and M. Lewis, eds. New York, NY: The Guilford Press.

Jacobs, R.A., M.I. Jordon, S.J. Nowland, and G.E. Hinton
1991 Adaptive mixtures of local experts. *Neural Computation* 3:79-87.

James, W.
1890 *Principles of Psychology*. New York, NY: Holt.

Janis, I.L.
1989 *Crucial Decisions: Leadership in Policymaking and Crisis Management*. New York, NY: The Free Press.

Janis, I.L., and L. Mann
1977 *Decision Making: A Psychological Analysis of Conflict, Choice and Commitment*. New York, NY: The Free Press.

Jin, Y., and R.E. Levitt
1993 I-AGENTS: Modeling organizational problem solving in multi-agent. *International Journal of Intelligent Systems in Accounting, Finance and Management* 2(4):247-270.
1996 The virtual design team: A computational model of project organizations. *Computational and Mathematical Organization Theory* 2(3):171-196.

Jochem, T.M., D.A. Pomerleau, and C.E. Thorpe
1995 Vision-Based Neural Network Road and Intersection Detection and Traversal. IEEE Conference on Intelligent Robots and Systems, Pittsburgh, PA, August 5-9.

John, B.E., and D.E. Kieras
1996 The GOMS family of user interface analysis techniques: Comparison and contrast. *ACM Transactions on Computer-Human Interaction* 3:320-351.

Johnson, E.J., J. Hershey, J. Meszaros, and H. Kunreuther

1993 Framing, probability distortions and insurance decisions. *Journal of Risk and Uncertainty* 7:35-51.

Johnson, T.R., J. Krems, and N.K. Amra

1994 A computational model of human abductive skill and its acquisition. Pp. 463-468 in *Proceedings of the Sixteenth Annual Conference of the Cognitive Science Society*, A. Ram and K. Eiselt, eds. Hillsdale, NJ: Lawrence Erlbaum Associates.

Johnson, W.L.

1994 Agents that learn to explain themselves. In *Proceedings of the Twelfth National Conference on Artificial Intelligence* (AAAI-94). August. Seattle, WA.

Johnson-Laird, P.N

1983 *Mental Models: Toward a Cognitive Science of Language, Inference, and Consciousness*. Cambridge, MA: Harvard University Press.

Jordan, M.I.

1990 Motor learning and the degrees of freedom problem. In *Attention and Performance, XIII*, M. Jeannerod, ed. Hillsdale, NJ: Lawrence Erlbaum.

Josephson, J.R., and S.G. Josephson

1994 *Abductive Inference: Computation, Philosophy, Technology*. New York, NY: Cambridge University Press.

Juslin, P.

1994 Well calibrated confidence judgments for general knowledge items, inferential recognition decisions and social predictions. In *Contributions to Decision Making*, J-P. Caverni, M. Bar-Hillel, F.H. Barron, and H. Jungerman, eds. Amsterdam, NL: North-Holland/ Elsevier.

Kaempf, G.L., G. Klein, M.L. Thordsen, and S. Wolf

1996 Decision making in complex naval command-and-control environments. *Human Factors* 38(2):220-231.

Kahneman, D.

1973 *Attention and Effort*. Englewood Cliffs, NJ: Prentice-Hall.

Kahneman, D., and A. Tversky

1979 Prospect theory: Analysis of decision making under risk. *Econometrica* 47:263-291.

1982 The psychology of preferences. *Scientific American* 246(1):160-173.

Kambhampati, S.

1995 AI planning: A prospectus on theory and applications. *ACM Computing Surveys* 27(3)September.

1997 Refinement planning as a unifying framework for plan synthesis. *AI Magazine*(Summer).

Karayiannis, N.B., and A.N. Venetsanopoulos

1993 *Artificial Neural Networks: Learning Algorithms, Performance Evaluation, and Applications*. Boston. MA: Kluwer Academic Publishers.

Karr, C.R.

1996 Unit level representation. In *Proceedings of the Defense Modeling and Simulation Office Workshop on Unit Level Behavior, August 7-8*. Alexandria, VA: Defense Modeling and Simulation Office.

Katz, D.

1960 The functional approach to the study of attitudes. *Public Opinion Quarterly* 24:163-204.

Katz, E.

1961 The social itinerary of technical change: Two studies on the diffusion of innovation. In *Studies of Innovation and Communication to the Public,* E. Katz, ed. Stanford, CA: Institute for Communication Research.

Kaufer, D., and K.M. Carley
1993 *Communication at a Distance: The Effects of Print on Socio-cultural Organization and Change*. Hillsdale, NJ: Lawrence Erlbaum.
Kawato, M., Y. Maeda, Y. Uno, and R. Suzuki
1990 Trajectory formation of arm movement by cascade neural network model based on minimum torque-change criterion. *Biological Cybernetics* 62:275-288.
Keeney, R., and H. Raiffa
1976 *Decisions with Multiple Objectives*. New York, NY: Wiley.
Kephart, J.O., B.A. Huberman, and T. Hogg
1992 Can predictive agents prevent chaos? Pp. 41-55 in *Economics and Cognitive Science*, P. Bourgine and B. Walliser, eds. Oxford, United Kingdom: Pergamon Press.
Kern, R.P.
1966 A Conceptual Model of Behavior Under Stress, With Implications for Combat Training. Human Resources Research Office. Washington, DC: George Washington University.
Kerstholt, J.H., and J.G.W. Raaijmakers
forthcoming Decision making in dynamic task environments. In *Decision Making: Cognitive Models and Explanations*, R. Crozier and O. Svenson, eds. London, England: Routledge.
Kieras, D.E., and D.E. Meyer
1996 The Epic architecture: Principles of operation. Published electronically at ftp.eecs.umich.edu:/people/kieras/EPICarch.ps.
1997 An overview of the EPIC architecture for cognition and performance with application to human-computer interaction. *Human-Computer Interaction* 12(4):391-438.
Kieras, D.E., S.D. Wood, and D.E. Meyer
1997 Predictive engineering models based on the EPIC architecture for a multimodal high-performance human-computer interaction task. *ACM Transactions on Computer-Human Interaction* 4(3):230-275.
Kieras, D.E., D.E. Meyer, S. Mueller, and T. Seymour
1998 Insights into Working Memory from the Perspective of the EPIC Architecture for Modeling Skilled Perceptual-Motor and Cognitive Human Performance. Technical Report No. 10 (TR-98/ONR-EPIC-10). Electrical Engineering and Computer Science Department, University of Michigan.
Kiesler, C., B. Collins, and N. Miller
1969 *Attitude Change*. New York, NY: Wiley.
Kiesler, S., J. Siegel, and T. McGuire
1984 Social psychological aspects of computer-mediated communication. *American Psychologist* 39:1123-1134.
Kilduff, M., and D. Krackhardt
1994 Bringing the individual back in: A structural analysis of the internal market for reputation in organizations. *Academy of Management Journal* 37:87-107.
Kinnear, K.E., ed.
1994 *Advances in Genetic Programming*. Cambridge, MA: MIT Press
Kirkpatrick, S., C.D. Gelatt, and M.P. Vecchi
1983 Optimization by simulated annealing. *Science* 220(4598):671-680.
Kirwan, B., and L.K. Ainsworth, eds.
1992 *A Guide to Task Analysis*. Bristol, PA: Taylor and Francis.
Kitayama, S., and H.R. Markus, eds.
1994 *Emotion and Culture*. Washington, DC: The American Psychological Association.
Klayman, J., and Y-W. Ha
1987 Confirmation, disconfirmation, and information in hypothesis testing. *Psychological Review* 94:211-228.

Kleijnen, J., and W. van Groenendaal
1992 *Simulation: A Statistical Perspective.* New York, NY: Wiley.
Klein, G.A.
1989 Recognition-primed decisions. Pp. 47-92 in *Advances in Man-Machine Systems Research*, W. Rouse, ed. Greenwich, CT: JAI.
1994 A recognition-primed decision (RPD) model of rapid decision making. In *Decision Making in Action: Models and Methods*, G.A. Klein, J. Orasanu, R. Calderwood, and C.E. Zsambok, eds. Norwood, NJ: Ablex Publishing Company.
1997 *Recognition-Primed Decision Making: Looking Back, Looking Forward.* Hillsdale, NJ: Lawrence Erlbaum.
Klein, G.A., R. Calderwood, and A. Clinton-Cirocco
1986 Rapid decision-making on the fire ground. In *Proceedings of the Human Factors Society 30th Annual Meeting.* Santa Monica, CA: Human Factors Society.
Kleinman, D.L.
1976 Solving the optimal attention allocation problem in manual control. *IEEE Transactions in Automatic Control* 21(6):815-821.
Kleinman, D.L., S. Baron, and W.H. Levison
1970 Optimal control model of human response, Part I: Theory and validation. *Automatica* 6:357-369.
1971 A control theoretic approach to manned-vehicle systems analysis. *IEEE Transactions on Automatic Control* 16:824-832.
Klimoski, R., and S. Mohammed
1994 Team mental model: Construct or metaphor? *Journal of Management* 20(2):403-437.
Koehler, J.J.
1996 The base rate fallacy reconsidered: Descriptive, normative and methodological challenges. *Behavioral and Brain Sciences* 19:1-54.
Koller, D., and J. Breese
1997 Belief Network and Decision-Theoretic Reasoning for Artificial Intelligence. AAAI Fourteenth National Conference on Artificial Intelligence, Providence, RI.
Kolodner, J.L.
1987 Extending problem solver capabilities through case-based inference. Pp. 167-178 in *Proceedings of the Fourth International Workshop on Machine Learning.* San Mateo, CA: Morgan Kaufmann.
Kontopoulos, K.M.
1993 Neural networks as a model of structure. Pp. 243-267 in *The Logics of Social Structure.* New York, NY: Cambridge University Press.
Konz, S.
1995 Predetermined time systems. Pp. 464-475 in *Work Design: Industrial Ergonomics, Fourth Edition.* Columbus, OH: Publishing Horizons, Inc.
Kornbrot, D.
1988 Random walk models of binary choice: The effect of deadlines in the presence of asymmetric payoffs. *Acta Psychologica* 69:109-127.
Koza, J.R.
1990 Genetic Programming: A Paradigm for Genetically Breeding Populations of Computer Programs to Solve Problems. TR STAN-CS-90-1314 (June). Department of Computer Science, Stanford University.
Knapp, A., and J.A. Anderson
1984 A signal averaging model for concept formation. *Journal of Experimental Psychology: Learning, Memory, and Cognition* 10:617-637.

Krackhardt, D.
1994 Graph theoretical dimensions of informal organizations. In *Computational Organization Theory*, K.M. Carley and M.J. Prietula, eds. Hillsdale, NJ: Lawrence Erlbaum.
forthcoming Organizational Viscosity and the Diffusion of Controversial Innovations. *Journal of Mathematical Sociology.*
Krackhardt, D., and D.J. Brass
1994 Intraorganizational networks: The micro side. Pp. 207-229 in *Advances in Social Networks Analysis*, S. Wasserman and J. Galaskiewiscz, eds. Thousand Oaks, CA: Sage Publications.
Krackhardt, D., J. Blythe, and C. McGrath
1994 KrackPlot 3.0: An improved network drawing program. *Connections* 17(2):53-55.
Kraus, M.K., D.J. Franceschini, T.R. Tolley, L.J. Napravnik, D.E. Mullally, and R.W. Franceschini
1996 CCTT SAF and ModSAF behavior integration techniques. Pp. 159-170 in *Proceedings of the Sixth Conference on Computer Generated Forces and Behavioral Representation.* Orlando, FL: Institute for Simulation and Training.
Krems, J., and T. Johnson
1995 Integration of anomalous data in multicausal explanations. Pp. 277-282 in *Proceedings of the Seventeenth Annual Conference of the Cognitive Science Society*, J.D. Moore and J.F. Lehman, eds. Mahwah, NJ: Lawrence Erlbaum.
Kristofferson, A.B.
1967 Attention and psychophysical time. Pp. 93-100 in *Attention and Performance*, A.F. Sanders, ed. Amsterdam, The Netherlands: North-Holland Publishing Co.
Kruschke, J.K.
1992 ALCOVE: An exemplar-based connectionist model of category learning. *Psychological Review* 99:22-24.
Kuhn, G.W.S.
1989 Ground Forces Battle Casualty Rate Patterns: The Empirical Evidence. Logistics Management Institute, McLean, VA.
Kuipers, B., A. Moskowitz, and J. Kassirer
1988 Critical decisions under uncertainty: Representation and structure. *Cognitive Science* 12(2):177-210.
Kuokka, D.R.
1991 MAX: A meta-reasoning architecture for "X." *SIGART Bulletin* 2:93-97.
Laird, J.
1996 Soar/IFOR: Intelligent Synthetic Agents. Presentation to National Research Council Panel on Modeling Human Behavior and Command Decision Making, October 17.
Laird, J.E., and P.S. Rosenbloom
1990 Integrating, execution, planning, and learning in Soar for external environments. Pp. 1022-1029 in *Proceedings of the Eighth National Conference on Artificial Intelligence.* Menlo Park, CA: American Association for Artificial Intelligence.
Laird, J.E., A. Newell, and P.S. Rosenbloom
1987 SOAR: An architecture for general intelligence. *Artificial Intelligence* 33:1-64.
Laird, J.E., K. Coulter, R. Jones, P. Kenny, F. Koss, and P. Nielsen
1997 Review of Soar/FWA Participation in STOW-97. Published electronically at http://ai.eecs.umich.edu/ifor/stow-review.html. November 7.
Lallement, Y., and B.E. John
1998 Cognitive architecture and modeling idiom: An examination of three models of Wicken's Task. *Proceedings of the Twentieth Annual Conference of the Cognitive Science Society* (August).
Laming, D.R.
1968 *Information Theory of Choice Reaction Time.* San Diego, CA: Academic Press.

Landauer, T.K., and S.T. Dumais
1997 A solution to Plates problem—The latent semantic analysis theory of acquisition, induction and representation of knowledge. *Psychological Review* 104(2):211-240.
Langley, P.
1996 *Elements of Machine Learning*. San Francisco, CA: Kaufman.
Langley, P., K.B. McKusick, J.A. Allen, W.F. Iba, and K. Thompson
1991 A design for the ICARUS architecture. *SIGART Bulletin* 2:104-109.
Lant, T.K.
1994 Computer simulations of organizations as experiential learning systems: Implications for organization theory. In *Computational Organization Theory*, K.M. Carley and M.J. Prietula, eds. Hillsdale, NJ: Lawrence Erlbaum.
Lant, T.L., and S.J. Mezias
1990 Managing discontinuous change: A simulation study of organizational learning and entrepreneurship. *Strategic Management Journal* 11:147-179.
Lant, T.L., and S.J. Mezias
1992 An organizational learning model of convergence and reorientation. *Organization Science* 3(1):47-71.
Latane, B.
1996 Dynamic social impact: The creation of culture by communication. *Journal of Communication* 46(4):13-25.
Latane, B., and M.J. Bourgeois
1996 Experimental evidence for dynamic social impact: The emergence of subcultures in electronic groups. *Journal of Communication* 46(4):35-47.
Latane, B., J.H. Liu, A. Nowak, M. Bonevento, and Z. Long
1995 Distance matters: Physical space and social impact. *Personality and Social Psychology Bulletin* 21(8):795-805.
Latorella, K.A.
1996a Investigating Interruptions: Implications for Flightdeck Performance. Unpublished doctoral dissertation. State University of New York at Buffalo, New York.
1996b Investigating interruptions: An example from the flightdeck. Pp. 249-253 in *Proceedings of the Human Factors and Ergonomics Society 40th Annual Meeting*. Santa Monica, CA: Human Factors and Ergonomics Society.
Laughery, K.R., and K.M. Corker
1997 Computer modeling and simulation of human/system performance. In *Handbook of Human Factors*, second edition, G. Salvendy, ed. New York, NY: John Wiley and Sons.
Lauritzen, S.L., and D.J. Spiegelhalter
1988 Local computation with probabilities in graphical structures and their applications to expert systems. *Journal of the Royal Statistical Society* B:50(2).
LaVine, N.D., S.D. Peters, and K.R. Laughery
1996 Methodology for Predicting and Applying Human Response to Environmental Stressors. Report DNA-TR-95-116. Defense Special Weapons Agency, Alexandria, VA.
LaVine, N.D., R.K. Laughery, B. Young, R. Kehlet
1993 Semi-automated forces (SAFOR) crew performance degradation. Pp. 405-415 in *Proceedings of the Third Conference on Computer Generated Forces and Behavioral Representation*, Orlando, FL, March 17-19.
Lawrence, P.R., and J. Lorsch
1967 *Organization and Environment: Managing Differentiation and Integration*. Cambridge, MA: Harvard University Press.
Lawson, K.E., and C.T. Butler
1995 Overview of Man-in-the-Loop Air-to-Air System Performance Evaluation Model (MIL-AASPEM II). D658-10485-1. The Boeing Company (September).

Lazarus, R.S.

1991 *Emotion and Adaptation*. New York, NY: Oxford University Press.

Leake, D.

1996 CBR in context: The present and future. In *Case-Based Reasoning: Experiences, Lessons, and Future Directions*. Menlo Park: AAAI Press.

Leary, M.R.

1982 Hindsight distortion and the 1980 presidential election. *Personality and Social Psychology Bulletin* 8:257-263.

LeDoux, J.E.

1987 Emotion. Pp. 419-460 in *The Nervous System V*. Bethesda, MD: American Physiological Society.

1989 Cognitive-emotional interactions in the brain. *Cognition and Emotion* 3(4):267-289.

1992 Brain mechanisms of emotion and emotional learning. *Current Opinion in Neurobiology* 2(2):191-197.

Lee, J.J., and P.A. Fishwick

1994 Real-time simulation-based planning for computer generated forces. *Simulation* 63(5):299-315.

Levin, D.S.

1991 *Introduction to Neural and Cognitive Modeling*. Hillsdale, NJ: Lawrence Erlbaum.

Levin, I.P., and G. J.Gaeth

1988 How consumers are affected by the framing of attribute information before and after consuming a product. *Journal of Consumer Research* 15:374-378.

Levitt, B., and J. March

1988 Organizational learning. *Annual Review of Sociology* 14:319-340.

Levitt, R.E., G.P. Cohen, J.C. Kunz, C.I. Nass, T. Christiansen, and Y. Jin

1994 The 'virtual design' team: Simulating how organization structure and information processing tools affect team performance. Pp. 1-18 in *Computational Organization Theory*, K.M. Carley and M.J. Prietula, eds. Hillsdale, NJ: Lawrence Erlbaum.

Lewin, K.

1935 *A Dynamic Theory of Personality*. New York, NY: McGraw-Hill.

Lewis, R.L.

1996 Interference in short-term memory: The magical number two (or three) in sentence processing. *Journal of Psycholinguistic Research* 25:93-115.

1997a Leaping off the garden path: Reanalysis and limited repair parsing. In *Reanalysis in Sentence Processing*, J. Fodor and F. Ferreira, eds. Boston: Kluwer.

1997b Specifying architectures for language processing: Process, control, and memory in parsing and interpretation. In *Architectures and Mechanisms for Language Processing*, M. Crocker, M. Pickering, and C. Clifton, eds. Cambridge University Press.

Libicki, M.C.

1995 *What is Information Warfare*? Washington, DC: U.S. Government. Printing Office.

Lichtenstein, S., and B. Fischhoff

1977 Do those who know more also know more about how much they know? The calibration of probability judgments. *Organizational Behavior and Human Performance* 20:159-183.

Lin, N., and R.S. Burt

1975 Differential effects of information channels in the process of innovation diffusion. *Social Forces* 54:256-274.

Lin, Z., and K.M. Carley

1993 Proactive and reactive: An analysis of the effect of agent style on organizational decision making performance. *International Journal of Intelligent Systems in Accounting, Finance and Management* 2(4):271-288.

1997a Organizational response: The cost performance tradeoff. *Management Science* 43(2):217-234.

1997b Organizational decision making and error in a dynamic task environment. *Journal of Mathematical Sociology* 22(2):125-150.

Lindsey, P.H., and D.A. Norman

1977 *Human Information Processing: An Introduction to Psychology*. San Diego, CA: Academic Press.

Link, S.W.

1992 *Wave Theory of Difference and Similarity*. Hillsdale, NJ: Erlbaum.

Liu, Y.

1996 Queueing network modeling of elementary mental processes. *Psychological Review* 103(1):116-136.

Liu, Y.

1997 Queueing network modeling of human performance of concurrent spatial and verbal tasks. *IEEE Transactions on Systems, Man, and Cybernetics* 27(2):195-207.

Lockett, J., and S. Archer

1997 *Human Behavior Modeling and Decision Making for Military Simulation*. Boulder, CO: MicroAnalysis and Design, Inc.

Logan, G.D.

1988 Toward an instance theory of automatization. *Psychological Review* 95:492-527.

1991 Automatizing alphabet arithmetic: I. Is extended practice necessary? *Journal of Experimental Psychology: Learning, Memory, and Cognition* 17:179-195.

1992 Shapes of reaction time distributions and shapes of learning curves: A test of instance theory of automaticity. *Journal of Experimental Psychology: Learning, Memory, and Cognition* 18:883-914.

Logan, G.D., and C. Bundesen

1997 Spatial effects in the partial report paradigm—A challenge for theories of visual-spatial attention. *Psychology of Learning* 35:243-282.

Logan, G.D., and S.T. Klapp

1991 Automatizing alphabet arithmetic: I. Is extended practice necessary to produce automaticity? *Journal of Experimental Psychology: Learning, Memory and Cognition* 17(3):478-496.

Logicon, Inc.

1989 A Model for the Acquisition of Complex Cognitive Skills. Technical Report, Logicon Project 1150. Logicon, Inc., San Diego, CA.

Loomes, G., and R. Sugden

1982 Regret theory: An alternative theory of rational choice under uncertainty. *Economic Journal* 92:805-824.

Lopes, L.L.

1987 Between hope and fear: The psychology of risk. *Advances in Experimental Social Psychology* 20:255-295.

Lu, Z., and A.H. Levis

1992 Coordination in distributed decision making. Pp. 891-897 in *IEEE International Conference on Systems, Man and Cybernetics, Volume I*. New York: Institute of Electrical and Electronic Engineers.

Luce, R.D.

1959 *Individual Choice Behavior*. New York, NY: Wiley.

Luce, R.D., and P.C. Fishburn

1991 Rank and sign dependent linear utility for finite first order gambles. *Journal of Risk and Uncertainty* 4:29-60.

Lussier, J.W., and D.J. Litavec

1992 Battalion Commanders' Survey: Tactical Commanders Development Course Feedback. Research Report 1628 (AD-A258 501). U.S. Army Research Institute for the Behavioral and Social Sciences, Alexandria, VA.

Lussier, J.W., R.E. Solick, and S.D. Keene

1992 Experimental Assessment of Problem Solving at the Combined Arms Services Staff School. Research Note 92-52. U.S. Army Research Institute for Behavioral and Social Sciences, Alexandria, VA.

MacCrimmon, K.R., and S. Larsson

1979 Utility theories: Axioms versus paradoxes. Pp. 333-409 in *Expected Utility Hypothesis and the Allais Paradox*, M. Allais and O. Hagen, eds. Dordrecht, Netherlands; Boston, MA: D. Reidel Publishing Co.

MacKay, D.J.C.

1997 *Information Theory, Probability, and Neural Networks*. Cambridge, England: Cambridge University Press.

MacLeod, C.M., and K. Dunbar

1988 Training and Stroop like interference: Evidence for a continuum of automaticity. *Journal of Experimental Psychology: Learning, Memory, and Cognition* 14:126-135.

MacMillan, J., S.E. Deutsch, and M.J. Young

1997 A Comparison of Alternatives for Automated Decision Support in a Multi-Tasking Environment. Human Factors and Ergonomics Society 41st Annual Meeting, Albuquerque, NM.

MacWhinney, B., and J. Leinbach

1991 Implementations are not conceptualizations: Revising the verb learning model. *Cognition* 40:121-153.

Macy, M.W.

1990 Learning theory and the logic of critical mass. *American Sociological Review* 55:809-826.

1991a Learning to cooperate: Stochastic and tacit collusion in social exchange. *American Journal of Sociology* 97(3): 808-843.

1991b Chains of cooperation: Threshold effects in collective action. *American Sociological Review* 56:730-747.

Majchrzak, A., and L. Gasser

1992 HITOP-A: A tool to facilitate interdisciplinary manufacturing systems design. *International Journal of Human Factors in Manufacturing* 2(3):255-276.

Malone, T.W.

1987 Modeling coordination in organizations and markets. *Management Science* 33:1317-1332.

Mann, M.

1988 States, war and capitalism. In *The MacMillan Student Encyclopedia of Sociology*, B. Blackwell, ed. London: MacMillan.

Manthey, M.

1990 Hierarchy and Emergence: A Computational View. 0106-0791, R90-25. Department of Mathematics and Computer Science, Institute for Electronic Systems, University of Aalborg, Denmark.

March, J., and R. Weissinger-Baylon, eds.

1986 *Ambiguity and Command: Organizational Perspectives on Military Decision Making*. Boston, MA: Pitman.

March, J.G.

1994 *A Primer on Decision Making: How Decisions Happen*. New York, NY: Free Press; Toronto; Maxwell MacMillan Canada.

1996 Exploration and exploitation in organizational learning. In *Organizational Learning*, M.D. Cohen and L.S. Sproull, eds. Thousand Oaks, CA: Sage.

March, J.G., and H.A. Simon

1958 *Organizations*. New York, NY: Wiley.

Marshall, S.P.

1995 Learning in Tactical Decision-Making Situations: A Hybrid Model of Schema Development. Presented at the ONR Hybrid Learning Kick-Off Meeting, Washington, DC, August 3-4.

Masuch, M., and P. LaPotin

1989 Beyond garbage cans: An AI model of organizational choice. *Administrative Science Quarterly* 34:38-67.

Maurer, M.

1996 *Coalition Command and Control: Key Considerations*. Washington, DC: U.S. Government Printing Office.

McClelland, J.L., and D.E. Rumelhart

1981 An interactive activation model of context effects in letter perception: Art 1. An account of basic findings. *Psychological Review* 88:375-407.

1985 Distributed memory and the representation of general and specific information. *Journal of Experimental Psychology: General* 114:159-188.

1986 *Parallel Distributed Processing: Explorations in the Microstructure of Cognition: Volume 2*. Cambridge, MA: MIT Press.

1988 *Explorations in Parallel Distributed Processing*. Cambridge, MA: MIT Press.

McDermott, D.

1978 Planning and acting. *Cognitive Science* 2(2):71-109.

McLeod, R.W., and M.J. Griffin

1989 A review of the effects of translational whole-body vibration on continuous manual control operations. *Journal of Sound and Vibration* 133:55-114.

McNeil, B.J., S.G. Pauker, H.C. Sox, and A. Tversky

1982 On the elicitation of preferences for alternative therapies. *New England Journal of Medicine* 306:1259-1262.

Mead, G.H.

1934 *Mind, Self, and Society*. Chicago, IL: University of Chicago Press.

Medin, D.L., and M.M. Shaffer

1978 Context theory of classification learning. *Psychological Review* 85:207-238.

Mellers, B., and K. Biagini

1994 Similarity and choice. *Psychological Review* 101:505-518.

Mellers, B.A., A. Schwartz, and A.D.J. Cooke

forthcoming Judgment and decision making. *Annual Review of Psychology*.

Mengshoel, O., and D. Wilkins

1997 Visualizing uncertainty in battlefield reasoning using belief networks. In *Proceedings of First Annual Symposium of Army Research Laboratory for Advanced and Interactive Displays*. Adelphi, MD: ARL Adelphi Laboratory.

Merritt, A.C., and R.L. Helmreich

1996 Culture in the cockpit: A multi-airline study of pilot attitudes and values. In *Proceedings of the Eighth International Symposium on Aviation Psychology*. Columbus OH: Ohio State University.

Mesterson-Gibbons, M.

1992 *Introduction to Game-Theoretic Modelling*. Redwood City, CA: Addison-Wesley.

Metcalf, J.
1986 Decision making and the Grenada rescue operation. Pp. 277-297 in *Ambiguity and Command: Organizational Perspectives on Military Decision Making*, J.G. March and R. Weissinger-Baylon, eds. Marshfield, MA: Pitman Publishing Inc.

Metlay, W.D., D. Liebling, N. Silverstein, A. Halatyn, A. Zimberg, and E. Richter
1985 Methodology for the Assessment of the Command Group Planning Process. Department of Psychology, Applied Research and Evaluation, Hofstra University, Hamstead, NY.

Meyer, D.E., and D.E. Kieras
1997a A computational theory of executive cognitive processes and multiple-task performance. Part 1. Basic mechanisms. *Psychological Review* 104(1):2-65.
1997b A computational theory of executive cognitive processes and multiple-task performance. Part 2. Accounts of psychological refractory-period phenomena. *Psychological Review* 104:749-791.

Miao, A.X., G. Zacharias, and S. Kao
1997 A computational situation assessment model for nuclear power plant operations. *IEEE Transactions on Systems, Man and Cybernetics. Part A: Systems and Humans* 27(6):728-742.

Milgram, P., R. van der Winngaart, H. Veerbeek, O.F. Bleeker, and B.V. Fokker
1984 Multi-crew model anlaytic assessment of landing performance and decision making demands. Pp. 373-399 in *Proceedings of 20th Annual Conference on Manual Control, Vol. II.* Moffett Field, CA: NASA Ames Research Center.

Mil-HDBK-759-A
1981 Chapter 9: Moderator Variables

Miller, C.R.
1985 *Modeling Soldier Dimension Variables in the Air Defense Artillery Mission Area.* Aberdeen Proving Ground, MD: TRASANA.

Miller, C.S., and J.E. Laird
1996 Accounting for graded performance within a discrete search framework. *Cognitive Science* 20:499-537.

Miller, G.A.
1956 The magical number seven, plus or minus two: Some limits in our capacity for processing information. *Psychological Review* 63:81-97.

Miller, N.E.
1944 Experimental studies of conflict. Pp. 431-465 in *Personality and the Behavior Disorders, Volume 1*, J. Mcv. Hunt, ed. New York, NY: The Ronald Press.

Miller, R.A., H.E. Pople, Jr., and J.D. Myers
1982 Internist-1: An experimental computer-based diagnostic consultant for general internal medicine. *New England Journal of Medicine* 307(8):468-476.

Minar, N., R. Burkhart, C. Langton, and M. Askenazi
1996 The Swarm Simulation System: A Toolkit for Building Multi-Agent Simulations. Working Paper 96-06-0 42. Santa Fe Institute.

Mineka, S., and S.K. Sutton
1992 Cognitive biases and the emotional disorders. *Psychological Science* 3(1):65-69.

Mischel, W.
1968 *Personality and Assessment.* New York, NY: Wiley.

Mitchell, T.M., J. Allen, P. Chalasani, J. Cheng, O. Etzioni, M. Ringuette, and J.C. Schlimmer
1991 Theo: A framework for self-improving systems. Pp. 325-355 in *Architectures for Intelligence,* K. VanLehn, ed. Hillsdale, NJ: Lawrence Erlbaum.

Modi, A.K., D.M. Steier, and A.W. Westerberg
1990 Learning to use approximations and abstractions in the design of chemical processes. In *Proceedings of the AAAI Workshop on Automatic Generation of Approximation and Abstractions*, July 30, 1990. Boston, MA.

Moffat, D.
1997 Personality parameters and programs. Pp. 120-165 in *Creating Personalities for Synthetic Actors: Towards Autonomous Personality Agents*, R. Trappl and P. Petta, eds. Lecture Notes in Artificial Intelligence 1195, Subseries of Lecture Notes in Computer Science. Berlin, Germany: Springer-Verlag.

Molm, L.D.
1978 Sex-role attitudes and the employment of married women: The direction of causality. *Sociological Quarterly* 19:522-533.

Moray, N., M. Dessouky, B. Kijowski, and R. Adapathya
1991 Strategic behavior, workload, and performance in task scheduling. *Human Factors* 33(6):607-629.

Moreland, R.L.
forth-coming Transactive memory in work groups and organizations. In *Shared Knowledge in Organizations*, L. Thompson, D. Messick, and J. Levine, eds. Mahwah, NJ: Lawrence Erlbaum.

Moreland, R.L., L. Argote, and R. Krishnan
1996 Social shared cognition at work: Transactive memory and group performance. Pp. 57-84 in *What's Social About Social Cognition? Research on Socially Shared Cognition in Small Groups*. Newbury Park, CA: Sage.

Morris, M.
1994 Epidemiology and Social Networks: Modeling Structured Diffusion. Pp. 26-52 in *Advances in Social Network Analysis*, S. Wasserman and J. Galskiewicz, eds. Thousand Oaks, CA: Sage.

Moses, Y.O., and M. Tennenholtz
1990 Artificial Social Systems. Part I, Basic Principles. Technical Report CS90-12. Department of Computer Science, Makhon Vaitsmann le-mada, Weizmann Institute of Science, Rehovot, Israel.

Moss, S., and B. Edmonds
1997 A formal preference-state model with qualitative market judgments. *Omega—The International Journal of Management Science* 25(2):155-169.

Moss, S., and O. Kuznetsova
1996 Modelling the process of market emergence. Pp. 125-138 in *Modelling and Analysing Economies in Transition*, J.W. Owsinski and Z. Nahorski, eds. Warsaw: MODEST.

Muller, J.
1997 Control architectures for autonomous and interacting agents: A survey. Pp. 1-21 in *Intelligent Agent Systems*, L. Cavedon, A. Rao, and W. Wobcke, eds. New York: Springer Verlag.

Mulgund, S., G. Rinkus, C. Illgen, and J. Friskie
1997 OLIPSA: On-Line Intelligent Processor for Situation Assessment. Second Annual Symposium and Exhibition on Situational Awareness in the Tactical Air Environment, Patuxent River, MD.

Mulgund, S.S., G.J. Rinkus, et al.
1996 A Situation Driven Workload Adaptive Pilot/Vehicle Interface. Charles River Analytics, Cambridge, MA.

Myerson, R.
1991 *Analysis and Conflict*. Cambridge, MA: Harvard University Press.

NASA Ames Research Center
1996 MIDAS Core Redesign: Release 1.0 Documentation. Unpublished report. NASA Ames Research Center, Moffett Field, CA.

Nass, C., Y. Moon, B.J. Fogg, B. Reeves, and C. Dryer
1997 *Can Computer Personalities Be Human Personalities?* Department of Communications. Stanford, CA: Stanford University.

Navon, D., and D. Gopher
1979 On the economy of the human processing system. *Psychological Review* 86(3):214-255.

Nelson, G.H., J.F. Lehman, and B.E. John
1994a Integrating cognitive capabilities in a real-time task. Pp. 353-358 in *Proceedings of the Sixteenth Annual Conference of the Cognitive Science Society*, August.
1994b Experiences in interruptible language processing. In *Proceedings of the 1994 AAAI Spring Symposium on Active Natural Language Processing*.

Nerb, J., J. Krems, and F.E. Ritter
1993 Rule learning and the power law: A computational model and empirical results. Pp. 765-770 in *Proceedings of the 15th Annual Conference of the Cognitive Science Society*. Boulder, CO. Hillsdale, New Jersey: LEA.

Newcomb, T.
1953 An approach to the study of communicative acts. *Psychological Review* 60:393-404.

Newell, A.
1990 *Unified Theories of Cognition*. Cambridge, MA: Harvard University Press.

Newell, A., and P.S. Rosenbloom
1981 Mechanisms of skill acquisition and the law of practice. In *Cognitive Skills and Their Acquisition*, J.R. Anderson, ed. Hillsdale, NJ: Erlbaum.

Newell, A., and H.A. Simon
1963 GPS, A program that simulates human thought. Pp. 279-293 in *Computers and Thought*, E.A. Feigenbaum and J. Feldman, eds. New York, NY: McGraw-Hill.

Ng, K-C., and B. Abramson
1990 Uncertainty management in expert systems. *IEEE Expert* 1990:29-48.

Nicoll, J.R., and D.M. Hsu
1995 A Search for Understanding: Analysis of Human Performance on Target Acquisition and Search Tasks Using Eyetracker Data. Institute of Defense Analysis Paper P-3036.

Nii, P.H.
1986a Blackboard systems: The blackboard model of problem solving and the evolution of blackboard architectures. Part One. *AI Magazine* 38-53.
1986b Blackboard systems: The blackboard model of problem solving and the evolution of blackboard architectures. Part Two. *AI Magazine* 82-106.

Nobel, P.A.
1996 Response Times in Recognition and Recall. Unpublished doctoral dissertation. Indiana University.

Noble, D.
1989 Schema-based knowledge elicitation for planning and situation assessment aids. *IEEE Transactions on Systems, Man, and Cybernetics* 19(3)May/June:473-482.

North, R.A., and V.A. Riley
1989 A predictive model of operator workload. Pp. 81-90 in *Application of Human Performance Models to System Design*, G.R. McMillan, D. Beevis, E. Salas, M. Strub, R. Sutton, and L. Van Breda, eds. New York, NY: Plenum Press.

Nosofsky, R.M., and T.J. Palmeri
1997 An exemplar-based random walk model of speeded classification. *Psychological Review* 104:266-308.

Ntuen, C.A., D.N. Mountjoy, M.J. Barnes, and L.P. Yarborough
1997 A battlefield reasoning system. In *Conference Proceedings of the First Annual Symposium of Advanced Displays and Interactive Displays Federated Laboratory, January 28-29*. Adelphi, MD: Army Research Laboratory.

Oatley, K., and P.N. Johnson-Laird
1987 Towards a cognitive theory of emotions. *Cognition and Emotion* 1:29-50.

Ogasawara, G.H., and S.J. Russell
1994 Real-time, decision-theoretic control of an autonomous underwater vehicle. Special issue of *CACM on Real-World Applications of Uncertain Reasoning*.
Ortony, A., J. Slack, and O. Stock
1992 *Communications from an Artificial Intelligence Perspective: Theoretical and Applied Issues*. New York, NY: Springer Verlag.
Ortony, A., J. Black, and O. Stock, eds.
1992 *Communication from an Artificial Intelligence Perspective*. New York, NY: Springer-Verlag.
Ortony, A., G.L. Clore, and A. Collins
1988 *The Cognitive Structure of Emotions*. New York, NY: Cambridge University Press.
Osgood, C., and P. Tannenbaum
1955 The principle of congruity in the prediction of attitude change. *Psychological Review* 62:42-55.
Osgood, C.E.
1967 *The Measurement of Meaning*. Champaign, IL: The University of Illinois Press.
Osgood, C., G. Succi, and P. Tannenbaum
1957 *The Measurement of Meaning*. Urbana, IL: The University of Illinois Press.
Padgham, L., and G. Taylor
1997 A system for modeling agents having emotion and personality. In *Intelligent Agent Systems*, L. Cavedon, A. Rao, and W. Wobcke, eds. Berlin, Germany: Springer-Verlag.
Palmer, J.
1994 Set-size effects in visual search: The effect of attention is independent of the stimulus for simple tasks. *Vision Research* 34:1703-1721.
Parasuraman, R., and D.F. Davies, eds.
1984 *Varieties of Attention*. Orlando, FL: Academic Press.
Parker, J.D., R.M. Bagby, and R.T. Joffe
1996 Validation of the biosocial model of personality: Confirmatory factor analysis of the tridimensional personality questionnaire. *Psychological Assessment* 8(2):139-144.
Parsons, K.C.
1996 Ergonomics assessment of thermal environments. In *Evaluation of Human Work*, J.R. Wilson and E.N. Corlett, eds. Bristol, PA: Taylor and Francis.
Pattipati, K.R., and D.L. Kleinman
1991 A review of the engineering models of information-processing and decision-making in multi-task supervisory control. Pp. 35-68 in *Multiple Task Performance*, D.L. Damos, ed. London: Taylor and Francis Ltd.
Payne, J.W., J.R. Bettman, and E.J. Johnson
1993 *The Adaptive Decision Maker*. Cambridge, England: Cambridge University Press.
Pearl, J.
1986 Fusion, progagation, and structuring in belief networks. *Artificial Intelligence* 29(3):241-288.
1988 *Probabilistic Reasoning in Intelligent Systems: Networks of Plausible Inference*, San Mateo, CA: Morgan Kaufmann.
Peck, V.A., and B.E. John
1992 Browser-Soar: A cognitive model of a highly interactive task. Pp. 165-172 in *Proceedings of CHI*. Monterey, CA, May 3-7. New York, NY: ACM.
Pennings, J.
1975 The relevance of the structural-contingency model for organizational effectiveness. *Administrative Science Quarterly* 20:393-410.
Pennington, D.C., D.R. Rutter, K. McKenna, and I.E. Morley
1980 Estimating the outcome of a pregnancy test: Women's judgment in foresight and hindsight. *British Journal of Social and Clinical Psychology* 79:317-323.

Persons, J.B., and E.B. Foa
1984 Processing of fearful and neutral information by obsessive-compulsives. *Behavior Research Therapy* 22:260-265.

Pete, A., K.R. Pattipati, and D.L. Kleinman
1993 Distributed detection in teams with partial information: A normative descriptive model. *IEEE Transactions on Systems, Man and Cybernetics* 23:1626-1648.
1994 Optimization of detection networks with multiple event structures. *IEEE Transactions on Automatic Control* 39:1702-1707.

Pew, R.W.
1995 The state of situation awareness measurement: Circa 1995. Pp. 7-15 in *Experimental Analysis and Measurement of Situation Awareness*, D.J. Garland and M.R. Endsley, eds. Daytona Beach, FL: Embry-Riddle Aeronautical University.

Pew, R.W., and A.S. Mavor, eds.
1997 *Representing Human Behavior in Military Simulations: Interim Report.* Panel on Modeling Human Behavior and Command Decision Making: Representations for Military Simulations. Washington, DC: National Academy Press.

Pfeffer, J., and G.R. Salancik
1978 *The External Control of Organizations: A Resource Dependency Perspective.* New York, NY: Harper and Row.

Pfeiffer, M.G., A.I. Siegel, S.E. Taylor, and L. Shuler, Jr.
1979 Background Data for the Human Performance in Continuous Operations. ARI Technical Report 386. U.S. Army Research Institute for the Behavioral and Social Sciences, Alexandria, VA. July.

Pinker, S., and A. Prince
1988 On language and connectionism: Analysis of a parallel distributed processing model of language acquisition. *Cognition* 28:73-193.

Plaut, D.C., J.L. McClelland, and M.S. Seidenberg
1996 Understanding word reading: Computational principles in quasi-regular domain. *Psychological Review* 103:56-115.

Plutchik, R., and H.R. Conte
1997 Circumplex models of personality and emotions. In *Circumplex Models of Personality and Emotions*, R. Plutchik and H.R. Conte, eds. Washington, DC: American Psychological Association.

Polk, T.A., and A. Newell
1995 Deduction as verbal reasoning. *Psychological Review* 102:533-566.

Polk, T.A., and P.S. Rosenbloom
1994 Task-independent constraints on a unified theory of cognition. In *Handbook of Neuropsychology, Volume 9*, F. Boller and J. Grafman, eds. Amsterdam, Netherlands: Elsevier.

Pomerleau, D.A., J. Gowdy, and C.E. Thorpe
1991 Combining artificial neural networks and symbolic processing for autonomous robot guidance. *Journal of Engineering Applications of Artificial Intelligence* 4(4):279-285.

Posner, M.I.
1980 Orienting of attention. *Quarterly Journal of Experimental Psychology* 32:3-25.

Pospisil, L.
1994 War and peace among the Kapaupu. In *Studying War: Anthropological Perspectives*, S.P. Reyna and R.E. Downs, eds. Langhorne, PA: Gordon and Breach.

Powell, W.W., and P.J. DiMaggio
1994 *The New Institutionalism in Organizational Analysis.* Chicago, IL: The University of Chicago Press.

Price, D.J. de S.
1965 Networks of scientific papers. *Science* 149:510-515.

Prosser, T.W.
1996a JWARS. In *Proceedings of the Defense Modeling and Simulation Office Workshop on Unit Level Behavior, August 7-8*. Arlington, VA: Defense Modeling and Simulation Office.
1996b JWARS Testbed White Paper: Intelligence Fusion. Joint Warfare System Office, Arlington, VA.

Quiggen, J.
1982 A theory of anticipated utility. *Journal of Economic Behavior and Organizations* 3:323-343.

Quillen, J.R.
1986 Induction of decision trees. *Machine Learning* 1:81-106.

Raaijmakers, J.G.W., and R.M. Shiffrin
1981 Search of associative memory. *Psychological Review* 88:93-134.

Raby, M., and C.D. Wickens
1994 Strategic workload management and decision biases in aviation. *The International Journal of Aviation Psychology* 4(3):211-240.

Ramsey, J.D., and S.J. Morrissey
1978 Isodecrement curves for task performance in hot environments. *Applied Ergonomics* 9:66-72.

Rao, A.S.
1996 AgentSpeak(L): Agents speak out in a logical computable language. In *Agents Breaking Away*, W. Van de Velde and J.W. Perram, eds. New York, NY: Springer-Verlag.

Rapoport, A.
1953 Spread of information through a population with socio-structural bias: I. Assumption of transitivity. II. Various models with partial transitivity. *Bulletin of Mathematical Biophysics* 15:523-546.

Ratches, J.A.
1976 Statistic performance model for thermal imaging systems. *Optical Engineering* 15(6):525-530.

Ratcliff, R.
1978 A theory of memory retrieval. *Psychological Review* 85:59-108.

Ratcliff, R., T. van Zandt, and G. McKoon
1998 Connectionistic and Diffusion Models of Reaction Time. Unpublished manuscript.

Reece, D.
1996 Computer controlled hostiles for TTES. In *Proceedings of the Defense Modeling and Simulation Office Individual Combatant Workshop, July 1-2*. Alexandria, VA: Defense Modeling and Simulation Office.

Reece, D.A., and P. Kelley
1996 An architecture for computer generated individual combatants. In *Sixth Conference on Computer Generated Forces and Behavioral Representations: Project Status Reports*, J. Clarkson, W.T. Noff, J.C. Peacock, and J.D. Norwood. Orlando, FL: University of Central Florida.

Reece, D.A., and R. Wirthlin
1996 Detection models for computer generated individual combatants. In *Proceedings of the Sixth Conference on Computer Generated Forces and Behavioral Representation, July 23-25*. Report IST-TR-96-18. Orlando, FL: Institute for Simulation and Training.

Reed, G.F.
1969 Under-inclusion—A characteristic of obsessional personality. *British Journal of Psychiatry* 115:781-785.

Reger, R.K., and A.S. Huff
1993 Strategic groups: A cognitive perspective. *Strategic Management Journal* 14:103-124.
Remy, P.A., and A.H. Levis
1988 On the generation of organizational architectures using Petri nets. Pp. 371-385 in *Advances in Petri Nets*. G. Rozenberg, ed. Berlin, West Germany: Springer-Verlag.
Restle, F.
1961 *Psychology of Judgment and Choice*. New York, NY: Wiley.
Reynolds, L.D., and N.G. Wilson
1968 *Scribes and Scholars: A Guide to the Transmission of Greek and Latin Literature*. Oxford: Clarendon Press.
Rice, R.E.
1984 *The New Media: Communication, Research, and Technology*. Beverly Hills, CA: Sage.
1994 Network Analysis and Computer Mediated Communication Systems. Pp. 167-206 in *Advances in Social Network Analysis*, S. Wasserman and J. Galskiewicz, eds. Thousand Oaks, CA: Sage.
Rice, R.E., and C. Aydin
1991 Attitudes toward new organizational technology: Network proximity as a mechanism for social information processing. *Administrative Science Quarterly* 39:219-244.
Rice, R.E., and D. Case
1983 Electronic message systems in the university: A description of use and utility. *Journal of Communication* 33:131-152.
Richards, T., and F. Giminez
1994 Cognitive Style and Strategic Choice. Working Paper #282. Center for Business Research, University of Manchester, United Kingdom.
Rieman, J., R.M. Young, and A. Howes
1996 A dual-space model of iteratively deepening exploratory learning. *International Journal of Human-Computer Studies* 44:743-775.
Ritter, F.E., and J.H. Larkin
1994 Developing process models as summaries of HCI action sequences. *Human Computer Interaction* 9(3-4):345-383.
Roberts, C.W.
1989 Other than counting words: A linguistic approach to content analysis. *Social Forces* 68:147-177.
Roberts, K.
1990 Some characteristics of one type of high reliability organizations. *Organization Science* 1(2):160-176.
Roberts, N.C., and K.A. Dotterway
1995 The Vincennes incident: Another player on the stage. *Defense Analysis* 11(1):31-45.
Robinson, J.A.
1965 A machine-oriented logic based on the resolution principle. *Journal of the Association for Computing Machinery* 12:23-41.
Rochlin, G.I.
1991 Iran Air Flight 655 and the USS Vincennes: Complex, large-scale military systems and the failure of control. In *Social Responses to Large Technical Systems: Control or Anticipation*, T.R. LaPorte, ed. Netherlands: Kluwer Academic Publishers.
Roethlisberger, F.J., and W.J. Dickson
1939 *Management and the Worker*. Cambridge, MA: Harvard University Press.
Rogers, E.M.
1983 *Diffusion of Innovations*. New York, NY: Free Press.

Rogers, W.H.
1996 Flight deck task management: A cognitive engineering analysis. Pp. 239-243 in *Proceedings of the Human Factors and Ergonomics Society 40th Annual Meeting*. Santa Monica, CA: Human Factors and Ergonomics Society.

Rosen, L.D., and M.M. Weil
1995 Computer anxiety: A cross-cultural comparison of students in ten countries. *Computers in Human Behavior* 11(4):45-64.

Rosenberg, M.
1956 Cognitive structure and attitudinal affect. *Journal of Abnormal and Social Psychology* 53:3667-3672.

Rosenbloom, P.S., A. Newell, and J.E. Laird
1991 Toward the knowledge level in Soar: The role of architecture in the use of knowledge. In *The Twentysecond Carnegie Mellon Symposium on Cognition*, K. Van Lehn, ed. Hillsdale, NJ: Lawrence Erlbaum Associates, Inc..

Ross, B.H.
1989 Some psychological results on case-based reasoning. In *Proceedings Case-Based Reasoning Workshop (DARPA II)*. San Mateo, CA: Morgan Kaufman.

Rumelhart, D.E.
1989 Toward a microstructural account of human reasoning. Pp. 298-312 in *Similarity and Analogical Reasoning*, S. Vosniadou and A. Ortony, eds. New York: Cambridge University Press.

Rumelhart, D.E., and J. McClelland
1986 *Parallel Distributed Processing: Explorations in the Microstructure of Cognition, Volume I.* Cambridge, MA: MIT Press.

Rumelhart, D.E., G.E. Hinton, and R.J. Williams
1986a Learning internal representations by error propagation. In *Parallel Distributed Processing: Explorations in the Microstructure of Cognition, Volume 1: Foundations*, D.E. Rumelhart and J.L. McClelland, eds. Cambridge, MA: MIT Press.

Rumelhart, D.E., J.L. McClelland, and the PDP Research Group, eds.
1986b *Parallel Distributed Processing: Fundamentals, Volume 1.* Cambridge, MA: MIT Press.

Rumelhart, D.E., P. Smolensky, J.L. McClelland, and G.E. Hinton
1986c Schemata and sequential thought processes in PDP models. Pp. 7-57 in *Parallel Distributed Processing: Explorations in the Microstructure of Cognition, Volume 2*, J.L. McClelland, D.E. Rumelhart, and the PDP Research Group, eds. Cambridge, MA: MIT Press.

Russell, P., and P. Norvig
1995 *Artificial Intelligence: A Modern Approach.* Englewood Cliffs, NJ: Prentice Hall.

Rutenbar, R.A.
1989 Simulated annealing algorithms: An overview. *IEEE Circuits and Devices Magazine* 5:12-26.

Ryan, M.
1997 Overview of International Simulation Advisory Group. In *Conference Proceedings of Sixth Annual Modeling and Simulation Briefing to Government and Industry, May 22-23.* Alexandria, VA: Defense Modeling and Simulation Office.

Sacerdoti, E.D.
1974 Planning in a hierarchy of abstraction spaces. *Artificial Intelligence* 5:115-135.
1975 The nonlinear nature of plans. Pp. 206-214 in *Proceedings of the Fourth International Joint Conference on Artificial Intelligence*, Menlo Park, CA.
1977 *A Structure for Plans and Behavior.* New York, NY: Elsevier-North Holland.

Salancik, G.R., and J. Pfeffer
1978 A social information professing approach to job attitudes and task design. *Administrative Science Quarterly* 23:224-253.

Sampson, S.F.
1968 *A novitiate in a period of change: An experimental and case study of relationships.* Ph.D. Dissertation. Cornell University.

Sarter, N.B., and D.D. Woods
1991 Situation awareness: a critical but ill-defined phenomenon. *International Journal of Aviation Psychology* 1:45-57.
1995 How in the world did we ever get into that mode? Mode error and awareness in supervisory control. *Human Factors* 37(1):5-19.

Schachter, D.L.
1987 Implicit memory: History and current status. *Journal of Experimental Psychology: Learning, Memory, and Cognition* 13:501-513.

Schlabach, J.
1997 RAVEN requirements development perspective. Personal communication, September.

Schlabach, J.L., C.C. Hayes, and D.E. Goldberg
1997 SHAKA-GA: A Genetic Algorithm for Generating and Analyzing Battlefield Courses of Action. White paper. University of Illinois, Champaign.

Schlegel, R.P., and D. DiTecco
1982 Attitudinal structures and the attitude-behavior relation. In *Consistency in Social Behavior*, M. Zanna, E. Higgins, and C. Herman, eds. Hillsdale, NJ: Erlbaum.

Schneider, W., and R.M. Shiffrin
1977 Controlled and automatic human information processing: I. Detection, search, and attention. *Psychological Review* 84:1-66.

Schneider, W., S.T. Dumais, and R.M. Shiffrin
1984 Automatic and control processing and attention. Pp. 1-27 in *Varieties of Attention*, R. Parasuraman and D.F. Davies, eds. Orlando, FL: Academic Press.

Schutte, P.C., and A.C. Trujillo
1996 Flight crew task management in non-normal situations. Pp. 244-248 in *Proceedings of the Human Factors and Ergonomics Society 40th Annual Meeting.* Santa Monica, CA: Human Factors and Ergonomics Society.

Schweikert, R., and B. Boruff
1986 Short-term memory capacity: Magic number seven or magic spell? *Journal of Experimental Psychology: Learning, Memory, and Cognition* 16:419-425.

Scott, W.R.
1981 *Organizations: Rational, Natural, and Open Systems.* Prentice Hall Inc.

Scribner, B.L, A.D. Smith, R.H. Baldwin, and R.L. Phillips
1986 Are smart tankers better? AFQT and military productivity. *Armed Forces and Society* 12(2)(Winter).

Seidenberg, M.S., and J.L. McClelland
1989 A distributed developmental model of word recognition and naming. *Psychological Review* 96:523-568.

Selfridge, O.G.
1959 Pandemonium: A paradigm for learning. Pp. 511-529 in *Symposium on the Mechanization of Thought Processes.* London, England: HM Stationery Office.

Serfaty, D., J. MacMillan, and J. Deckert
1991 Towards a Theory of Tactical Decision-Making Expertise. TR-496-1. Alphatech, Burlington, MA.

Serpe, R.
1987 Stability and change in self: A structural symbolic interactionist explanation. *Social Psychology Quarterly* 50:44-55.

1988 The cerebral self: Thinking, planning and worrying about identity relevant activity. In *Self and Society: A Social-Cognitive Approach*, J. Howard and P. Callero, eds. Cambridge, England: Cambridge University Press.

Servan-Schreiber, E.
1991 The Competitive Chunking Theory: Models of Perception, Learning, and Memory. Ph.D. dissertation. Carnegie Mellon University, Pittsburgh, PA.

Shah, P.P.
1995 Who are Our Social Referents? A Network Perspective to Determine the Referent Other. Paper presented at the Academy of Management. Strategic Management Faculty, University of Minnesota.

Shankar, R.
1991 Z-Scheduler: Integrating theories on scheduling behavior into a computational model. Pp. 1219-1223 in *Proceedings of the IEEE International Conference on Systems, Man, and Cybernetics*, Charlottesville, VA.

Shapiro, K.L., and J.E. Raymond
1994 Temporal allocation of visual attention: Inhibition or interference. In *Inhibitory Processes in Attention, Memory and Language*, D. Dagenbach and T. Carr, eds. New York, NY: Academic Press.

Sharkey, A.J.C., and N.E. Sharkey
1995 Cognitive modeling: Psychology and connectionism. In *The Handbook of Brain Theory and Neural Networks*, M.A. Arbib, ed. Cambridge, MA: MIT Press.

Sher, K.J., R.O. Frost, M. Kushner, and T.M. Crews
1989 Memory deficits in compulsive checkers: Replication and extension in a clinical sample. *Behavior Research and Therapy* 27(1):65-69.

Sherif, M., and C. Hovland
1961 *Social Judgment*. New Haven, CT: Yale Press.

Sherif, M., C. Sherif, and R. Nebergall
1965 *Attitude and Attitude Change*. Philadelphia, PA: Saunders.

Sherman, P.J., and R.L. Helmreich
1996 Attitudes toward automation: The effect of national culture. In *Proceedings of the 8th Ohio State Symposium on Aviation Psychology*. Columbus, OH: The Ohio State University.

Shiffrin, R.M.
1988 Attention. Pp. 739-811 in *Stevens' Handbook of Experimental Psychology, Volume 2: Learning and Cognition*, R.A. Atkinson, R.J. Herrnstein, G. Lindsay, and R.D. Luce, eds. New York, NY: John Wiley and Sons.

Shiffrin, R.M., and G.T. Gardner
1972 Visual processing capacity and attentional control. *Journal of Experimental Psychology* 93:72-83.

Shiffrin, R.M., and N. Lightfoot
1997 Perceptual learning of alphanumeric-like characters. *The Psychology of Learning and Motivation* 36:45-80.

Shiffrin, R.M., and W. Schneider
1977 Controlled and automatic human information processing: II. Perceptual learning, automatic attending, and a general theory. *Psychological Review* 84:127-190.

Shiffrin, R.M., and M. Steyvers
1997 A model for recognition memory: REM–retrieving effectively from memory. *Psychonomic Bulletin and Review* 4:145-166.

Shively, J.
1997 Overview of MIDAS situation awareness modeling effort. Presentation at NASA Ames Research Center, Moffett Field, CA, November.

Shoham, Y., and M. Tennenholtz
1994 Co-Learning and the Evolution of Social Activity. STAN-CS-TR-94-1511 Report. Department of Computer Science, Stanford University, CA.
Siegel, A.I., and J.J. Wolf
1969 *Man-Machine Simulation Models.* New York, NY: Wiley-Interscience.
Siegel, A.I., M.G. Pfeiffer, F. Kopstein, and J.J. Wolf
1980 Human Performance in Continuous Operations: Volume III, Technical Documentation. ARI Research Product 80-4C, AD-A088319. U.S. Army Research Institute for the Behavioral and Social Sciences, Alexandria, VA. March.
Siegel, A.I., J.J. Wolf, and A.M. Schorn
1981 Human Performance in Continuous Operations: Description of a Simulation Model and User's Manual for Evaluation of Performance Degradation. ARI Technical Report 505, AD-A101950. U.S. Army Research Institute for the Behavioral and Social Sciences, Alexandria, VA. January.
Simon, H.A.
1947 *Administrative Behavior.* New York, NY: Free Press.
1967 Motivational and emotional controls of cognition. *Psychological Review* 74:29-39.
Sloman, A.
1997 Personalities for synthetic actors: Current issues and some perspectives. In *Synthetic Actors: Towards Autonomous Personality Agents*, R. Trappl and P. Petta, eds. New York, NY: Springer.
Smith, B.R., S.W. Tyler, and K.M. Corker
1996 Man-Machine Integration Design and Analysis System. Presentation to National Research Council Panel on Modeling Human Behavior and Command Decision Making. Washington, D.C., November 18.
Smith, M.
1973 Political attitudes. In *Handbook of Political Psychology*, J. Knutson, ed. San Francisco: Jossey-Bass.
Smith, M., J. Bruner, and R. White
1956 *Opinions and Personality.* New York, NY: Wiley.
Smith, P.L.
1996 Psychophysically principled models of visual reaction time. *Psychological Review* 102:567-593.
Sperling, G.
1960 The information available in brief visual presentations. *Psychological Monographs* 74(11):Whole No. 498.
Sperling, G., and E. Weichselgartner
1995 Episodic theory of the dynamics of spatial attention. *Psychological Review* 102(3):503-532.
Sperry, R.W.
1995 The riddle of consciousness and the changing scientific worldview. *Journal of Humanistic Psychology* 35(2):7-33.
Spick, M.
1988 *The Ace Factor.* Annapolis, MD: Naval Institute Press.
Spranca, M., E. Minsk, and J. Baron
1991 Omission and commission in judgment and choice. *Journal of Experimental Psychology* 21:76-105.
Sproull, L., and S. Kiesler
1986 Reducing social context cues: Electronic mail in organizational communication. *Management Science* 32:1492-1512.

1991 *Connections: New Ways of Working in the Networked Organization.* Cambridge, MA: MIT Press.

Stanfill, C., and D. Waltz
1986 Toward memory-based reasoning. *Communication of the ACM* 29(12):??.

Staw, B.M.
1976 Knee-deep in the big muddy: A study of escalating commitment to a chosen course of action. *Organizational Behavior and Human Performance* 16:27-44.

Staw, B.M., and S.G. Barsade
1993 Affect and managerial performance—A test of the sadder but wiser vs. happier-and-smarter hypothesis. *Administrative Science Quarterly* 38(2):304-331.

Staw, B.M., and H. Hoang
1995 Sunk costs in the NBA: Why draft order affects playing time and survival in professional basketball. *Administrative Science Quarterly* 40:474-494.

Staw, B.M., and J. Ross
1989 Understanding behavior in escalation situations. *Science* 246(4927):216-220.

Stefik, M.J.
1981a Planning with constraints (MOLGEN: Part 1). *Artificial Intelligence* 16:111-140.
1981b Planning and metaplanning (MOGEN: Part 2). *Artificial Intelligence* 16:141-170.

Stevens, W.K., and L. Parish
1996 Representation of unit level behavior in the naval simulation system (NSS). In *Proceedings of the Defense Modeling and Simulation Workshop on Unit Level Behavior, August 7-8.* Alexandria, VA: Defense Modeling and Simulation Office.

Stiffler, D.R.
1988 Graduate level situation awareness. *USAF Fighter Weapons Review* (Summer):115-120.

Stigler, J.W., R.A. Schweder, and L. Herdt, eds.
1990 *Cultural Psychology.* Cambridge, United Kingdom: Cambridge University Press.

Stites, J.
1994 Complexity. *Omni* 16(8):42-52.

Stone, M.
1960 Models for reaction time. *Psychometrika* 25:251-260.

Stryker, S.
1980 *Symbolic Interactionism.* Menlo Park, CA: Benjamin Cummings.

Stryker, S., and R. Serpe
1982 Commitment, identity salience, and role behavior. In *Personality, Roles and Social Behavior*, W. Ickes and E. Knowles, eds. Berlin, Germany: Springer-Verlag.

Sullivan, K., and T. Kida
1995 The effect of multiple reference points and prior gains and losses on managers' risky decision making. *Organizational Behavior and Human Decision Processes* 64(1):76-83.

Sun, R.
1995 A Hybrid Learning Model for Sequential Decision Making. Presented at the ONR Hybrid Learning Kick-Off Meeting, Washington, DC, August 3-4.

Sun, R., and F. Alexandre
1997 *Connectionistic-Symbolic Integration.* Hillsdale, NJ: Lawrence Erlbaum.

Sussman, G.J.
1975 *A Computer of Skill Acquisition.* New York, NY: American Elsevier.

Sutton, R.S.
1992 Special issue on reinforcement learning. *Machine Learning* 8:1-395.

Tajfel, H.
1982 Social psychology of intergroup relations. *Annual Review of Psychology* 33:1-39.

Tambe, M.
1996a Executing team plans in dynamic, multi-agent domains. In *AAAI FALL Symposium on Plan Execution: Problems and Issues.*
1996b Teamwork in real-world, dynamic environments. *In Proceedings of the International Conference on Multi-agent Systems (ICMAS).*
1996c Tracking dynamic team activity. In *Proceedings of the National Conference on Artificial Intelligence*, August.
1997 Agent architectures for flexible, practical teamwork. In *Proceedings of the American Association of Artificial Intelligence.*
Tambe, M., W.L. Johnson, R.M. Jones, F. Koss, J.E. Laird, P.S. Rosenbloom, and K. Schwamb
1995 Intelligent agents for interactived simulation environments. *AI Magazine* (Spring):15-37.
Tang, Z., K. Pattipati, and D. Kleinman
1991 An algorithm for determining the decision thresholds in a distributed detection problem. *IEEE Transactions on Systems, Man, and Cybernetics* 21(1):231-237.
Tate, A.
1977 Generating project networks. Pp. 888-893 in *Proceedings of the Fifth International Joint Conference on Artificial Intelligence*, Menlo Park, CA.
Tate, A., J. Hendler, and M. Drummond, eds.
1990 A review of AI planning techniques. In *Readings in Planning*. San Mateo, CA: Morgan Kaufmann Publishers, Inc.
Taube, J.S., and H.L. Burton
1995 Head direction cell activity monitored in a novel environment and during a cue conflict situation. *Journal of Neurophysiology* 74(5):1953-1971.
Thagard, P.
1989 Explanatory coherence. *Behavioral and Brain Sciences* 12(3):435-502.
Theeuwes, J.
1994 Endogenous and exogenous control of visual selection. *Perception* 23(4):429-440.
Thomas, E.M.
1994 Management of violence among Iu/wasi of Nyae Nyae. In *Studying War: Anthropological Perspectives*, S.P. Reyna and R.E. Downs, eds. Langhorne, PA: Gordon and Breach.
Thompson, J.D.
1967 *Organizations in Action.* New York, NY: McGraw-Hill.
Thorsden, M., J. Galushka, G.A. Klein, S. Young, and C.P. Brezovic
1989 Knowledge Elicitation Study of Military Planning. Technical Report 876. Army Research Institute, Alexandria, VA.
Thorngen, B.
1977 Silent actors: Communication networks for development. In *The Social Impact of the Telephone*, I. del Sola Pool, ed.. Cambridge, MA: MIT Press.
Thurstone, L.L.
1959 *The Measurement of Values.* Chicago, IL: Chicago University Press.
Tolcott, M.A., F.F. Marvin, and P.E. Lehner
1989 Expert decisionmaking in evolving situations. *IEEE Transactions on Systems, Man and Cybernetics* 19(3):606-615.
Touretzky, D.S.
1995 Connectionist and symbolic representations. In *The Handbook of Brain Theory and Neural Networks*, M.A. Arbib, ed. Cambridge, MA: MIT Press.
Touretzky, D.S., and G.E. Hinton
1988 A distributed connectionist production system. *Cognitive Science* 12:423-466.
Treisman, A., and S. Sato
1990 Conjunction search revisited. *Journal of Experimental Psychology: Human Perception and Performance* 16(3):459-478.

Treisman, A.M.
1969 Strategies and models of selective attention. *Psychological Review* 76:282-299.
Tulga, M.K., and T.B. Sheridan
1980 Dynamic decisions and work load in multitask supervisory control. *IEEE Transactions on Systems, Man, and Cybernetics* 10(5):217-231.
Tulving, E.
1972 *Episodic and Semantic Memory*. New York, NY: Academic Press.
Tversky, A.
1972 Elimination by aspects: A theory of choice. *Psychological Review* 79:281-289.
Tversky, A., and D. Kahneman
1980 *Causal Schemes in Judgments Under Uncertainty*. New York, NY: Cambridge University Press.
1992 Advances in prospect theory: Cumulative representation of uncertainty. *Journal of Risk and Uncertainty* 5:297-323.
U.S. Army
1997 *Knowledge and Speed: The Annual Report of the Army After Next Project*. Washington, DC: U.S. Army Chief of Staff.
U.S. Army Command and General Staff College
1993 *The Tactical Decisionmaking Process*. Fort Leavenworth, KS: U.S. Army Command and General Staff College.
U.S. Congress
1989 Iran Airflight 655 Compensation Hearings before the Defense Policy Panel of the Committee on Armed Services. House of Representatives, Second Session, held on August 3 and 4, September 9, and October 6. Washington, DC: U.S. Government Printing Office.
U.S. Department of the Army
1980 *Technical Bulletin, Medical 507*. Occupational and Environmental Health, Prevention, Treatment and Control of Heat Injury. July. Washington, DC: Department of the Army.
1983 Chapter 9: Moderator Variables
1993a *Field Manual 22-9, Soldier Performance in Continuous Operations*. December.
1993b *FM 100-5 Operations*. Headquarters. Fort Monroe, VA: U.S. Department of the Army.
U.S. Department of Defense
1981 *Human Engineering Design Criteria for Military Systems, Equipment and Facilities: Mil-STD 1472D*. Washington, DC: U.S. Department of Defense.
1994 Joint Pub 1-02. *Department of Defense Dictionary of Military and Associated Terms*. s.v. command and control, p. 77.
1995 *Modeling and Simulation (M&S) Master Plan*. Training and Doctrine Command. October. Washington, DC: Department of Defense.
1996 *Verification, Validation and Accreditation (VV&A) Recommended Practices Guide*. Office of the Director of Defense Research and Engineering. Washington, DC: Defense Modeling and Simulation Office.
Utgoff, P.E.
1989 Incremental induction of decision trees. *Machine Learning* 4:161-186.
Valente, T.
1995 *Network Models of the Diffusion of Innovations*. Cresskill, NJ: Hampton Press.
Van Creveld, M.L.
1985 *Command in War*. Cambridge, MA: Harvard University Press.
VanLehn, K.
1996 Cognitive skill acquisition. *Annual Review of Psychology* 47:513-539.
Van Nostrand, S.J.
1986 Model Effectiveness as a Function of Personnel. CAA-SR-86-34. U.S. Army Concepts Analysis Agency, Bethesda, MD.

Van Schie, E.C.M., and J. van der Plight
1995 Influencing risk preference in decision making: The effects of framing and salience. *Organizational Behavior and Human Decision Processes* 63(3):264-275.
Vere, S., and T. Bickmore
1990 A basic agent. *Computational Intelligence* 6:41-60.
von der Malsburg, C.
1973 Self-organization of orientation sensitive cells in the striate cortex. *Kybernetic* 14:85-100.
Von Neumann, J., and O. Morgenstern
1947 *Theory of Games and Economic Behavior*. **??**: Princeton University Press.
von Winterfeldt, D., and W. Edwards
1986 *Decision Analysis and Behavioral Research*. Cambridge, United Kingdom: Cambridge University Press.
Walden, R.S., and W.B. Rouse
1978 A queueing model of pilot decisionmaking in a multitask flight management situation. *IEEE Transactions on Systems, Man, and Cybernetics* 8(12):867-874.
Waldinger, R.
1977 Achieving several goals simultaneously. In *Machine Intelligence 8*, R. Waldinger, ed. Chichester, England: Ellis Norwood Limited.
Wallace, J.R.
1982 Gideon Criterion: The Effects of Selection Criteria on Soldier Capability and Battle Results. USA Recruiting Command, Fort Sheridan, IL. January.
Wallsten, T.S.
1995 Time pressure and payoff effects on multidimensional probabilistic inference. Pp. 167-179 in *Time Pressure and Stress in Human Judgment*, O. Svenson and J. Maule, eds. New York, NY: Plenum Press.
Walsh, J.P.
1995 Managerial and organizational cognition: Notes from a trip down memory lane. *Organizational Science* 6(3):280-321.
Wasserman, P.D.
1989 *Neural Computing: Theory and Practice*. New York, NY: Van Nostrand Reinhold.
1993 *Advanced Methods in Neural Computing*. New York, NY: Van Nostrand.
Wasserman, S., and K. Faust
1994 *Social Network Analysis: Methods and Applications*. Cambridge, United Kingdom: Cambridge University Press.
Watson, R., J. Barry, and R. Sandza
1988 A case of human error. *Newsweek* (August 15):18-21.
Weber, M., and C. Camerer
1987 Recent developments in modeling preferences under risk. *Specktrum* 9:129-151.
Wegner, D. M.
1995 Transactive memory: A contemporary analysis of the group mind. Pp. 185-208 in *Theories of Group Behavior*, B. Mullen and G.R. Goethals, eds. New York, NY: Springer-Verlag.
Wegner, D. M., Erber, R., and Raymond, P.
1991 Transactive memory in close relationships. *Journal of Personality and Social Psychology* 61:923-929.
Weick, K.E., and K.A. Roberts
1993 Collective mind in organizations: Heedful interrelating on flight decks. *Administrative Science Quarterly* 38:357-381.
Weismeyer, M.D.
1992 An Operator-Based Model of Human Covert Visual Attention. Ph.D. thesis. Electrical Engineering and Computer Science Department, University of Michigan.

Wellman, B.
1991 Network analysis: Some basic principles. *Sociological Theory* 1:155-200.
Wellman, B., and S. Berkowitz
1988 *Social Structures: A Network Approach*. Norwood, NJ: Ablex. (Reprinted 1997.)
Wellman, B., J. Salaff, D. Dimitrova, L. Garton, M. Gulia, and C. Haythornthwaite
1996 Computer Networks as Social Networks: Virtual Community, Computer Supported Co-operative Work and Telework. *Annual Review of Sociology* 22: 213-238.
White, H.
1989 Learning in artificial neural networks: A statistical perspective. *Neural Computation* 1:425-464.
White, H.C., S.A. Boorman, and R.L. Breiger
1976 Social structure from multiple networks. I. Blockmodels of roles and positions. *American Journal of Sociology* 81:730-780.
Whittaker, J.
1967 Resolution and the communication discrepancy issue in attitude change. In *Attitude, Ego-Involvement and Change*, C. Sherif and M. Sherif, eds. New York, NY: Wiley.
Wickens, C.D.
1984 Processing resources in attention. Pp. 63-102 in *Varieties of Attention*, R. Parasuraman and D.A. Davies, eds. Orlando, FL: Academic Press.
1989 Resource management and timesharing. In *Human Performance Models for Computer-Aided Engineering*, J.I. Elkind, S.K. Card, J. Hochberg, and B.M. Huey, eds. Washington, DC: National Academy Press.
1992 *Engineering Psychology and Human Performance*, second edition. New York, NY: HarperCollins.
Wickens, C.D., A.S. Mavor, and J.P. McGee, eds.
1997 *Flight to the Future: Human Factors in Air Traffic Control*. Panel on Human Factors in Air Traffic Control Automation. Washington, DC: National Academy Press.
Wigdor, A.K., and B.F. Green, Jr., eds.
1991 *Performance Assessment for the Workplace: Volume 1*. Committee on the Performance of Military Personnel. Washington, DC: National Academy Press.
Wilensky, R.
1981 Meta-planning: Representing and using knowledge about planning in problem solving and natural language understanding. *Cognitive Science: Incorporating the Journal Cognition and Brain Theory: A Multidisciplinary Journal (CSc)* 5(3):197-233.
Wilkins, D.
1988 *Practical Planning*. San Francisco, CA: Morgan Kaufmann.
Wilkins, D.E.
1984 Domain-independent planning: Representation and plan generation. *Artificial Intelligence* 22:269-301.
Williams, J.M.G., F.N. Watts, C. MacLeod, and A. Mathews
1997 *Cognitive Psychology and Emotional Disorders*. (revised edition) New York, NY: John Wiley.
Woldron, A.N.
1997 The art of shi. *The New Republic* 2/6:36-40.
Wolfe, J.M.
1994 Guided search 2.0: A revised model of visual search. *Psychonomic Bulletin of Review* 1:202-238.
Yaari, M.E.
1987 The dual theory of choice under uncertainty. *Econometrica* 55:95-115.

Yantis, S.

1993 Stimulus-driven attentional capture. *Current Directions in Psychological Science* 2:156-161.

Yates, J.

1990 What is theory? A response to Springer. *Cognition* 36(1):91-96.

Young, R.M., and R. Lewis

forthcoming The Soar cognitive architecture and human working memory. In *Models of Working Memory: Mechanisms of Active Maintenance and Executive Control*, A. Miyake and P. Shah, eds. ??: Cambridge University Press.

Zacharias, G.L., S. Baron, and R. Muralidharan

1981 A supervisory control model of the AAA crew. Pp. 301-306 in *Proceedings of the 17th Conference on Manual Control*, Los Angeles, CA, June.

Zacharias, G.L., A.X. Miao, C. Illgen, and J.M. Yara

1996 SAMPLE: Situation Awareness Model for Pilot-in-the-Loop Evaluation. Final Report R95192, December 12. Charles River Analytics, Cambridge, MA.

Zacharias, G.L., A.X. Miao, A. Kalkan, and S.-P. Kao

1994 Operator-based metric for nuclear operations automation assessment. Pp. 181-205 in *22nd Water Reactor Safety Information Briefing*. NUREG/CP-0140, Volume 1. Washington: U.S., Nuclear Regulatory Commission.

Zacharias, G., A.X. Miao, E.W. Riley, and R.K. Osgood

1992 Situation Awareness Metric for Cockpit Configuration Evaluation. Final AL-TR-1993-0042, November. Wright-Patterson Air Force Base, OH.

Zacharias, G., A. Miao, C. Illgen, J. Yara, and G. Siouris

1996 SAMPLE: Situation Awareness Model for Pilot in the Loop Evaluation. First Annual Conference on Situation Awareness in the Tactical Air Environment, Naval Air Warfare Center, Patuxent River, MD.

Zachary, W., J.M. Ryder, and J.H. Hicinbothom

forthcoming Cognitive task analysis and modeling of decision making in complex environments. In *Decision Making Under Stress: Implications for Training and Simulation*, J. Cannon-Bower and E. Salas, eds. Washington, DC: American Psychological Association.

Zachary, W., J.-C. Le Mentec, and J. Ryder

1996 Interface agents in complex systems. In *Human Interaction With Complex Systems: Conceptual Principles and Design Practice*, C.N. Ntuen and E.H. Park, eds. Kluwer Academic Publishers.

Zachary, W., J. Ryder, L. Ross, and M.Z. Weiland

1992 Intelligent computer-human interaction in real-time, multi-tasking process control and monitoring systems. In *Human Factors in Design for Manufacturability*, M. Helander and M. Nagamachi, eds. New York, NY: Taylor and Francis.

Zadeh, L.A.

1986 A simple view of the Dempster-Shafer theory of evidence and its implication for the rule of combination. *The AI Magazine* (Summer):85-90.

1973 Outline of a new approach to the analysis of complex systems and decision processes. *IEEE Transactions on Systems, Man, and Cybernetics*. SMC-3(1):28-44.

1997 The roles of fuzzy logic and soft computing in the conception, design and deployment of intelligent systems. In *Software Agents and Soft Computing*, H.S. Nwana and A. Nader, eds. Berlin, Germany: Springer-Verlag.

Zellenberg, M.

1996 On the importance of what might have been. In *Psychological Perspectives on Regret and Decision Making*. Amsterdam, NL: The Faculty of Psychology of the University of Amsterdam.

Appendix

Biographical Sketches

RICHARD W. PEW (Chair) is principal scientist at BBN Technologies, a unit of GTE Internetworking, in Cambridge, Massachusetts. He holds a bachelors degree in electrical engineering from Cornell University (1956), a master of arts degree in psychology from Harvard University (1960), and a Ph.D. in psychology with a specialization in engineering psychology from the University of Michigan (1963). He has 30 years of experience in human factors, human performance, and experimental psychology as they relate to systems design and development. Throughout his career he has been involved in the development and utilization of human performance models and in the conduct of experimental and field studies of human performance in applied settings. He spent 11 years on the faculty of the Psychology Department at Michigan, where he was involved in human performance teaching, research and consulting before moving to BBN in 1974. His current research interests include the impact of automation on human performance, human-computer interaction, and human performance modeling. Dr. Pew was the first chairman of the National Research Council Committee on Human Factors and has been president of the Human Factors Society and president of Division 21 of the American Psychological Association, the division concerned with engineering psychology. He has also been chairman of the Biosciences Panel of the Air Force Scientific Advisory Board. Dr. Pew has authored more than 60 publications, including book chapters, articles, and technical reports.

JEROME BUSEMEYER received his Ph.D. from the University of South Carolina, 1979. He is a past president of the Society of Mathematical Psychology and is currently a member of the National Institute of Mental Health Perception and Cognition Review Committee. He is also a member of the editorial boards of the *Journal of Mathematical Psychology*, the *Psychological Bulletin*, and the *Journal of Experimental Psychology: Learning, Memory, and Cognition.* His research concerns dynamic, emotional, and cognitive models of judgment and decision making; neural network models of function learning, interpolation, and extrapolation; methodology for comparing and testing complex models of behavior; and measurement theory with error-contaminated data.

KATHLEEN M. CARLEY is currently an associate professor of sociology and organizations at Carnegie Mellon University. She received an S.B. in political science and an S.B. in economics from Massachusetts Institute of Technology and a Ph.D. in sociology from Harvard University. Her current research is in the areas of organizational design and adaptation, computational organization theory, social and organizational networks, evolution of social networks, communication and technology, social theory, communication and diffusion of information, statistical and computational techniques for analyzing social networks and their evolution over time, and computer-assisted textual analysis techniques for coding mental models. Dr. Carley is a founding coeditor of the journal *Computational and Mathematical Organization Theory.*

TERRY CONNOLLY is professor of management and policy at the University of Arizona. He has taught at the Georgia Institute of Technology, the University of Illinois, and the University of Chicago since completing his Ph.D. at Northwestern University in organization behavior and systems theory. His undergraduate degree was in electrical engineering (Manchester) and his M.A. in sociology (Northwestern). His primary research interests are in judgment and decision making, and he has published widely on these topics. He serves on several editorial boards, including *Organizational Behavior and Human Decision Processes*, the *Journal of Behavioral Decision Making*, and the *Administrative Science Quarterly.* Dr. Connolly is a fellow of the American Psychological Society and past president of the Judgment and Decision Making Society, and he serves on the Committee on Human Factors of the National Research Council.

JOHN R. CORSON is a retired U.S. Army infantry colonel and currently president of JRC Research & Analysis L.L.C. He is a consultant to the U.S. Army Operational Test and Evaluation Command, where he has provided operational evaluation support for the Army's Advanced Warfighting Experiments and assessment of the Army Experimental Force for the Army of the 21st Century. He has over 30 years of experience in senior-level management, training and training development, operations research/systems analysis, and operational test and evaluation. His professional experience includes 2 years as chief operating officer of Integrated Visual Learning, a commercial joint venture company; 4 years as vice president/senior program manager for a professional technical support defense contractor corporation; and 2 years as deputy commander of the U.S. Army Operational Test and Evaluation Agency. He has a B.Sc. in industrial management from Drexel University (1961) and an MBA from The Ohio State University (1972). He previously served as a National Research Council panel member on the Panel on Human Factors in the Design of Tactical Display Systems for the Individual Soldier.

KENNETH H. FUNK received a B.A. in biology from Taylor University in 1975 and an M.S. and Ph.D. in industrial and systems engineering from The Ohio State University in 1977 and 1980, respectively. He is currently with the Industrial and Manufacturing Engineering Department of Oregon State University, where he teaches courses in human factors engineering, industrial engineering, system safety, and artificial intelligence. His research interests include human factors engineering (especially aviation human factors), applied artificial intelligence, and social implications of technology. He is a member of the Human Factors and Ergonomics Society, the Association of Aviation Psychologists, and the Institute of Electrical and Electronics Engineers.

BONNIE E. JOHN is an associate professor in the Human-Computer Interaction Institute and the Departments of Psychology and Computer Science at Carnegie Mellon University. She develops engineering models of human performance that aid in the design of computer systems. She also studies human computer interaction usability evaluation methods to understand their usefulness, effectiveness, learnability, and usability. She received a bachelor of engineering degree in mechanical engineering from The Cooper Union in 1977 and an MS in mechanical engineering from Stanford in 1978. She worked at Bell Laboratories from 1977 to 1983, designing data and telecommunications systems. After taking courses in human factors at Stevens Institute of Technology, she left Bell Labs to get a masters degree (1984) and Ph.D. (1988) in cognitive psychology at Carnegie Mellon University. She received a National Science Foundation Young Investigator Award in 1994 and is currently serving on the National Research Council's Committee on Human Factors.

JERRY S. KIDD is senior adviser for the Committee on Human Factors and its various projects. He received a Ph.D. from Northwestern University in social psychology in 1956; he then joined RAND Corporation to help on a project to simulate air defense operations. He left RAND in late 1956 to join the staff at the Laboratory of Aviation Psychology at Ohio State University. There he worked under Paul Fitts and George Briggs until 1962, when he joined the staff of AAI, Incorporated, north of Baltimore, Maryland. In 1964, he moved to the National Science Foundation as program director for special projects. He joined the faculty of the College of Library and Information Services at the University of Maryland in 1967 and retired in 1992.

ANNE MAVOR is study director for the Panel on Modeling Human Behavior and Command Decision Making, and director of the Committee on Human Factors and the Committee on Techniques for the Enhancement of Human Performance: Occupational Analysis. Her previous work as a National Research Council senior staff officer has included a study of human factors in air traffic control automation, a study of human factors considerations in tactical display for soldiers, a study of scientific and technological challenges of virtual reality, a study of emerging needs and opportunities for human factors research, a study of modeling cost and performance of military enlistment, a review of federally sponsored education research activities, and a study to evaluate performance appraisal for merit pay. For the past 25 years her work has concentrated on human factors, cognitive psychology, and information system design. Prior to joining the National Research Council she worked for the Essex Corporation, a human factors research firm, and served as a consultant to the College Board. She has an M.S. in experimental psychology from Purdue University.

RICHARD M. SHIFFRIN, a member of the National Academy of Sciences and the American Academy of Arts and Sciences, is currently a professor of psychology, the Luther Dana Waterman Research Professor, and director of the Cognitive Science Program at Indiana University. He is an experimental psychologist who specializes in mathematical and computer simulation models of cognition. He is best known for his research and theories of short-term and long-term memory and of attention and automatism. Although his research lies primarily in basic processes of cognition, many of his Ph.D. students have gone on to work in applied research areas of human factors and information science and technology in academia, business, and the military. Dr. Shiffrin received a B.A. in mathematics from Yale University in 1964 and a Ph.D. in experimental and mathematical psychology from Stanford University in 1968, and has been on the faculty at Indiana University since 1968.

GREG L. ZACHARIAS is principal scientist and director of research at Charles River Analytics Inc. Since cofounding the company in 1983, he has led efforts in developing behavioral representations of human information processing, situation awareness, and decision making in a variety of complex task environments, including piloted flight, nuclear power plant operations, and military decision making. As a senior scientist at Bolt Beranek and Newman from 1977 to 1983, Dr. Zacharias developed and applied models of visual and motion cueing for piloted flight control applications and was a codeveloper of a family of computational models currently used in evaluating multi-operator performance in multitask environments. Earlier, as a research engineer at the C.S. Draper Laboratory, he focused on pilot/vehicle interface design issues, building on a previous Air Force assignment at the National Aeronautics and Space Administration's Johnson Spacecraft Center, where he was responsible for preliminary design definition of the space shuttle reentry flight control system. Dr. Zacharias obtained a Ph.D. in aeronautics and astronautics from Massachusetts Institute of Technology (MIT) in 1977, is a research affiliate of the Man-Vehicle Laboratory at MIT, and is a member of the Committee on Human Factors of the National Research Council.

Index

A

F

J

N

O

P

Q

R

S

U

V

W